NUMERICAL SOLUTION OF INTEGRAL EQUATIONS

EDITED BY

L. M. DELVES

UNIVERSITY OF LIVERPOOL

AND

J. WALSH

UNIVERSITY OF MANCHESTER

CLARENDON PRESS · OXFORD

1974

Oxford University Press, Ely House London W. 1

GLASGOW NEW YORK TORONTO MELBOURNE WELLINGTON
CAPE TOWN IBADAN NAIROBI DAR ES SALAAM LUSAKA ADDIS ABABA
DELHI BOMBAY CALCUTTA MADRAS KARACHI LAHORE DACCA
KUALA LUMPUR SINGAPORE HONG KONG TOKYO

ISBN 0 19 853342 X

Printed photolitho in Great Britain by
J. W. Arrowsmith Ltd., Bristol

PREFACE

The aim of this volume is to give an introduction to the principal problems and methods in the numerical solution of integral equations, together with some theoretical background, and a number of applications. The basic methods are treated in some detail, and recent developments are also discussed and compared, with full lists of references for further study.

Part I (Chapters 1 - 5) deals with mathematical preliminaries, in the theory of integral equations, in numerical analysis, function spaces and approximation theory. Part II (Chapters 6 - 18) considers the main numerical methods used for the different types of integral equations, treating general classes of problems rather than particular cases. Chapter 6 gives the direct approach to the numerical solution of Fredholm equations of the second kind, using quadrature formulae. Chapters 7, 8 and 9 consider methods based on expansions of various kinds, with the solution obtained by variational or Galerkin methods, or by collocation. Chapter 10 applies quadrature and variational methods to the solution of the eigenvalue problem. Chapters 11 and 12 deal with Volterra equations of the first and second kinds, with a discussion of the stability of step-by-step methods. Chapter 13 considers the problem of Fredholm equations of the first kind, and discusses the ill-conditioned nature of the problem and the associated numerical difficulties. Chapter 14 surveys the wide field of integro-differential equations, with a summary of recent theoretical work and a brief description of some particular examples. Chapters 15, 16 and 17 treat the general problem of nonlinear integral equations, with a study of the convergence of iterative methods, and error estimation. Part II ends with a discussion in Chapter 18 of library programs for the practical solution of certain classes of integral equations.

In Part III (Chapters 19 - 25) some fields of application of integral equations are surveyed. Integral equations often arise in the solution of potential problems and similar partial differential equations, and this is illustrated by examples from diffraction problems, water waves, scattering in quantum mechanics, and conformal mapping. The practical difficulties of handling three-dimensional problems are discussed, and an example is given of the use of finite elements for the Helmholtz equation.

This collection of papers is based on the material presented at a joint Summer School in July 1973, organized by the Department of Mathematics, University of Manchester, and the Department of Computational and Statistical Science, University of Liverpool. The detailed programme was drawn up principally by C.T.H. Baker, in collaboration with the other contributors. We acknowledge with thanks the careful and skilful typing of the text, which was done by Kendal Anderson, Pat McMunn and Maureen Parker at Manchester, Doreen Manley and Norma Wilson at Liverpool.

L.M. Delves

J. Walsh

CONTENTS

PART 1. BACKGROUND THEORY

PART 3. APPLICATIONS

CONTRIBUTORS

Professor I. Barrodale (Chapters 5, 8)

Mathematics Department, University of Victoria,
Victoria, B.C. Canada.

Mr. A. Burton (Chapter 21)

Mr. G.F. Miller (Chapters 13, 18)

Dr. G.T. Symm (Chapters 20, 24)

Division of Numerical Analysis and Computation,
National Physical Laboratory, Teddington, England.

Dr. D.F. Mayers (Chapter 6)

Oxford University Computing Laboratory,
19 Parks Road, Oxford, England.

Professor L.B. Rall (Chapters 15, 16, 17)

Mathematics Research Center, University of Wisconsin,
Madison, Wisconsin 53706, U.S.A.

Dr. C.T.H. Baker (Chapters 7, 10, 12, 14)

Dr. I. Gladwell (Chapter 3)

Professor F. Ursell (Chapters 1, 22)

Dr. J. Walsh (Chapter 2)

Department of Mathematics, The University,
Manchester M13 9PL, England.

Professor L.M. Delves (Chapters 4, 9, 25)

Dr. R. Wait (Chapter 23)

Department of Computational and Statistical Science,
The University, Liverpool L69 3BX, England.

PART 1

BACKGROUND THEORY

CHAPTER 1 INTRODUCTION TO THE THEORY OF LINEAR INTEGRAL EQUATIONS

F. Ursell

University of Manchester

1. Definitions and Examples

The following are examples of linear integral equations

$$\int_a^b f(y)\ K(x, y)dy = g(x), \qquad c < x < d, \tag{1.1}$$

$$f(x) = g(x) + \lambda\int_a^b K(x, y)\ f(y)dy, \qquad a < x < b, \tag{1.2}$$

or
$$f(x) = g(x) + \int_a^b K(x, y, \lambda)f(y)dy, \qquad a < x < b. \tag{1.3}$$

Here $K(x, y)$ and $g(x)$ are known functions, and λ is a complex-valued parameter. The unknown function is $f(x)$. The equations are to hold over a prescribed range of x (which, in the case of (1.2) and (1.3), is the same as the range of integration). The functions $K(x, y)$ in (1.1), and $K(x, y)$ (or $\lambda K(x, y)$) in (1.2), and $K(x, y, \lambda)$ in (1.3) are the kernels of the corresponding integral equations. Equation (1.1) is an equation of the first kind, and equations (1.2), (1.3) are equations of the second kind.

When $g(x) \equiv 0$ in (1.1), i.e. when

$$\int_a^b f(y)\ K(x, y)dy = 0, \tag{1.4}$$

we say that (1.4) is a homogeneous equation of the first kind. Similarly the equations

$$f(x) = \lambda\int K(x, y)f(y)dy \tag{1.5}$$

and

$$f(x) = \int K(x, y, \lambda)f(y)dy \tag{1.6}$$

are homogeneous equations of the second kind.

Examples of equations of the first kind.

(i) $\sqrt{\frac{2}{\pi}}\int_0^\infty f(y)\cos xy\ dy = g(x), \qquad 0 < x < \infty.$

Solution: $f(y) = \sqrt{\frac{2}{\pi}} \int_0^\infty g(x)\cos xy\, dx, \ 0 < y < \infty.$

(Fourier cosine transform, 1811)

(ii) $\int_o^x \frac{f(y)dy}{(x-y)^{\frac{1}{2}}} = g(x).$

Solution: $f(y) = \frac{1}{\pi}\frac{d}{dy}\left(\int_o^y g(x) \frac{dx}{(y-x)^{\frac{1}{2}}}\right).$

(Abel's equation, 1823)

Example of an equation of the second kind.

The differential equation $w'' + p(x)w = q(x)$ with the two-point boundary condition $w(0) = 0$, $w(1) = 0$ can be transformed into an integral equation of the second kind:

$$w(x) = F_1(x) + \int_o^1 K(x, y)p(y)w(y)dy,$$

where

$$K(x, y) = \begin{cases} \frac{y(1-x)}{1}, & y \lesssim x, \\ \frac{x(1-y)}{1}, & x \lesssim y, \end{cases}$$

and

$$F_1(x) = -\int_o^1 K(x, y)q(y)dy.$$

Note that the integral equation is equivalent to the differential equation together with its boundary conditions.

An important special case arises when $K(x, y) = 0$ for $y > x$. This type is called a Volterra equation.

Thus,

$$f(x) = g(x) + \lambda \int_a^x K(x, y)f(y)dy$$

is a Volterra equation of the second kind.

The equation $w'' + p(x)w = q(x)$ with the one-point boundary condition $w(0) = c_0$, $w'(0) = c_1$ can be reduced to this form:

$$w(x) = \int_o^x (x-y)q(y)dy - \int_o^x (x-y)p(y)w(y)dy + c_0 + c_1 x.$$

2. Solution by Iteration. The Liouville-Neumann Series

Equations of the second kind can be solved by iteration when the kernel is sufficiently small (or when the kernel is a Volterra kernel). Define the iterated kernels

$$K^{(n)}(x, y) = \int_a^b K(x, z)\, K^{(n-1)}(z, y)dz, \; n \geq 2.$$

It is easy to show that

$$K^{(m+n)}(x, y) = \int_a^b K^{(m)}(x, z)\, K^{(n)}(z, y)dz$$

for all integral m and n. Also write

$$Kg = \int_a^b K(x, y)g(y)dy, \; K^n g = \int_a^b K^{(n)}(x, y)g(y)dy.$$

Then the iterative solution of

$$f = g + \lambda Kf$$

is the Liouville-Neumann series

$$f = g + \lambda Kg + \lambda^2 K^2 g + \dots .$$

For instance, when g and K are continuous, the series can be shown to converge when $|\lambda|\,(b-a) \max_{x,y} |K(x, y)| < 1$. This condition is sufficient but not necessary.

To show that it is not necessary, consider the case

$$K(x, y) = \cos x \sin y, \quad 0 \leq x, y \leq \pi.$$

Then $K^{(2)}(x, y) = \int_0^\pi \cos x.\, \sin z \cos z.\, \sin y\, dz = 0$, and all the iterated kernels vanish. In this case, $f = g + \lambda Kg$,

i.e. $f(x) = g(x) + \lambda \cos x \int_0^\pi g(y) \sin y\, dy.$

The Volterra equation of the second kind can always be solved by iteration. We have

$$K^{(n)}(x, y) = \int_y^x K(x, z)K^{(n-1)}(z, y)dz.$$

It is easy to prove by induction that

$$|K^{(n)}(x, y)| \leq \frac{(x-y)^{n-1}}{(n-1)!} (\max |K|)^n;$$

thus the Liouville-Neumann series is comparable to an exponential series, and is therefore convergent.

The Liouville-Neumann series is fundamental in the application of integral equations to asymptotics. The problem is reduced to an integral equation with a small kernel; then the Liouville-Neumann series gives a rigorous estimate of the residual $f(x) - g(x)$.

Note that there is no Liouville-Neumann series for equations of the first kind.

3. Existence of Solutions. The Fredholm Alternative

Equations of the second kind have an existence theory even when the kernel is not small. Suppose, to fix ideas, that the range of integration is finite, and that the known functions $g(x)$ and $K(x, y)$ are continuous. (These conditions can be much relaxed.) Then we have the

Fredholm Alternative for equations of the second kind.

Either: λ is a regular value, then the equation

$$f(x) = g(x) + \lambda \int_a^b K(x, y)\, f(y)dy \tag{3.1}$$

has a unique solution for any arbitrary function $g(x)$;

Or: λ is a characteristic value of (3.1), then the homogeneous equation

$$f(x) = \lambda \int_a^b K(x, y)\, f(y)dy \tag{3.2}$$

has a finite number (l, say) of linearly independent solutions $\phi_1(x), \ldots, \phi_l(x)$. In this case the transposed homogeneous equation

$$\psi(x) = \lambda \int_a^b K(y, x)\psi(y)dy$$

also has l solutions $\psi_1(x), \ldots, \psi_l(x)$; and the equation (3.1) has a solution if and only if $g(x)$ is orthogonal to all the ψ_j, i.e. if and only if

$$\int_a^b g(x)\psi_j(x)dx = 0, \quad j = 1, \ldots \; l.$$

This solution is clearly not unique since we can add any linear combination of the ϕ_j's. (Note the analogy with the theory of the finite-dimensional matrix equation $(I - \lambda A)\underline{x} = 0$.) For an outline of a proof see §4 below.

The Fredholm Alternative holds also for the equation

$$f(x) = g(x) + \int K(x, y, \lambda) f(y) dy, \tag{3.3}$$

for we have not used the fact that λ is a factor outside the integral sign.

The Fredholm Alternative may be put into another form:

If the homogeneous equation (3.2) has the unique solution $f \equiv 0$, then (3.1) is solvable for arbitrary g (and the solution must evidently be unique). In other words, uniqueness implies existence.

The Fredholm Alternative does <u>not</u> in general apply to equations of the <u>first kind</u>; in this case, uniqueness need not imply existence. Counter-example: Consider the kernel $\frac{1}{x+y+1}$, $0 \leq x, y \leq 1$.

The equation $\int_0^1 \frac{f(y)dy}{x+y+1} = 0$ has no continuous solution other than $f(y) \equiv 0$. (It follows that the characteristic functions of the kernel form a complete set over (0, 1).) Nevertheless the equation $\int_0^1 \frac{f(y)dy}{x+y+1} = 1$ has no continuous solution.

Proof: Consider $F(z) \equiv \int_0^1 \frac{f(y)dy}{z+y+1}$, which is evidently an analytic function of z in the z-plane cut from $z = -2$ to $z = -1$. (In particular, $F(z)$ is regular when $0 < x < 1$.) Then we have, by a well-known result,

$$F(-1 - \xi \pm i\eta) = \int_0^1 \frac{\phi(y)dy}{y-\xi\pm i\eta}$$

$$\rightarrow \int_0^1 \frac{\phi(y)dy}{y-\xi} \pm i\pi\phi(\xi) \quad \text{as} \quad \eta \rightarrow +0.$$

(The integral in the last equation is to be interpreted as a Cauchy principal value.) It follows that the discontinuity across the cut is $2i\pi\phi(\xi)$.

Suppose now that $F(x) = 0$ when $0 < x < 1$. It follows by analytic continuation that $F(z) \equiv 0$ everywhere in the cut plane. Thus there is no discontinuity across the cut, i.e. $f(\xi) \equiv 0$.

Thus $F(x) = 0$ when $0 < x < 1$ implies $f(x) = 0$ when $0 < x < 1$. (Continuity of $f(x)$ has been assumed but evidently the argument is applicable to integrable functions. It follows in this case that $f(x) = 0$ almost everywhere.)

Now consider the integral equation of the first kind

$$F(x) \equiv \int_0^1 \frac{f(y)dy}{x+y+1} = 1.$$

We have just seen that this has at most one solution. In fact it has no solution, for $F(z) - 1$ vanishes on $(0, 1)$ and therefore everywhere in the cut plane. But evidently $F(z) - 1 \to -1$ as $z \to +\infty$, a contradiction. Thus the Fredholm Alternative does not hold in this case.

In this example we see that the continuous function $f(x)$ is mapped into the analytic function $\int_0^1 \frac{f(y)dy}{x+y+1}$ which cannot in general be equal to an arbitrary continuous function $g(x)$. This example shows that equations of the first kind cannot be expected to have solutions unless the kernel is singular.

4. Proof of the Fredholm Alternative for Equations of the Second Kind.

If $K(x, y)$ is degenerate, i.e. if it is a finite sum of the form

$$K_d(x, y) = \sum_{k=1}^{n} X_k(x)\, Y_k(y),$$

then the integral equation is equivalent to a system of linear equations in n unknowns and can be treated by linear algebra. (See e.g. Riesz-Nagy, 1955, §70.) If K is not degenerate, approximate it by a degenerate kernel,

$$K(x, y) = \sum_{k=1}^{n} X_k(x)\, Y_k(y) + h(x, y),$$

where $h(x, y)$ is small; e.g. when $K(x, y)$ is continuous, choose n so large that $|\lambda|(b-a) \max|h(x, y)| < 1$. Now apply the resolvent kernel for $h(x, y)$. Then the integral equation is transformed into an equation with a degenerate kernel, and the Fredholm Alternative follows. (Riesz-Nagy, §71.)

Fredholm's method of proof was different; he showed that the solution is

$$f(x) = g(x) + \lambda \int_a^b R(x, y, \lambda) g(y) dy,$$

where $R(x, y, \lambda) = \frac{D(x,y,\lambda)}{D(\lambda)}$ is the quotient of two entire functions convergent for all λ,

$$D(x, y, \lambda) = K(x, y) + \sum_1^\infty (-1)^n d_n(x, y) \frac{\lambda^n}{n!},$$

$$D(\lambda) = 1 + \sum_1^\infty (-1)^n d_n \frac{\lambda^n}{n!},$$

where $d_n = \int \cdots \int \begin{vmatrix} K(t_1, t_1) \dots K(t_1, t_n) \\ \dots\dots\dots\dots \\ K(t_n, t_1) \dots K(t_n, t_n) \end{vmatrix} dt_1 \dots dt_n,$

$$d_n(x, y) = \int \cdots \int \begin{vmatrix} K(x, y) & K(x, t_1) \dots K(x, t_n) \\ K(t_1, y) & \vdots \\ \dots\dots & \dots\dots\dots \\ K(t_n, y) & \dots\dots K(t_n, t_n) \end{vmatrix} dt_1 \dots dt_n.$$

Fredholm was able to prove convergence of these power series for all λ. The homogeneous equation has a non-trivial solution for $\lambda = \lambda_0$ if and only if $D(\lambda_0) = 0$. Fredholm's formulae have so far found only a few applications, either analytical or numerical, apart from providing a foundation for the theory.

5. Characteristic Values and Characteristic Functions

Consider the homogeneous equation of the second kind

$$f(x) = \lambda \int_a^b K(x, y) f(y) dy.$$

Fredholm theory shows that this has a non-trivial solution if and only if λ is a characteristic value (i.e., a value at which the Fredholm denominator $D(\lambda)$ vanishes). If $K(x, y)$ is continuous (or belongs to L_2, or has other suitable regularity properties) we know from Fredholm theory that the set $\{\lambda_n\}$ of characteristic values (c.v.'s) is denumerable; the corresponding characteristic functions (c.f.'s) are denoted by $\phi_n(x)$. There may be several linearly independent c.f.'s belonging to the same c.v.

An integral equation need not have any c.v.'s, e.g. Volterra equations (for which $K(x, y) = 0$ when $y > x$) have no c.v.'s as we have seen. In general it is difficult to decide whether a given kernel has a c.v. The best-known results relate to Hermitian kernels ($K(x, y) = \overline{K(y, x)}$), of which real symmetric kernels are a special case. They are as follows.

(i) Hermitian kernels have at least one c.v.

(ii) The c.v.'s λ_n are all real.

(iii) The c.f.'s $\phi_m(x)$, $\phi_n(x)$ belonging to distinct c.v.'s λ_m, λ_n are orthogonal, i.e. $\int_a^b \overline{\phi_m(x)}.\phi_n(x)dx = 0$.

(iv) There are at most finitely many linearly independent c.f.'s belonging to a given c.v. λ_n. These may be arranged to be orthogonal. If this is done then all distinct c.f.'s are orthogonal.

(v) For theoretical work it is often convenient to suppose that the c.f.'s are normalized

$$\int_a^b |\phi_n(x)|^2 dx = 1.$$

A set of functions such that

$$\int_a^b \overline{\phi_m(x)}.\phi_n(x)dx = \delta_{mn} = \begin{cases} 1, & m = n \\ 0, & m \neq n \end{cases}$$

is said to be orthonormal (orthogonal and normalized).

Henceforth we shall consider real-valued symmetric kernels rather than complex-valued Hermitian kernels. The theory for the latter is a straightforward extension of the theory for the former.

A real symmetric kernel need not have more than one c.v.

Example: $K(x, y) = \cos x.\cos y$, $0 \leqslant x, y \leqslant \pi$.

Thus $\phi(x) = \lambda \cos x \int_0^\pi \cos y.\ \phi(y)dy$, i.e. $\phi(x)$ is a multiple of $\cos x$.

Apply $\int_0^\pi \ldots \cos x\ dx$. Then

$$\int_0^\pi \phi(x)\cos x\ dx = \lambda\left(\int_0^\pi \cos^2 x\ dx\right)\left(\int_0^\pi \phi(y)\cos y\ dy\right)$$

$$= \tfrac{1}{2}\pi\lambda \int_0^\pi \phi(x)\cos x\ dx.$$

We conclude that

$$\lambda = \frac{2}{\pi}, \qquad \phi(x) = \sqrt{\frac{2}{\pi}} \cos x \qquad \text{(normalized)}.$$

In practice it would not be useful to normalize.

The example $K(x, y) = \cos x.\cos y$ shows that the set of c.f.'s will in general be incomplete, i.e. an arbitrary continuous function $F(x)$ cannot be approximated arbitrarily closely by sums $\sum c_n \phi_n(x)$. A test for completeness is that the equation

$$\int K(x, y)\, f(y) dy = 0$$

has no solution other than $f(y) \equiv 0$. (We have shown this in §3 above for the kernel $\frac{1}{x+y+1}$ in the interval $(0, 1)$.)

Expansion theorem. Functions $G(x)$ of the form

$$G(x) = \int_a^b K(x, y)\, g(y) dy$$

can be expanded in terms of the c.f.'s $\phi_n(x)$ of the symmetric kernel $K(x, y)$.

Important remark: Little is known about the asymptotic behaviour of the c.v.'s λ_n and the c.f.'s $\phi_n(x)$ as $n \to \infty$.

6. Singular Values and Singular Functions

Suppose now that the kernel K is not symmetric. Then we can form the symmetric kernels

$$[K]^2(x, y) = \int K(x, z) K(y, z) dz \tag{6.1}$$

and

$$[K^*]^2(x, y) = \int K(z, x) K(z, y) dz, \tag{6.2}$$

and these have characteristic functions and (positive) characteristic values. Denote the c.f.'s of (6.1) by ϕ_n, and the c.v.'s by μ_n^2. Then it is easy to see that

$$\psi_n(x) = \mu_n \int K(z, x)\, \phi_n(z) dz$$

is a c.f. of (6.2) belonging to the same c.v. μ_n^2. Similarly, if ψ_n is a c.f. of (6.2), then

$$\phi_n(x) = \mu_n \int K(x, z)\, \psi_n(z) dz$$

is a c.f. of (6.1). It is easy to show that the ψ_n are orthonormal if the ϕ_n are orthonormal. The functions ϕ_n and ψ_n are the singular functions of K belonging to the singular value μ_n.

References

There are many good accounts of the theory of integral equations, in monographs, in text-books on mathematical analysis, in works on partial differential equations etc. Many of these are listed by Cochran (1972). Only a few are given here.

COCHRAN, J.A. (1972) Analysis of Linear Integral Equations. New York: McGraw Hill.

(The mathematical theory, with references to most of the recent literature.)

GOURSAT, E. (1964) Course of Mathematical Analysis. English ed., Vol. 3, Part 2. New York: Dover.

(Gives the classical theory in detail, and is still of use.)

HELLINGER, E. and TOEPLITZ, O. (1927) Integralgleichungen und Gleichungen mit unendlich vielen Unbekannten. Encycl. der math. Wissenschaften, Vol. II C13. Leipzig: Teubner (reprinted by Chelsea, New York, 1953).

(A full and readable survey of the literature up to 1925.)

HOCHSTADT, H. (1973) Integral Equations, New York: Wiley-Interscience.

(A readable brief account, with many examples and exercises.)

RIESZ, F. and SZ-NAGY, B. (1955) Functional Analysis. New York: Ungar.

(A masterly account of functional analysis, including integral equations, illustrated by many applications.)

SCHMEIDLER, W. (1950) Integralgleichungen mit Anwendungen in Physik und Technik, Leipzig: Akademie-Verlag.

(Theory and applications, with an appendix listing all the examples treated in the book.)

SMIRNOV, V.I. (1964) Course of Higher Mathematics, Vol. IV. London: Pergamon Press.

(An account of the linear theory, with many applications.)

SMITHIES, F. (1962) Integral Equations. Cambridge University Press.

(A careful account of the theory.)

CHAPTER 2 QUADRATURE AND FOURIER SERIES

J. Walsh

University of Manchester

1. Simple Quadrature Rules

We shall consider methods for approximating the value of a definite integral which is assumed to exist in the sense of Riemann. A simple rule for approximate integration (or quadrature) has the following form:

$$\int_a^b f(x)dx \simeq w_1 f(x_1) + w_2 f(x_2) + \ldots + w_n f(x_n), \qquad (1.1)$$

i.e. the integral is represented by a weighted sum of values of the integrand at a finite number of points x_i. For straightforward integration these points are usually taken to lie within the interval $[a, b]$, but for certain applications, e.g. Volterra integral equations, it may be convenient to take some of them outside $[a, b]$. Various classes of formulae of the type (1.1) may be obtained by imposing different conditions on x_i and w_i, e.g. that the points should be equally spaced, or that certain of the weights should take prescribed values, etc. When these conditions have been met, the remaining free parameters are chosen to make the formula as accurate as possible, by making it exact for successively higher powers of x. We substitute $f(x) = 1, x, x^2, \ldots$ in (1.1), and equate the left and right hand sides to give a system of equations for the unknown x_i, w_i. If these equations (which are nonlinear in general) have a real solution, we obtain a formula which is exact for all polynomials of a certain degree k say, depending on the number of free parameters.

Such formulae are useful in cases where $f(x)$ can be accurately represented by a polynomial, but they are unsatisfactory when $f(x)$ has a singularity in the range; special methods for this case are discussed in §3. When $f(x)$ is nonsingular, it may be approximated either by a single high-order polynomial over the whole interval, or by a set of low-order polynomials over sub-intervals. The latter is often more convenient for numerical work; it means that we use simple quadrature rules with a small number of parameters, and apply them repeatedly.

We now summarize the principal formulae of type (1.1), with their error terms. Let us write $I(f)$ for the integral $\int_a^b f(x)dx$. The error depends on some high derivative of $f(x)$ at a point ξ or η in $[a, b]$, where we assume that the derivative occurring is continuous. For the repeated forms, the interval $[a, b]$ is divided into n equal steps of length h, so that $nh = b - a$, and the formula is applied over sub-intervals. The first three formulae use equally-spaced points, and we let $x_i = a + i\,h$.

(i) <u>Mid-point rule</u>

$$\left.\begin{aligned} &\text{Single interval:}\quad I(f) = (b-a)f\{\tfrac{1}{2}(a+b)\} + E_1, \\ &\qquad \text{where the error}\quad E_1 = \frac{1}{24}(b-a)^3 f''(\xi). \end{aligned}\right\} \qquad (1.2)$$

$$\left.\begin{aligned} &\text{Repeated form:}\quad I(f) = h\sum_{i=1}^{n} f(x_{i-\frac{1}{2}}) + R_1, \\ &\qquad \text{where}\quad R_1 = \frac{1}{24}(b-a)h^2 f''(\eta). \end{aligned}\right\} \qquad (1.3)$$

(ii) <u>Trapezium rule</u>

$$\left.\begin{aligned} &\text{Single interval:}\quad I(f) = \tfrac{1}{2}(b-a)\{f(a) + f(b)\} + E_2, \\ &\qquad \text{where}\quad E_2 = -\frac{1}{12}(b-a)^3 f''(\xi). \end{aligned}\right\} \qquad (1.4)$$

Repeated form:

$$\left.\begin{aligned} &I(f) = \tfrac{1}{2}h\,\{f(a) + 2\sum_{i=1}^{n-1} f(x_i) + f(b)\} + R_2, \\ &\text{where}\quad R_2 = -\frac{1}{12}(b-a)h^2 f''(\eta). \end{aligned}\right\} \qquad (1.5)$$

(iii) <u>Simpson's rule</u>

Single interval:

$$\left.\begin{aligned} &I(f) = \frac{1}{6}(b-a)\,\{f(a) + 4f\left[\tfrac{1}{2}(a+b)\right] + f(b)\} + E_3, \\ &\text{where}\quad E_3 = -\frac{1}{2880}(b-a)^5 f^{iv}(\xi). \end{aligned}\right\} \qquad (1.6)$$

Repeated form:

$$\left.\begin{aligned} &I(f) = \frac{1}{3}h\,\{f(a) + 4\sum_{i=1}^{\frac{1}{2}n} f(x_{2i-1}) + 2\sum_{i=1}^{\frac{1}{2}n-1} f(x_{2i}) + f(b)\} + R_3, \\ &\text{where}\quad R_3 = -\frac{1}{180}(b-a)h^4 f^{iv}(\eta). \end{aligned}\right\} \qquad (1.7)$$

In formula (1.7), n must be even. The above formulae all use equally-spaced points, and higher-order formulae of the same type may be derived (Newton-Cotes formulae). However, the first three are the most important in practice. (The "Three-Eighths" rule is also useful in special circumstances, see Davis and Rabinowitz, 1967, p.30.)

Another class of formulae is obtained when we remove the condition that the points should be equally spaced. These are the formulae for Gaussian quadrature and its variants. The values of x_i and w_i have to be computed for each n, and they are extensively tabulated in Stroud and Secrest (1966). We describe briefly three important forms.

(iv) Simple Gauss rule

Single interval: Formula as in (1), with x_i, w_i chosen to give an exact result for a polynomial of degree $2n-1$.

Error term $$E_4 = K_4 (b - a)^{2n+1} f^{(2n)}(\xi), \tag{1.8}$$

where K_4 depends on n, but not on a and b (details in Davis and Rabinowitz, 1967, §2.7).

Repeated form: If the interval is subdivided as before into n subintervals of width h, and the Gauss rule is applied over each, the error term becomes

$$R_4 = K_4 (b - a) h^{2n} f^{(2n)}(\eta). \tag{1.9}$$

(v) Radau quadrature

The form of the rule is as in (1.1), with x_1 fixed at the left end-point, and the rule is exact for polynomials of degree $2n - 2$.

$$\left.\begin{aligned} I(f) &= \frac{b - a}{n^2} f(a) + \sum_{i=2}^{n} w_i f(x_i) + E_5, \\ \text{where } E_5 &= K_5 (b - a)^{2n} f^{(2n-1)}(\xi). \end{aligned}\right\} \tag{1.10}$$

There is a corresponding formula using $f(b)$ instead of $f(a)$.

(vi) Lobatto quadrature

The form is again as in (1.1), with x_1 and x_n fixed at the two end-points. The rule is exact for polynomials of degree $2n - 3$.

$$\left.\begin{aligned} I(f) &= \frac{b - a}{n(n-1)} \{f(a) + f(b)\} + \sum_{i=2}^{n-1} w_i f(x_i) + E_6, \\ \text{where } E_6 &= K_6 (b - a)^{2n-1} f^{(2n-2)}(\xi). \end{aligned}\right\} \tag{1.11}$$

The last two methods may also be used in a repeated form, with obvious modifications.

2. Convergence and Correction Terms

For functions with the requisite degree of continuity, the derivative occurring in the error term is bounded in $[a, b]$. So the results in §1 show that we have convergence of the approximate result to the exact value of the integral for repeated rules, as $h \to 0$, and they also give the rate of convergence. If the function has a lower degree of continuity than that indicated, we can still use the approximation, but the convergence will generally be poorer. For any degree of continuity, the error term may be obtained in the form of an integral by using Peano's theorem (Davis, 1963, p.70). In solving integral equations, it is quite common to have a kernel with a simple discontinuity in the gradient at some point, and this is best handled by a fairly simple method such as the trapezium rule.

For smooth functions it is possible to obtain high accuracy by using a low-order quadrature rule with correction terms. The best-known example is the Euler-Maclaurin formula. Regarded as a method of integration, this expresses the error of the trapezium rule in terms of a series of derivatives of $f(x)$. Let $T(f, h)$ represent the trapezium rule approximation to $I(f)$ at interval h,

$$T(f, h) = \tfrac{1}{2}h \left\{f(a) + 2 \sum_{i=1}^{n-1} f(a + ih) + f(b)\right\}. \tag{2.1}$$

Then we can show that

$$\left.\begin{aligned} I(f) = T(f, h) &- \frac{1}{12} h^2\{f'(b) - f'(a)\} + \frac{1}{720} h^4\{f'''(b) - f'''(a)\} \\ &+ \dots - \frac{B_{2m}}{(2m)!} h^{2m}\{f^{(2m-1)}(b) - f^{(2m-1)}(a)\} + R, \end{aligned}\right\} \tag{2.2}$$

where the remainder R is given by

$$R = -\frac{B_{2m+2}}{(2m+2)!} h^{2m+2}(b - a) f^{(2m+2)}(\xi), \tag{2.3}$$

and the coefficients B_i are the Bernouilli numbers. The condition required on $f(x)$ is that the highest derivative occurring should be continuous.

The Euler-Maclaurin formula may be applied directly in certain cases, for example if the higher derivatives are known explicitly, or if we have

$$f^{(2k-1)}(a) = f^{(2k-1)}(b), \quad k = 1, 2, \dots, m. \tag{2.4}$$

In the latter case, the trapezium rule can give very high accuracy. An alternative form which is often useful in practice is Gregory's formula, obtained by replacing the derivatives in (2.2) by forward and backward differences. We write $f_0 = f(a)$, $f_n = f(b)$,

$$\Delta f_0 = f(a + h) - f(a), \quad \nabla f_n = f(b) - f(b - h), \text{ etc.} \tag{2.5}$$

Then we have

$$\left. \begin{aligned} I(f) = T(f, h) &- \frac{1}{12} h(\nabla f_n - \Delta f_0) - \frac{1}{24} h(\nabla^2 f_n + \Delta^2 f_0) \\ &- \frac{19}{720} h(\nabla^3 f_n - \Delta^3 f_0) + \ldots \end{aligned} \right\} \tag{2.6}$$

The function values involved in the correction terms are inside the range of integration, if we do not go beyond the nth differences. The formula may be used to estimate the accuracy of the trapezium rule, for smooth functions, or to provide a difference correction in solving integral equations.

Another application of the Euler-Maclaurin formula is in deriving the method of Romberg extrapolation. Let us write (2.2) as

$$T(f, h) = I(f) + Ah^2 + Bh^4 + \ldots + O(h^{2m+2}), \tag{2.7}$$

assuming the appropriate derivatives are continuous. If we calculate the trapezium rule approximation for different values of h, we can use this result to eliminate successive error terms from the expansion. Thus, from $T(f, h)$ and $T(f, \frac{1}{2}h)$ we can obtain

$$J(f, h) = \frac{4}{3} T(f, \tfrac{1}{2}h) - \frac{1}{3} T(f, h), \tag{2.8}$$

which has an error of $O(h^4)$. Similarly, the quantity

$$V(f, h) = \frac{16}{15} J(f, \tfrac{1}{2}h) - \frac{1}{15} J(f, h) \tag{2.9}$$

has error of order h^6, etc. This method of successive extrapolation is convenient and efficient for automatic computation, provided the integrands are sufficiently smooth. The interval need not be halved at each stage of extrapolation; any two values of h can be used, with an appropriate modification in (2.8) and (2.9). But halving is most economical in practice, because the function values used in calculating $T(f, h)$ are also needed for $T(f, \frac{1}{2}h)$, etc.

3. Special Quadrature Formulae

So far we have considered the integration of smooth functions, which can be approximated by polynomials. In cases where the integrand is singular, we modify the approach of §1 and consider formulae of the type

$$\int_a^b p(x)f(x)dx \simeq w_1f(x_1) + w_2f(x_2) +\ldots+ w_nf(x_n). \tag{3.1}$$

We suppose that the integral exists, and that the integrand can be expressed as the product of a singular function $p(x)$ and a smooth function $f(x)$. If the integrals

$$\int_a^b x^k p(x)dx, \qquad k = 0, 1, 2,\ldots,$$

can be evaluated, we proceed as before, choosing the parameters w_i, x_i to make (3.1) exact when $f(x)$ is a polynomial of a certain degree. The degree will depend on the number of free parameters, as in the previous quadrature rules of Gauss type.

An example of such a formula is the Gauss-Chebyshev rule

$$\left.\begin{aligned} &\int_{-1}^{+1} \frac{f(x)}{\sqrt{1-x^2}}\,dx = \frac{\pi}{n}\sum_{i=1}^{n} f\{\cos(i-\tfrac{1}{2})\tfrac{\pi}{n}\} + E_7, \\ &\text{where} \\ &\qquad E_7 = \frac{2\pi}{2^{2n}(2n)!}\,f^{(2n)}(\xi). \end{aligned}\right\} \tag{3.2}$$

This is exact for polynomials of degree $2n - 1$. Similar formulae can be obtained for the range $[a, b]$, whenever $p(x)$ has the form $(x - a)^\alpha\ (b - x)^\beta$, with $\alpha, \beta > -1$ (Hildebrand, 1956, §8.9).

In the application to integral equations, the kernel is a function of two variables, and we have to consider integrals of the type $\int_a^b K(x, y)f(y)dy$. The function $K(x, y)$, which is known, may have a singularity, and $f(y)$ is an unknown function which is the solution of the integral equation. To take a specific example (Atkinson, 1971, p.104), suppose

$$K(x, y) = G(x, y)|x - y|^\alpha, \tag{3.3}$$

where $G(x, y)$ is smooth. If we use linear interpolation for the non-singular part of the integrand, the integral is approximated by

$$\left.\begin{aligned}\sum_{i=1}^{n}\int_{y_{i-1}}^{y_i}\left\{\left(\frac{y_i-y}{h}\right)G(x,\ y_{i-1})f(y_{i-1})+\left(\frac{y-y_{i-1}}{h}\right)G(x,\ y_i)f(y_i)\right\}\\ \times\ |x-y|^{\alpha}dy,\end{aligned}\right\}\qquad(3.4)$$

where $y_i = a + ih$, and $n\,h = b - a$. We evaluate (3.4) at the points $x_k = a + kh$ to obtain linear combinations of the values of $f(y_i)$, of the form $\sum_i W_{ki} f(y_i)$, which are then substituted into the integral equation. A typical term of the integral in (3.4) has the form

$$\int_{y_{i-1}}^{y_i}\left(\frac{y-y_{i-1}}{h}\right)\ |x-y|^{\alpha}dy \qquad (3.5)$$

which certainly exists if $\alpha > -1$. Putting $y = y_{i-1} + ht$, $x = x_k = a + kh$, (3.5) becomes

$$h^{\alpha+1}\int_0^1 t|t-(i-k-1)|^{\alpha}dt. \qquad (3.6)$$

This integral may be calculated for all relevant values of $(i - k)$, and it is then easy to obtain the weights W_{ki}. The calculation is particularly simple in this case, because the singular function is symmetric in x and y, but the same method may be used for more general kernels.

4. Construction of Fourier Series

We turn now to a rather different problem, the representation of a function $f(x)$ as a Fourier series. We consider the Fourier series not only in the classical sense, as an expansion in terms of sine and cosine functions, but also in the general sense where the expansion functions are any orthogonal system. Suppose the functions $\phi_k(x)$, $k = 0, 1, 2,\ldots$ are orthogonal with respect to the inner product defined by

$$(u,\ v) = \int_a^b w(x)u(x)v(x)dx, \qquad (4.1)$$

where $w(x)$ is a non-negative weight function on $[a, b]$. Then $(\phi_i, \phi_j) = 0$, $i \neq j$, and $(\phi_i, \phi_i) = d_i$ say (where $d_i = 1$ if the functions are normalized). Any function $f(x)$ defined on $[a, b]$ may be formally expanded in terms of the $\phi_k(x)$ as follows

$$f(x) = \sum_{k=0}^{\infty} \alpha_k \phi_k(x), \tag{4.2}$$

where $$\alpha_k = \frac{1}{d_k} \int_a^b w(x) f(x) \phi_k(x) dx. \tag{4.3}$$

We assume that $f(x)$ is such that the integrals in (4.3) exist. The problem of the convergence of the series is mentioned later, but it is clear that the system $\{\phi_k(x)\}$ must have a completeness property if it is to be capable of representing a general function. Without going into the theory (for which see Davis, 1963, Ch.11), we quote the result that the sequence $\{\phi_k(x)\}$ is complete in a space with an inner product such as (4.1) if we have $(f, \phi_k) = 0$ for all k if and only if $f(x) \equiv 0$. The powers of x are complete in this sense if $[a, b]$ is a finite interval, for functions which are continuous in $[a, b]$. (It is known from Weierstrass' theorem that any continuous function over a finite interval may be approximated uniformly by a polynomial of sufficiently high degree.) The simple powers x^k, $k = 0, 1, 2, \ldots$, are of course not orthogonal with respect to the inner product (4.1), but it is easy to construct polynomials which are. Provided the integrals $\int_a^b x^k w(x) dx$ exist for all k, we can use the Gram-Schmidt process to orthogonalize the system $\{x^k\}$, but more convenient methods are available for practical work. Polynomials which are orthogonal with respect to an inner product such as (4.1) have a number of other properties which enable us to construct them easily; perhaps the most useful is the fact that they satisfy a three-term recurrence relation, as illustrated in examples (a) and (b) below.

We now give the definitions and some basic properties of three important orthogonal systems.

(a) Legendre polynomials are obtained by taking $\phi_k(x) = P_k(x)$, a polynomial of degree k, with $a = -1$, $b = 1$, $w(x) = 1$ in (4.1). With the scaling condition $P_k(1) = 1$, we have $P_0(x) = 1$, $P_1(x) = x$, and

$$(k + 1)P_{k+1}(x) - (2k + 1)P_k(x) + k\,P_{k-1}(x) = 0, \tag{4.4}$$

from which successive polynomials may be calculated. The normalizing factor d_k is equal to $2/(2k+1)$.

(b) Chebyshev polynomials are obtained by taking $\phi_k(x) = T_k(x)$, a polynomial of degree k, with $a = -1$, $b = 1$, $w(x) = 1/\sqrt{1-x^2}$ in (4.1). They satisfy the relation

$$T_{k+1}(x) - 2xT_k(x) + T_{k-1}(x) = 0, \tag{4.5}$$

with $T_0(x) = 1$, $T_1(x) = x$. The normalizing factor $d_k = \pi/2$ for $k > 0$, and $d_0 = \pi$.

c) Classical Fourier series are obtained by taking the interval $[-\pi, +\pi]$, with the set $\{\phi_k(x)\}$ as 1, cos x, sin x, cos 2x, sin 2x,... The weighting function $w(x) = 1$. The Fourier series is given formally by

$$\left.\begin{aligned} f(x) &= \frac{1}{2}\alpha_0 + \sum_{k=1}^{\infty} (\alpha_k \cos kx + \beta_k \sin kx), \\ \text{where} \quad \alpha_k &= \frac{1}{\pi}\int_{-\pi}^{+\pi} f(x)\cos kx\, dx. \\ \beta_k &= \frac{1}{\pi}\int_{-\pi}^{+\pi} f(x)\sin kx\, dx. \end{aligned}\right\} \tag{4.6}$$

The theory and results given above for orthogonal expansions may be exactly paralleled by a theory for the approximation of $f(x)$ over a discrete set of points. The integration in (4.1) and (4.2) is replaced by summation, so that if the points are x_j, $j = 0, 1,\ldots, N$, the inner product corresponding to (4.1) is defined by

$$(u, v) = \sum_{k=0}^{n} w_k\, u(x_k)v(x_k). \tag{4.7}$$

We may define orthogonal systems over $\{x_k\}$ just as before, and the expansion is exactly analogous to (4.2), except that it is finite. The results may be useful in two ways, either for direct curve-fitting of discrete data, or for obtaining an approximation to the results in the continuous case. For example the Chebyshev expansion may be approximated in the following way (Clenshaw, 1962). The infinite Chebyshev series is given by

$$\left.\begin{aligned} f(x) &= \frac{1}{2}\alpha_0 + \alpha_1 T_1(x) + \alpha_2 T_2(x) + \ldots, \\ \text{where } \alpha_k &= \frac{2}{\pi}\int_{-1}^{+1} f(x)T_k(x)/\sqrt{1-x^2}\,dx\,. \end{aligned}\right\} \tag{4.8}$$

Consider the quantities

$$c_k = \frac{2}{N}\left\{\frac{1}{2}f(x_0)T_k(x_0) + f(x_1)T_k(x_1) + \ldots + \frac{1}{2}f(x_N)T_k(x_N)\right\} \tag{4.9}$$

where the first and last terms are halved, and where $x_j = \cos(\pi j/N)$. If the Chebyshev series converges rapidly, the c_k are good approximations to the exact coefficients α_k, and we can show that

$$c_k = \alpha_k + \alpha_{2N-k} + \alpha_{2N+k} + \alpha_{4N-k} + \ldots \tag{4.10}$$

In numerical work, the summation is generally easier to perform than the exact integration, and in fact it corresponds to a method of approximating the integral in (4.8).

5. Convergence of Fourier Series

To discuss convergence of the series in (4.2), we have to define our measure of distance, or norm. Suppose we take the norm to be

$$\|u - v\| = \sqrt{(u - v,\ u - v)}, \tag{5.1}$$

where the functions $u(x)$, $v(x)$ are such that the required integrals exist. If the generalized Fourier series converges to $f(x)$ in this norm, we have

$$\left\|f(x) - \sum_{k=0}^{n} \alpha_k\phi_k(x)\right\| \to 0 \quad \text{as} \quad n \to \infty, \tag{5.2}$$

where the α_k are given by (4.3). Convergence in this sense is easier to prove than pointwise convergence; in general it follows if the system $\{\phi_k(x)\}$ is complete, and if $f(x)$ belongs to a complete function space. However, this is not strong enough for many cases of interest.

In practical work, uniform convergence is usually of more value than convergence in the sense of (5.2), and the results which can be obtained depend on the class of functions to which $f(x)$ belongs. Thus the Fourier expansion (4.6) is periodic, so it cannot converge to $f(x)$ at all points unless we have $f(-\pi) = f(+\pi)$. With this condition, the series is uniformly and absolutely convergent to $f(x)$ in $[-\pi, +\pi]$ if $f(x)$ is continuous and $f'(x)$ is piecewise continuous. It can be shown in this case that the coefficients α_k, β_k decrease at least as fast as $1/k$ as $k \to \infty$. For higher degrees of smoothness we have faster convergence, and many other results can be obtained about the convergence for functions of different types.

For the expansion of $f(x)$ in Legendre polynomials, the convergence results are similar. We have uniform and absolute convergence in the open interval $(-1, +1)$ if $f(x)$ is continuous and $f'(x)$ is piecewise continuous. Since the Legendre polynomials are simply linear combinations of the powers of x, we would expect to be able to represent <u>any</u> continuous function uniformly on $(-1, +1)$, by the Weierstrass theorem. If the function is analytic in some region in the complex plane which contains the segment $[-1, +1]$, we can obtain expressions for the rate of convergence of the coefficients (Davis, 1963, §12.4). The Chebyshev expansion of $f(x)$ is essentially the Fourier cosine expansion for the range $[0, \pi]$, as we see by putting $x = \cos\theta$ in (4.8), and its convergence properties are similar to those of classical Fourier theory. Periodicity is not required in this case, because the interval is only half the full range. A detailed discussion of the order of convergence of the Fourier coefficients α_k as $k \to \infty$ is given by Mead and Delves (1973) for some practically important expansions, and their results can be used to obtain bounds for the rate of pointwise convergence.

<u>References</u>

ATKINSON, K.E. (1971) A survey of numerical methods for the solution of Fredholm integral equations of the second kind. Department of Mathematics, Indiana University.

CLENSHAW, C.W. (1962) <u>Chebyshev Series for Mathematical Functions</u>. N.P.L. Math. Tables Vol. 5, H.M. Stationery Office.

DAVIS, P.J. (1963) <u>Interpolation and Approximation</u>. Blaisdell.

DAVIS, P.J. and RABINOWITZ, P. (1967) <u>Numerical Integration</u>. Blaisdell.

HILDEBRAND, F.B. (1956) <u>Introduction to Numerical Analysis</u>. McGraw Hill.

MEAD, K.O. and DELVES, L.M. (1973) On the convergence rate of generalised Fourier expansions. <u>J. Inst. Maths. and its Appl</u>. (to appear).

STROUD, A.H. and SECREST, D. (1966) <u>Gaussian Quadrature Formulas</u> Prentice-Hall.

CHAPTER 3 NUMERICAL LINEAR ALGEBRA

I. Gladwell
University of Manchester

1. Introduction. Systems of Linear Equations

In this chapter we discuss the solution of linear systems of equations and corresponding error bounds, linear least squares solutions, the problem of determining the rank of a matrix and the solution of simple and generalised algebraic eigenproblems and corresponding error bounds.

Many of these topics have been treated in considerable detail elsewhere and we shall state well-known results without reference. For those to whom the material is new we recommend Forsythe and Moler (1967), Fox (1964), Walsh (1966, chs. 2 and 3) and Wilkinson (1965). Many procedures implementing the methods we describe are given in Wilkinson and Reinsch (1971) and Forsythe and Moler (1967).

Consider the solution of

$$A\underline{x} = \underline{b} \tag{1.1}$$

where A is an $n \times n$ nonsingular matrix and $\underline{b}$ is an n-component vector. A numerical solution for (1.1) is usually obtained by Gaussian Elimination with partial pivoting - a process which is mathematically equivalent to finding a permutation matrix P, an upper triangular matrix U and a unit lower triangular matrix L such that

$$PA = LU, \tag{1.2}$$

and then solving the equations

$$L\underline{z} = P\underline{b}, \quad U\underline{x} = \underline{z} \tag{1.3}$$

for $\underline{x}$. Choosing P determines a strategy of interchanging the rows of A, and finding L and U is the "elimination" process. It is always possible to find matrices P, L and U satisfying (1.2) when exact arithmetic is employed. The matrix P is usually chosen (and hence the pivoting strategy determined) to make the elimination process numerically stable. A procedure for Gaussian elimination with partial pivoting is given in Wilkinson and Reinsch (1971, I/7).

For many systems of type (1.1), to perform the elimination (1.2) stably we must scale (equilibrate) the rows and columns of A in some suitable way before searching for P, L and U in (1.2). We usually choose diagonal matrices D_1 and D_2 so that

$$D_1 A D_2 = B \tag{1.4}$$

has a suitable form, then solve

$$B\underline{y} = D_1\underline{b} \tag{1.5}$$

for $\underline{y} = D_2^{-1}\underline{x}$, and thus determine $\underline{x}$.

A further refinement of Gaussian elimination is the use of complete pivoting, that is the interchange of both rows and columns of A during the elimination process. The complete pivoting technique applied to an equilibrated matrix B is mathematically equivalent to finding suitable permutation matrices P_1 and P_2 such that

$$P_1 B P_2 = L_1 U_1 \tag{1.6}$$

and modifying the equations (1.3) accordingly. In Wilkinson (1965, ch. 4) it is shown that with a suitable equilibration technique and complete pivoting one can achieve better bounds on the rounding error than with partial pivoting. However, it is generally believed that the error bounds obtainable for partial pivoting are pessimistic, and since complete pivoting is computationally expensive, partial pivoting is preferred in most applications. The practical effects of various pivoting and equilibration strategies are discussed in Wilkinson (1965), Van der Sluis (1969), Curtis and Reid (1972) and Kreifelts (1972).

A matrix A is symmetric if $A = A^T$ (that is, $a_{ij} = a_{ji}$, i, j = 1, 2,..., n) and positive definite if $\underline{x}^T A\underline{x} > 0$ for all $\underline{x} \neq \underline{0}$. [Equivalently, A is positive definite if all its eigenvalues are positive.] Gaussian elimination for positive definite symmetric matrices is a stable process without any pivotal strategy (Wilkinson, 1965). However, in this case we exploit symmetry in order to economise on computer storage, and use a Cholesky decomposition. To solve equation (1.1), we compute a lower triangular matrix L such that

$$A = LL^T, \tag{1.7}$$

and solve

$$L\underline{y} = \underline{b},\ L^T\underline{x} = \underline{y} \tag{1.8}$$

for $\underline{x}$. Alternatively, we might choose a unit lower triangular matrix

L_1 and a diagonal matrix D such that

$$A = L_1 D L_1^T, \tag{1.9}$$

and modify equations (1.8) accordingly. The choice (1.9) has better stability properties than (1.8) for some applications (Peters & Wilkinson, 1970). There is a procedure implementing the Cholesky decomposition in Wilkinson and Reinsch (1971, I/1). When A is symmetric but not positive definite, the Cholesky decomposition breaks down. However many other methods which exploit the symmetry of A have been devised; Bunch and Parlett (1971) give a survey of these methods and Parlett and Reid (1970) implement one of them in a procedure.

It can be shown that the process of Gaussian elimination with partial pivoting for solving equation (1.1) in floating-point arithmetic is mathematically equivalent to solving

$$(A + \delta A)\underline{x} = \underline{b} \tag{1.10}$$

for $\underline{x}$, where δA is a "small" matrix for which the following bound may be obtained (Forsythe and Moler, 1967, §21), $\|\delta A\| \leq h(n)\beta^{-t}\alpha$. Here $\|B\| = \|B\|_\infty = \max_{\|\underline{x}\|_\infty = 1} \|B\underline{x}\|_\infty$, where $\|\underline{x}\|_\infty = \max_{1 \leq i \leq n} |x_i|$, for any matrix B, β is the base of the arithmetic, t is the number of significant floating-point digits used, and α is the magnitude of the largest element in modulus occurring during the Gaussian elimination process. [It is easy to see that

$$\|B\| = \max_{1 \leq i \leq n} \sum_{j=1}^{n} |b_{ij}|.]$$

As Forsythe and Moler (1967) remark, if $n\beta^{-t} < .01$ then $h(n) = 1.01(n^3 + 3n^2)$, and in practice $\alpha < 8\|A\|$ almost always. Hence we can estimate $\|\delta A\|$.

Now if $\underline{x}$ satisfies equation (1.1) and $\hat{\underline{x}}$ satisfies equation (1.10) then

$$(A + \delta A)(\hat{\underline{x}} - \underline{x}) = -\delta A\underline{x}.$$

Since A is non-singular

$$A + \delta A = A(I + A^{-1}\delta A),$$

hence $A + \delta A$ is non-singular if

$$\|A^{-1}\| \, \|\delta A\| < 1 \tag{1.11}$$

(Wilkinson, 1965). Hence assuming that (1.11) is true, we have

$$\hat{\underline{x}} - \underline{x} = - (I + A^{-1}\delta A)^{-1}A^{-1}\delta A\underline{x}.$$

Now $\|(I + A^{-1}\delta A)^{-1}\| \leq 1/(1 - \|A^{-1}\| \, \|\delta A\|)$,

and so $\|\hat{\underline{x}} - \underline{x}\| \leq \dfrac{\|A^{-1}\| \, \|\delta A\| \, \|\underline{x}\|}{1 - \|A^{-1}\| \, \|\delta A\|}$.

Defining the condition number $K(A) = \|A\| \, \|A^{-1}\|$, we have

$$\frac{\|\hat{\underline{x}} - \underline{x}\|}{\|\underline{x}\|} \leq \frac{K(A) \, \|\delta A\|/\|A\|}{1 - K(A) \, \|\delta A\|/\|A\|}, \tag{1.12}$$

which provides a relative bound on the error in terms of the condition number of A and the "size" of δA relative to A. The bound is large when $K(A)$ is large and this (with a suitable scaling of A) corresponds to the case where $\|A^{-1}\|$ is large, that is A is "almost singular".

This same error bound (1.12) can provide some additional insight into the solution of the special equations which arise when solving linear and nonlinear Fredholm integral equations of the second kind. We seek solutions f of equations of the type

$$f - \lambda K f = g, \tag{1.13}$$

where K is an integral operator. Many approximations for this problem lead to equations

$$\underline{f} - \lambda H \underline{f} = \underline{g} \tag{1.14}$$

for a solution vector $\underline{f}$ where H is a matrix (this type of approximation is used at each iteration when solving a nonlinear equation). Hence in the bound (1.12) we may set $A = I - \lambda H$ when solving equation (1.14). Now if $\mu = 1/\lambda$ is close to an eigenvalue of the integral equation

$$\mu f - K f = 0, \tag{1.15}$$

then presuming that H is a good representation of K (as it would be if it were to be used say for finding the eigenvalues of K), it is likely that μ will be close to an eigenvalue of H, hence $\|A^{-1}\| = \|(\mu I - H)^{-1}\|$ will be large and the error $\hat{\underline{x}} - \underline{x}$ may be relatively large.

Let us return to the problem of solving equation (1.1). By Gaussian elimination with partial pivoting we obtain an approximate solution $\underline{x}_0$. We can form the residual $\underline{r}_0 = \underline{b} - A\underline{x}_0$, and if we now solve $A\,\delta\underline{x}_0 = \underline{r}_0$ we see that $\underline{x}_1 = \underline{x}_0 + \delta\underline{x}_0$ will be an exact solution if we compute $\delta\underline{x}_0$ exactly. In practice we cannot do this and we only obtain $\underline{x}_1$ approximately. This approximation can then be used to find a new solution $\underline{x}_2$ and so on. This algorithm is usually known as iterative refinement (or improvement) and is implemented in practice as follows.

Find P, L and U such that PA = LU and solve $LU\underline{x}_0 = P\underline{b}$ for $\underline{x}_0$.

For i = 0, 1, 2,... perform the following steps

(i) Form $\underline{r}_i = \underline{b} - A\underline{x}_i$, accumulating inner products double length to avoid cancellation.

(ii) Solve $A\delta\underline{x}_i = \underline{r}_i$ using the factors L and U already computed.

(iii) Form $\underline{x}_{i+1} = \underline{x}_i + \delta\underline{x}_i$. If $\|\delta\underline{x}_i\|$ is small enough then accept $\underline{x}_{i+1}$ as the correct answer and stop, otherwise return to (i).

Note that the solution in (ii) requires little computation compared with the original elimination. We remark on some features of iterative refinement.

(a) If $\dfrac{\|\delta A\|}{\|A\|} K(A) < \frac{1}{2}$, then the iterative refinement procedure will converge (Forsythe and Moler, 1967, p.110).

(b) As a useful working rule, we may assume that if all iterates $\underline{x}_r$ for $r \geq R$ agree to single precision then $\underline{x}_R$ is the exact solution rounded to single precision. [It is actually possible to find a counter-example.]

(c) If $\underline{x}_0$ and $\underline{x}_1$ agree to p binary digits, we may expect $\underline{x}_{r-1}$ and $\underline{x}_r$ to agree to about pr digits. The size of p is a measure of the condition of the matrix A - the smaller p is, the worse the condition of A.

A procedure incorporating iterative refinement is given in

Wilkinson and Reinsch (1971, I/7), and a procedure implementing a similar process for the Cholesky decomposition is given in the same reference (I/2).

Another method for obtaining a solution of equation (1.1) is to use the orthogonal decomposition

$$AP = QU , \tag{1.16}$$

where P is a permutation matrix giving interchange of columns, Q is an orthogonal matrix (that is $Q^{-1} = Q^T$) and U is an upper triangular matrix. Wilkinson (1965, p.233 ff.) describes an algorithm for finding the matrices P, Q, U in (1.16) using Householder transformations, and other techniques based on Givens rotations or Gram-Schmidt orthogonalisation are also available. To solve equation (1.1) using the decomposition (1.16) we find $\underline{x}$ from

$$\underline{y} = Q^T\underline{b}, \; U\underline{z} = \underline{y}, \; \underline{x} = P\underline{z}. \tag{1.17}$$

Wilkinson (1965), p.236) obtains a very satisfactory bound for the error in the decomposition (1.16) for Householder transformations, which would lead one to suspect that this process has better numerical stability properties than Gaussian elimination. However, Gaussian elimination is usually preferred in practice, because it involves about half as much work as the decomposition (1.16), and partial pivoting generally gives satisfactory stability.

2. Linear Least Squares Problems and the Determination of the Rank of a Matrix

Certain techniques for computing the numerical solution of an integral equation involve the solution of linear algebraic least squares problems. (In particular, these arise in the solution of Fredholm equations of the first kind.)

If $\|\underline{x}\|_2 = \sum_i x_i^2$, the (minimal) least squares problem can be expressed as: "Find the set of vectors X such that $\underline{x} \in X$ minimises

$$F(\underline{x}) = \|A\underline{x} - \underline{b}\|_2, \tag{2.1}$$

where A is an $m \times n$ matrix ($m \geq n$) and $\underline{b}$ is an m-component vector, then select the vector $\underline{x} \in X$ such that $\|\underline{x}\|_2$ is minimised (that is the vector of shortest length). If rank(A) = n, X consists of just one vector and this is the unique least squares solution. Peters and

Wilkinson (1970) give a comprehensive discussion of this problem and Wilkinson & Reinsch (1971, I/8, I/9, I/10) give procedures implementing various methods of solution.

We see that if rank(A) = n, the solution satisfies

$$A^T A\underline{x} = A^T\underline{b} \tag{2.2}$$

(the normal equations). If rank(A) = r < n, the matrix A^TA is singular and equation (2.2) does not have a unique solution. If A^TA is non-singular, the solution of equation (2.2) to obtain $\underline{x}$ is a superficially attractive method, but the matrix A^TA is often ill-conditioned as may be seen from the following example. Consider the matrix of rank two

$$A = \begin{bmatrix} 1 & 1 \\ \beta & 0 \\ 0 & \beta \end{bmatrix}, \quad \text{for which} \quad A^TA = \begin{bmatrix} 1+\beta^2 & 1 \\ 1 & 1+\beta^2 \end{bmatrix}.$$

We see that if β is small, $1+\beta^2$ rounds to 1 in machine arithmetic and the rank of A^TA appears to be 1, so the numerical solution of the normal equations will fail.

To alleviate such difficulties we usually use methods which operate on the original matrix A rather than on A^TA. If rank(A) = n, it can be shown (Peters and Wilkinson, 1970, p.313) that there exists a decomposition

$$AP = QU, \tag{2.3}$$

where P is an n x n permutation matrix, Q comprises the first n columns of an m x m orthogonal matrix and U is an n x n upper triangular matrix. The solution $\underline{x}$ is given by

$$\underline{x} = PU^{-1}Q^T\underline{b}. \tag{2.4}$$

To compute the decomposition (2.3), we can use Householder transformations in a similar way to the computation of the decomposition (1.16). Using suitable column interchanges to ensure the numerical stability of the process (thus determining P), we choose an orthogonal matrix $\tilde{Q}^T$ so that

$$\tilde{Q}^TAP = \begin{bmatrix} U \\ \hline 0 \end{bmatrix}, \tag{2.5}$$

then Q in (2.3) comprises the first n columns of $\tilde{Q}$. Note that if

$$\tilde{Q}^T \underline{b} = \left[\begin{array}{c} \underline{c} \\ \hline \underline{d} \end{array}\right] \tag{2.6}$$

where $\underline{c}$ has n components, then

$$Q^T \underline{b} = \underline{c} \tag{2.7}$$

and $||\underline{d}||_2$ is the length of the residual remaining after solving the least squares problem, so that

$$||\underline{d}||_2 = \|A\underline{x} - \underline{b}\|_2 \tag{2.8}$$

is a measure of how accurately $\underline{b}$ can be represented as a linear combination of the columns of A. If rank(A) = r < n, the solution of the least squares problem becomes more complicated. Peters and Wilkinson (1970) and Hanson and Lawson (1969) discuss this problem in detail.

Instead of using an orthogonal decomposition such as (2.3), it is possible to solve the least squares problem by using a decomposition similar to (1.2) into upper and lower triangular factors. The expression which replaces (2.4) in this situation is much more complicated and this method is not recommended.

Having obtained an approximate least squares solution, we can improve the solution by iterative refinement. It would seem that a useful algorithm could be devised similar to the one described in the previous section with Q replacing L; however Golub and Wilkinson (1966) show that if $||\underline{d}||_2$ in (2.8) is large enough this process of iterative refinement will not converge to correct working accuracy. Björck and Golub (1967) have implemented a modification of the iterative refinement procedure which can be shown to converge.

We have assumed above that the rank of the matrix A in (1.1) or (2.1) is known. In some situations this is not so and we may wish to determine the rank numerically. Since the rank of a matrix does not vary continuously with its elements, this is strictly impossible, because rounding errors can change the rank. However there are some commonly used methods for estimating the rank which we describe. In the formation of the decomposition AP = LU (equation (1.2)) we generate the rows and columns of U and L in a sequence of steps; at the ith step an important quantity calculated is the pivot u_{ii} which remains unchanged in subsequent steps. If the pivot is zero the matrix A is singular, that is it has rank less than n (since $|\det(A)| = |\det(U)| = |\prod_{i=1}^{n} u_{ii}|$), and the elimination process breaks down (though the decomposition exists and can be completed). In practice, zero pivots occur infrequently but small pivots which usually indicate singularity or near singularity of A may occur. In any case, small pivots in an equilibrated matrix A usually indicate poor numerical stability properties. It is conventional to test the pivots against a tolerance which is dependent on the size of $\|A\|$ and the working accuracy of the machine, in order to determine the rank of A, but whatever tolerance is used this is not an infallible technique (Wilkinson, 1965). If rank determination is a serious problem, it may be better to use an orthogonal decomposition $AP = QU$ and to inspect the diagonal elements of A, which play a similar role to the pivots in Gaussian elimination ($|\det(A)| = |\prod_{i=1}^{n} u_{ii}|$, since $\det(Q) = 1$). If one is still unsure of the rank, one can employ the singular value decomposition. Forsythe & Moler (1967, pp.5-7) show that any $m \times n$ matrix A can be written in the form

$$A = U\Sigma V^T , \tag{2.9}$$

where

$$U^TU = VV^T = I_n, \quad \Sigma = \underset{i}{\operatorname{diag}}(\sigma_i) \tag{2.10}$$

and

$$\sigma_1 \geq \sigma_2 \geq \ldots \geq \sigma_n \geq 0. \tag{2.11}$$

This is the singular value decomposition where the values $\sigma_i (i = 1, 2,\ldots, n)$ are the singular values of A; since $A^TAV = V\Sigma^2$, the values $\sigma_i^2 (i = 1, 2,\ldots, n)$ are the eigenvalues of A^TA. If the rank of A is r then

$$\sigma_{r+1} = \sigma_{r+2} = \ldots = \sigma_n = 0. \tag{2.12}$$

Golub and Reinsch have implemented a stable method for finding the singular value decomposition (Wilkinson and Reinsch, 1971, I/10), and the safest method at present for determining the rank of a matrix would seem to be to test the singular values against a small tolerance. The singular value decomposition can also be used to find the least-squares solution. If we compute the singular value decomposition of A and hence determine the rank of A to be r, say, then we set $\sigma_{r+1} = \sigma_{r+2} = \ldots = \sigma_n = 0$ and define

$$\Sigma^+ = \operatorname*{diag}_i(\sigma_i^+), \tag{2.13}$$

where

$$\sigma_i^+ = \begin{cases} 1/\sigma_i, & \sigma_i > 0 \\ 0, & \sigma_i = 0 \end{cases} .$$

The least squares solution is given by

$$\underline{x} = V\Sigma^+U^T\underline{b}. \tag{2.14}$$

This technique is recommended for computing the least squares solution if all other techniques fail to produce reliable answers (Wilkinson and Reinsch, 1971, p.7).

3. The Algebraic Eigenvalue Problem

We consider the problem of finding the eigenvalues λ and corresponding non-null eigenvectors $\underline{x}$ such that

$$A\underline{x} = \lambda\underline{x} , \tag{3.1}$$

and later we also study the more general eigenproblem

$$A\underline{x} = \lambda B\underline{x} , \tag{3.2}$$

where A and B are real square matrices of order n.

The eigenvalues λ of equation (3.1) satisfy the equation

$$f_n(\lambda) \equiv \det(A - \lambda I) = 0, \tag{3.3}$$

and solving (3.3) is mathematically equivalent to determining the roots of the characteristic polynomial $f_n(\lambda)$, which can be obtained by expanding the determinant in (3.3), and is of degree n. Hence A has n eigenvalues, adding algebraic multiplicities, which may be real or may occur in complex conjugate pairs. To each eigenvalue λ there corresponds a set of linearly independent eigenvectors and the rank of this set is the geometric multiplicity of λ, which is less than or equal to the algebraic multiplicity. We shall assume, as is generally the case in practice, that there is a complete set of eigenvectors for A. (If the set is not complete, there is a set of principal vectors (Wilkinson, 1965; Forsythe and Moler, 1967) which together with the eigenvectors span the space of n-component vectors.) Computationally, it may be difficult to establish in general whether A has a complete set of eigenvectors, but in the case when A is symmetric (which we can ensure for integral equations if the kernel is symmetric), the eigenvalues are real and there is a complete set of eigenvectors which can be made orthonormal, so that $\underline{x}_i^T \underline{x}_j = \delta_{ij}$ $(i, j = 1, 2, \ldots, n)$. If Q is the matrix whose i-th column is $\underline{x}_i$ and $\Lambda = \mathrm{diag}(\lambda_i)$, then in this case $AQ = Q\Lambda$ and Q is orthogonal $(Q^T = Q^{-1})$.

The methods used for solving problem (3.1) depend on a variety of factors including whether A is symmetric, whether any eigenvectors are required and how many eigenvalues are required. Methods based on finding the roots of the characteristic polynomial $f_n(\lambda)$ are notoriously unreliable, since the coefficients of the polynomial have to be determined in rounded arithmetic, and the roots of polynomials are generally sensitive to small perturbations in the coefficients. It is possible to apply standard root-finding techniques to equation (3.3) directly (using, say, Gaussian elimination to evaluate the determinant), but this is only recommended for tracing the path of an eigenvalue when the matrix A depends on a parameter.

In the case where we wish to compute the largest eigenvalue (in magnitude) of A and its corresponding eigenvector, we can apply the <u>power method</u>. Let $\lambda_1, \lambda_2, \ldots, \lambda_n$ and $\underline{x}_1, \underline{x}_2, \ldots, \underline{x}_n$ be the eigenvalues and corresponding eigenvectors of the real matrix A, and let

$$|\lambda_1| > |\lambda_2| \geq |\lambda_3| \geq \ldots \geq |\lambda_n| \qquad (3.4)$$

(assuming λ_1 is real). For any vector $\underline{y} = \sum_{i=1}^{n} c_i \underline{x}_i$, we may form $\underline{z}^{(r)} = A^r \underline{y}$ by the algorithm

$$\underline{z}^{(0)} = \underline{y}, \ \underline{z}^{(i)} = A\underline{z}^{(i-1)} \ (i = 1, 2, \ldots, r). \qquad (3.5)$$

Then

$$\underline{z}^{(r)} = \lambda_1^r \left\{ c_1 \underline{x}_1 + \sum_{i=2}^{n} c_i \left(\frac{\lambda_i}{\lambda_1}\right)^r \underline{x}_i \right\}, \qquad (3.6)$$

and from (3.4), it is clear that $\underline{z}^{(r)} \to \lambda_1^r c_1 \underline{x}_1$ as $r \to \infty$. Because we are working in machine arithmetic we will always have $c_1 \neq 0$ in practice and so, for r large enough, a comparison of successive iterates $\underline{z}^{(r)}$ will give an approximation for λ_1 and a unnormalised approximation for $\underline{x}_1$. In practice we must scale the iterates $\underline{z}^{(r)}$ in order to avoid overflow or underflow. An extension of this algorithm to the computation of eigenvalues of equal magnitude and complex conjugate pairs of eigenvalues is given by Fox (1964) and Wilkinson (1965, ch.9).

If, after calculating λ_1 and $\underline{x}_1$, we wish to determine the next eigenvalue λ_2 and its corresponding eigenvector $\underline{x}_2$, we can proceed by constructing a new eigenproblem with the same eigenvalues and eigenvectors as A, except that λ_1 is replaced by zero. We can then apply the power method to the new eigenproblem to calculate λ_2 and $\underline{x}_2$. The technique for modifying the eigenproblem is called deflation and a comprehensive description and analysis of the available methods is given by Wilkinson (1965, ch.9).

As Wilkinson (1965) shows, the power method combined with any stable form of deflation can be an effective technique for finding a few of the largest eigenvalues and corresponding eigenvectors of a matrix. An alternative approach to the problem is to use simultaneous iteration, in which a number of vectors are iterated using the power method. At each iteration the vectors are modified so that they remain linearly independent (as they must be if they are to represent the eigenvectors). Wilkinson (1965, p.602) describes this process in the form due to Bauer (who called the method treppen-iteration), and he demonstrates theoretical convergence for a particular implementation. However, he remarks that the method is fraught with practical difficulties. Recently, there has been much interest in simultaneous iteration for symmetric matrices; Rutishauser has implemented his own variant in a procedure (Wilkinson and Reinsch, 1971, II/9), and Jennings (1967) has also contributed useful ideas. It seems, however, that the effectiveness of the method still depends largely on the details of the implementation.

The methods we have described above may be the only feasible ones if the matrix A in (3.1) is full and very large. For moderate sized matrices we might consider methods where we operate directly on the matrix A, rather than on special vectors, with the aim of reducing the matrix to a form whose eigenvalues are easy to obtain. To do this, we use similarity transformations. If H is a non-singular matrix we say that $B = H^{-1} AH$ is similar to A, and it is easy to see that B has the same eigenvalues as A. In most applications the matrix H is chosen to be orthogonal, and Wilkinson (1965, p.126 ff.) shows that the commonly used orthogonal matrices, namely those derived from Householder transformations and plane rotations, lead to numerically stable algorithms. It is possible to choose such orthogonal matrices so that, under the similarity transformation, the matrix B is upper Hessenberg (that is, $b_{ij} = 0$, $j < i-1$) for a general A. If A is symmetric then B is symmetric and tridiagonal ($b_{ij} = 0$, $|i-j| > 1$). In the symmetric case we can then locate the largest roots accurately by the method of Sturm sequences. The tridiagonal matrix B has the form

$$B = \begin{bmatrix} \alpha_1 & \beta_1 & & & \\ \beta_1 & \alpha_2 & \beta_2 & & \\ & \cdots & & & \\ & & \cdots & & \\ & & & \cdots & \\ & & & & \beta_{n-1} \\ & & & \beta_{n-1} & \alpha_n \end{bmatrix} \qquad (3.7)$$

Let J_r be the leading principal minor of B of order r, and let $f_r(\lambda) = \det(J_r - \lambda I)$. We define $f_0(\lambda) = 1$ and define $f_r(\lambda)$ to have the opposite sign to $f_{r-1}(\lambda)$ when $f_r(\lambda) = 0$; then we have the following property (Wilkinson, 1965, p.300):

> The number of agreements in sign of consecutive members of the sequence $f_0(\lambda), f_1(\lambda), \ldots, f_n(\lambda)$ is equal to the number of eigenvalues of B which are greater than λ.

This property and the observation that we can generate the sequence $f_0(\lambda), f_1(\lambda), \ldots, f_n(\lambda)$ for any value of λ by the recurrence relation

$$f_i(\lambda) = (\alpha_i - \lambda) f_{i-1}(\lambda) - \beta_i^2 f_{i-2}(\lambda), \; f_{-1} = 0, \; f_0 = 1, \quad (3.8)$$

form the basis for many programs for finding the eigenvalues of a symmetric matrix. A modified form of the recurrence relation (3.8) is used in the procedure in Wilkinson and Reinsch (1971, II/5) and a procedure for reducing a general matrix to tridiagonal form is also given (II/2).

A general method for calculating all the eigenvalues of a matrix is the QR algorithm. Basically, each step of the algorithm proceeds as follows. Given A_s, form the orthogonal decomposition $A_s = Q_sR_s$ (as in the decomposition (1.16) with $R = U$) and then form $A_{s+1} = R_sQ_s$. Then $A_{s+1} = Q_s^TA_sQ_s$, and if $A_0 = A$, $A_s (s = 1, 2,\dots)$ has the same eigenvalues as A. It can be shown that if the eigenvalues of A are distinct then the iterates A_s tend to upper triangular form with the eigenvalues of A on the diagonal. In Wilkinson (1965, ch.8), a comprehensive description of the computational process is given, including the effects of shifts of origin, multiple eigenvalues and complex conjugate pairs of eigenvalues. It is shown that the QR algorithm preserves the upper Hessenberg form, or the tridiagonal form in the case when A is symmetric. We usually reduce the matrix A by similarity transformations to such a condensed form before applying the QR algorithm, in order to save computation. A procedure for reducing a general matrix to upper Hessenberg form is given in Wilkinson and Reinsch (1971, II/13) and procedures are also given for the QR algorithm (II/3, II/4, II/6, II/14, II/15). The procedure in II/6 will calculate the largest few eigenvalues of a real symmetric matrix but our experience shows that it must be used with caution, since it is mainly useful when the eigenvalues are well-separated.

If the eigenvectors of A are also required then the most commonly used technique is the method of inverse iteration. This method is a variant of the power method and is described and analysed in detail by Wilkinson (1965, pp.619-633). The iteration takes the form

$$\underline{y}_{s+1} = (A-qI)^{-1}\underline{y}_s \tag{3.9}$$

with $\underline{y}_0$ arbitrary and q an approximation to an eigenvalue. If q is close to an eigenvalue, the iteration (3.9) converges rapidly to the corresponding eigenvector in general (though the details of the implementation are important). Many of the algorithms mentioned above use inverse iteration implicitly or explicitly for calculating the eigenvectors, and in Wilkinson and Reinsch (1971, II/18) there is a new analysis and a procedure which is strongly recommended.

When an approximate eigenvalue λ and corresponding approximate eigenvector $\underline{x}$ have been found, we may wish to bound the error in terms of the residual

$$\underline{\eta} = A\underline{x} - \lambda\underline{x}. \tag{3.10}$$

If H is a matrix whose columns are the normalised eigenvectors of A and $\lambda_1, \lambda_2, \ldots, \lambda_n$ are the eigenvalues, we can show (Wilkinson, 1965, p.171) that

$$\min_i |\lambda - \lambda_i| \leq k \, \|\underline{\eta}\|_2 / \|\underline{x}\|_2, \tag{3.11}$$

where

$$k = \|H\|_2 \, \|H^{-1}\|_2 \geq 1 \tag{3.12}$$

is the spectral condition number of H. When A is symmetric, H is orthonormal and $k = 1$, and symmetric matrices have well-conditioned eigenproblems in this sense. For general matrices, if $\underline{y}_i$ and $\underline{x}_i$ are left- and right- eigenvectors corresponding to λ_i we have

$$A\underline{x}_i = \lambda_i\underline{x}_i, \quad A^T\underline{y}_i = \lambda_i\underline{y}_i \ . \tag{3.13}$$

If
$$s_i = \underline{y}_i^T\underline{x}_i / (\|\underline{x}_i\|_2 \, \|\underline{y}_i\|_2), \tag{3.14}$$

Wilkinson (1965, p.88) shows that $|s_i^{-1}| \leq k \leq \sum_{i=1}^{n} |s_i^{-1}|$. Hence the error bound (3.11) can be expected to be large when corresponding left- and right- eigenvectors of A are almost orthogonal. (If A is symmetric these eigenvectors are the same, and no problem arises.)

The above bounds are useful for the insight they provide when it is known that A has a complete set of eigenvectors. However, we can derive a bound for a simple eigenvalue without this knowledge. Let λ_i be a simple eigenvalue and let $\underline{x}_i$ and $\underline{y}_i$ be corresponding right- and left- eigenvectors, then, if λ is an approximation to λ_i, we define $\underline{\eta}$ by $\underline{\eta} = A\underline{x}_i - \lambda\underline{x}_i$. Premultiplying by $\underline{y}_i^T$ we obtain

$$\lambda_i - \lambda = \underline{y}_i^T\underline{\eta} / \underline{y}_i^T\underline{x}_i,$$

and hence

$$|\lambda_i - \lambda| \leq \|\underline{n}\|_2/(\|\underline{x}_i\|_2 s_i). \qquad (3.15)$$

We now turn to the generalised eigenproblem $A\underline{x} = \lambda B\underline{x}$. If we assume first that B is non-singular (the case B singular and A non-singular is similar), we can write $C = B^{-1}A$ and solve the eigenproblem for C by one of the techniques described above. Though this may sometimes be a good technique, it is not recommended in general for two reasons. First the eigenvalues of the computed matrix C may be considerably perturbed from the eigenvalues of (3.2) if B^{-1} is poorly determined, and, secondly, if A and B have any special property such as symmetry, this may be lost in the formation of C.

Iterative techniques such as the power method, inverse iteration and simultaneous iteration may be generalised directly for the eigenproblem (3.2), and details are given in Peters and Wilkinson (1970). If B is symmetric and positive definite we can convert the generalised problem (3.2) to a standard problem (3.1) as follows. Using the Cholesky decomposition (1.7) we find a lower triangular matrix L such that $B = LL^T$; then the problem $A\underline{x} = \lambda B\underline{x}$ is converted to the form $C\underline{y} = \lambda\underline{y}$ where

$$C = L^{-1}AL^{-T}, \quad \underline{y} = L^T\underline{x} . \qquad (3.16)$$

A procedure for forming C is given in Wilkinson and Reinsch (1971, II/10). If B has eigenvalues near zero then the matrix C in (3.16) may be ill-determined and in this case one of the methods described below should be used. If B is positive semi-definite (that is, it has some zero eigenvalues) and the number of zero eigenvalues is known, it is possible (Peters and Wilkinson, 1970, p.480) to use a modified form of the transformation (3.16). Fix and Heiberger (1972) describe a generalisation of (3.16) for positive semi-definite symmetric B where the rank of B is not known in advance. Their method, unlike the one mentioned above, distinguishes between eigenvalues which are well- or ill-conditioned with respect to perturbations of A and B.

A general method for solving the eigenproblem (3.2) is the QZ algorithm as described by Moler and Stewart (1971). This algorithm is an extension of the QR algorithm, and reduces to it when $B = I$. It is based on the observation that there exist orthogonal matrices Q and Z such that both QAZ and QBZ are upper triangular, and hence

$A\underline{x} = \lambda B\underline{x}$ can be converted to $QAZ\underline{y} = \lambda QBZ\underline{y}$, from which the eigenvalues can easily be extracted. The algorithm proceeds as follows. First, using a generalisation of the Householder reduction to upper Hessenberg form described above, we simultaneously reduce A to upper Hessenberg form and B to upper triangular form. Next, we apply a generalisation of the QR algorithm which preserves the form of B while extracting the eigenvalues, and finally the eigenvectors are recovered from the triangular matrices thus obtained.

References

BJÖRCK, A. and GOLUB, G. (1967) Iterative refinement of linear least square solutions by Householder transformation. BIT (Nord. Tidskr. Inf. Behandl.) Vol. 7, pp.322-337.

BUNCH, J.R. and PARLETT, B.N. (1971) Direct methods for solving symmetric indefinite systems of linear equations. SIAM J. numer. Anal. Vol. 8, pp.639-655.

CURTIS, A.R. and REID, J.K. (1972) On the automatic scaling of matrices for Gaussian elimination. J. Inst. Math. and its Appl. Vol. 10, pp.118-124.

FIX, G. and HEIBERGER, R.M. (1972) An algorithm for the ill-conditioned generalised eigenvalue problem. SIAM J. numer. Anal. Vol. 9, pp.78-88.

FORSYTHE, G. and MOLER, C.B. (1967) Computer Solution of Linear Algebraic Systems. Prentice-Hall. Englewood Cliffs, New Jersey.

FOX, L. (1964) An Introduction to Numerical Linear Algebra. Clarendon Press, Oxford.

GOLUB, G.H. and WILKINSON, J.H. (1966) Note on the iterative refinement of least squares solution. Num. Math. Vol. 9, pp.139-148.

HANSON, R.J. and LAWSON, C.L. (1969). Extensions and applications of the Householder algorithm for solving linear least squares problems. Maths. Comput. Vol. 23, pp.787-812.

JENNINGS, A. (1967) A direct iteration method of obtaining latent roots and vectors of a symmetric matrix. Proc. Camb. phil. Soc. math. phys. Sci. Vol. 63, pp.755-765.

KREIFELTS, Th. (1972) Uber Pivot-Strategien bei der Losung linearer Gleichungssysteme. Computing, Vol. 10, pp.167-175.

MOLER, C.B. and STEWART, G.W. (1971) An algorithm for the generalised matrix eigenvalue problem $Ax = \lambda Bx$. Tech. Rept CNA-32. Centre for

Numerical Analysis, Austin, Texas.

PARLETT, B.N. and REID, J.K. (1970) On the solution of a system of linear equations whose matrix is symmetric but not definite. BIT (Nord. Tidskr. Inf. Behandl.) Vol. 10, pp.386-397.

PETERS, G. and WILKINSON, J.H. (1970) The least squares problem and pseudo-inverses. Comput. J. Vol. 13, pp.309-316.

PETERS, G. and WILKINSON, J.H. (1970a) $Ax = \lambda Bx$ and the generalized eigenproblem. SIAM J. numer. Anal. Vol. 7, pp.479-492.

VAN DER SLUIS, A. (1970) Condition, equilibration and pivoting in linear algebraic systems. Num. Math. Vol. 15, pp.74-86.

WALSH, J.E. (ed.) (1966) Numerical Analysis: an Introduction. Academic Press: London.

WILKINSON, J.H. (1965) The Algebraic Eigenvalue Problem. Clarendon Press: Oxford.

WILKINSON, J.H. and REINSCH, C. (1971) Handbook for Automatic Computation. Volume II. Linear Algebra. Springer-Verlag: Berlin.

CHAPTER 4 FUNCTION SPACES AND LINEAR OPERATORS

L.M. Delves
University of Liverpool

Motivation : Why talk about Function Spaces?

In the background of any numerical solution of an equation or a set of equations lies (we hope) an analysis that assures us that the solution we seek exists, and is perhaps even unique. Some of the conditions necessary for this to be so may be obvious; others, less so. For example, we saw in Chapter I that the existence and uniqueness of solutions of the linear Fredholm equation

$$f(x) = g(x) + \lambda \int_a^b K(x,y)\, f(y)dy \tag{1.1}$$

depended partly on the question of the existence of nontrivial solutions of the homogeneous equation

$$f(x) = \lambda \int_a^b K(x,y)\, f(y)dy \tag{1.2}$$

(which we might have guessed); and partly on what set of solutions we decide to call admissible (which we might not have guessed). To take a concrete example, the equation

$$\left.\begin{aligned} f(x) &= g(x) + \lambda \int_a^x K(x,y)\, f(y)dy \\ K(x,y) &= y^{x-y} \quad 0 < y \leqslant x \leqslant 1 \\ &= 0 \quad\quad\; y = 0 \end{aligned}\right\} \tag{1.3}$$

has an unique ℓ^2 solution f_1 for any ℓ^2 function g and all λ. But if f_1 is a solution, it is easy to show that, for $\lambda = 1$ and independent of g, so is

$$f_2 = f_1 + \bar{f} \tag{1.4}$$

where

$$\begin{aligned} \bar{f}(x) &= x^{x-1} \quad 0 < x \leqslant 1 \\ &= 0 \quad\quad\;\; x = 0. \end{aligned} \tag{1.5}$$

Of course, $\bar{f}(x)$, and hence $f_2(x)$, is not an ℓ^2 function; and this difference is crucial.

Examples like this suggest that the theory be phrased in a way which concentrates attention wholly on members of some "allowed" class of functions, those with other properties then not entering at all: we are led naturally to consider linear function (vector) spaces, and in this chapter we give a very brief introduction to some of the basic concepts appropriate to a discussion of integral equations. The discussion inevitably represents largely a collection of definitions; but we try to motivate these in the context for which they are intended. Proofs of the theorems, as well as many other results, can be found in for example Taylor (1957), and we refer to this book for brevity as T. To read this chapter is no substitute for studying the subject; but it may serve as a reference or as an introduction to the ideas involved.

2. Some Relevant Vector Spaces

We recall first:

Definition 1

A vector space (linear space, function space) X is a set of elements $\{x, y, \dots\}$ referred to as points, or vectors, which is closed under addition ($x+y = z \in X$) and under multiplication by a (real or complex) scalar α ($\alpha x = y \in X$). For a fuller statement of these axioms, see (T, p.9).

Examples:

The spaces of interest here have as elements functions of one (or more) variables defined over an interval $[a,b]$:

a) The space $C[a,b]$ has as elements all functions $x(s)$ defined and continuous on $a \leq s \leq b$; addition is defined by $(x_1 + x_2)(s) = x_1(s) + x_2(s)$ and the axioms above obviously then apply.

b) The space $L^p[a,b]$.

Let L^p be the class of all functions x of the variable s such that $x(s)$ is defined for all $s \in [a,b]$ except possibly for a set of measure zero; is measurable; and such that $\int_a^b |x(s)|^p \, ds$ exists (in the Lebesgue sense). Then L^p is a vector space if equality is interpreted to mean "equality almost everywhere"; that is,+

+ We cannot expect to use the usual definitions of equality here; for if we did, then the zero element z would be such that $z(s)=0$ everywhere. But then $x+(-x)=z$ would not be true for all s if x were not defined everywhere.

$$x = y \Rightarrow x(s) = y(s) \text{ almost everywhere, } a \leq s \leq b$$

c) l^p.

This space (for $p \geq 1$) is defined to be the set of all sequences
$= \{\zeta_n\}$ such that

$$\sum_{n=1}^{\infty} |\zeta_n|^p < \infty$$

The connection between l^p (especially l^2) and L^p (L^2) becomes
pparent when we consider orthonormal (generalised Fourier) expansions
n L^2; the ζ_n then represent the expansion coefficients. Of these
paces, the case p=2 is the most important.

The axioms embodied in definition 1 by themselves are too
eneral to be very useful; we obtain more interesting spaces by
dding further properties of the elements we propose to use. Since
e want to talk about convergence, some notion of size is appropriate:

efinition 2

A norm on a vector space X is a positive real-valued function
$||.||$ with the following properties (T. p.83):

$$\begin{aligned} &\text{(i)} \quad ||x|| \neq 0 \text{ if } x \neq 0; \ ||0|| = 0 \\ &\text{(ii)} \quad ||\alpha x|| = |\alpha| \, ||x|| \qquad (2.1) \\ &\text{(iii)} \quad ||x_1 + x_2|| \leq ||x_1|| + ||x_2|| \quad \text{(triangle inequality).} \end{aligned}$$

if property (i) does not hold, we say that $||.||$ is a semi norm).
f $||x||$ is defined for every element of X, we say that X is a normed
ector space. Often, functions (norms) satisfying (2.1) can be
efined in many ways for a given vector space. Appropriate norms for
he spaces we have introduced are:

(a) $C[a,b]$: (i) $||x|| = \max_{a \leq s \leq b} |x(s)|$

$$\text{or (ii) } ||x|| = \int_a^b |x(s)| \, ds \qquad (2.2a)$$

(b) $L^p(a,b): \|x\| = \left\{ \int_a^b |x(s)|^p ds \right\}^{\frac{1}{p}}$ (2.2b)

(c) $l^p \quad : \|x\| = \left\{ \sum_{n=1}^{\infty} |\zeta|^p \right\}^{\frac{1}{p}}$ (2.2c)

With the concept of a norm we can talk about the convergence of an infinite sequence of elements:-

Definition 3 Convergence (Strong convergence).

Let $\{x_n\}$ be an (infinite) sequence of elements in a normed space X, and let $x \in X$ be such that $\lim_{n\to\infty} \|x_n - x\| = 0$. Then $\{x_n\}$ converges (strongly) to x.

For any convergent sequence we have for any n,m:-

$$\|x_n - x_m\| \leq \|(x_n - x) + (x - x_m)\| \leq \|x_n - x\| + \|x_m - x\| \underset{n,m\to\infty}{\rightarrow} 0 \tag{2.3}$$

Definition 4 Cauchy Sequence (T. p.74)

A sequence $\{x_n\}$ such that $\lim_{n,m\to\infty} \|x_n - x_m\| = 0$ is called Cauchy. If $\{x_n\}$ is convergent, (2.3) shows that it is Cauchy; but the converse is not necessarily true, because the limit x may not be in X. (As an example, the limit of a sequence of continuous functions may not be continuous). We therefore distinguish between spaces in which the limit point x is guaranteed to exist, and spaces in which there is no such guarantee.

Definition 5 Complete Space; Banach Space

A normed vector space X is complete if every Cauchy sequence $\{x_n\}$ is convergent. A complete normed vector space is termed a Banach space.

It is obviously very convenient to work with a complete space. Recall that much of classical analysis consists in proving that limits exist; there is none of that in a complete space.+

+ No wonder that functional analysis is sometimes referred to as "soft" analysis.

Whether a vector space is complete may depend on the norm chosen for it: the three spaces defined above $l^{(p)}$ and $L^{(p)}(a,b)$ are complete with the choice of norm (2.2b), (2.2c) while C[a,b] is complete with norm (i) but incomplete with norm (ii). It is therefore fortunate that we can always complete a space by adjoining additional elements to it; and this completion makes no essential difference to the properties of the space. The process is sufficiently important in principle that we describe briefly here the construction used. (T. pp.74,75)

Let us consider any two Cauchy sequences $\{x_n\},\{x_n'\}$ in X. If

$$||x_n - x_n'|| \underset{n}{\rightarrow} 0 \tag{2.4}$$

we say that the sequences are equivalent, and write $x_n \sim x_n'$. This process divides all possible Cauchy sequences into equivalence classes. We may denote each equivalence class by $\{x_n\}$, where $\{x_n\}$ is any Cauchy sequence in the class. Define the space $\overline{X}$ to be the set of all such equivalence classes. This clearly defines a vector space if addition is defined in the obvious way:-

$$\{x_n\} + \{y_n\} = \{x_n + y_n\} \tag{2.5}$$

and we may similarly define a norm:-

$$||\{x_n\}|| = \{||x_n||\} \tag{2.6}$$

Moreover, there is a very close correspondence between the elements $y=\{x_n\}$ of $\overline{X}$, and x of X. For to each x we can associate the element in $\overline{X}$ which contains the sequence $\{x,x,x,x \ldots\ldots\}$. Let $\overline{X}_0$ be the set of all such elements; clearly from (2.5), (2.6) $\overline{X}_0$ and X are isometric, (T. p.74) and can essentially be identified. Moreover, it can be shown (T. p.75) that X is complete (which is the point of the exercise) and that $\overline{X}_0$ is dense in $\overline{X}$; that is, given any element y in X and any $\epsilon>0$, we can find an element y_0 in $\overline{X}_0$ such that $||y_0-y||< \epsilon$. This then implies that any element in $\overline{X}$ can be approximated arbitrarily closely by one in X; so that the process of completion makes no essential difference to X.

Finally in this section, we consider still more structured spaces. These are the inner product spaces. (T. p.106)

Definition 6 (T. p.106)

An inner product is a scalar valued (that is, real or (in general) complex) function of two elements x,y with the following properties:

(i) $(x_1, x_2 + x_3) = (x_1, x_2) + (x_1, x_3)$

(ii) $(x_1, x_2) = (x_2, x_1)^*$ where * denotes complex conjugate

(iii) $(x_1, \alpha x_2) = \alpha(x_1, x_2)$

(iv) $(x_1, x_1) > 0,\ x \neq 0;\ (0,0) = 0$

Definition 7 (T.p.106)

An inner product space is a vector space X in which (x,y) is defined for each pair of elements x,y in X.

Examples For the spaces C(a,b), $\ell^2(a,b)$, we may choose to write

$$(x,y) = \int_a^b x^*(s)\, y(s)\, ds \qquad (2.7a)$$

while for l^2 we may choose

$$(x,y) = \sum_{n=1}^{\infty} x_n^* y_n \qquad (2.7b)$$

An inner product space may be complete or incomplete.

Definition 8 (T. p.118)

A Hilbert space is a complete inner product space

Comment An incomplete inner product space is sometimes called a pre-Hilbert space. It may be completed in the same way as an incomplete normed space. The spaces l^2 and $L^2(a,b)$ are complete inner product spaces with the definitions (2.7a),(2.7b). The subspace M of L^2 composed of functions which are continuous on [a,b],

.s incomplete.

Any inner product space may be viewed as a normed space; we ɪay define

$$||x|| = (x,x)^{\frac{1}{2}}$$

3. Basic Properties of Hilbert Spaces

In this next section we spell out briefly various properties of a Hilbert space which are useful in other chapters. Not all of these properties depend on the space being complete. The most basic is

Theorem 1 (Schwartz's inequality) (T. p.107)

$$|(x,y)|^2 \leq (x,x)\,(y,y) \qquad (3.1)$$

This property ensures, for example, that the inner products (2.7a) or (2.7b) are finite for each pair of elements x,y in L^2 or l^2 respectively.

A number of important properties concern orthonormal sets:-

Definition 9

The set $\{x_i\}$ is orthonormal iff $(x_i,x_j) = \delta_{ij}$.

Examples:

The set $\left\{\sqrt{\frac{2}{\pi}}\ \mathrm{Sin}\ ix,\ i=1,2,\ldots\right\}$ is orthonormal in $C[0,\pi]$ and $L^2(0,\pi)$

So is the set $\left\{\sqrt{\frac{2}{\pi}}\ \mathrm{Sin}\ ix,\ i=10,11,\ldots\right\}$.

The first of these two sets is complete in either of the spaces; the second is not. (See below). The following theorem does not assume completeness.

Theorem 2 (T.p.109)

If $\{x_i,\ i=1,\ldots,n\}$ is orthonormal, then for any element x

$$\sum_{i=1}^{n} |(x,x_i)|^2 \leq ||x||^2 \qquad \text{(Bessel's inequality)} \quad (3.2)$$

and for any x,y

$$\sum_{i=1}^{N} (x,x_i)(x_i,y) \leq ||x|| \; ||y|| \qquad \text{(Parseval's inequality)} \quad (3.3)$$

We are usually interested in orthonormal sets as sequences in which to expand a given element; that is, we attempt an expansion of the form

$$x = \sum_{i} a_i \, x_i \qquad \left\{x_i\right\} \text{ orthonormal} \qquad (3.4)$$

The conditions under which this expansion converges to x are given in the following.

Definition 10 (T. p.114)

Let M be a subspace of X. Suppose that $\left\{x_i\right\}$ is an orthonormal sequence such that the statement

$$(x, x_i) = 0 \;\forall\, i, \quad x \in M$$

implies x = 0. Then $\left\{x_i\right\}$ is said to be complete in M, and is referred to as a complete orthonormal sequence (c.o.n.s.) in M.

Examples:

See above under definition (9).

Theorem 3

If $\left\{x_i\right\}$ is a c.o.n.s. in M, $x \in M$ and we set $a_i=(x_i,x)$ in (3.4), then the series in (3.4) converges to x. Further, equality holds in the two relations of Theorem 2.

Definition 11

Let the Hilbert space X be such that it contains at least one c.o.n.s. Then it is said to be separable.

We need only consider separable spaces in these notes. If a space is separable, every c.o.n.s. in it has the same number of elements (rather, the same cardinality).

4. Linear Operators on Hilbert and Banach Spaces

So far we have considered mainly properties associated with the elements of a space. We obviously need to consider also functions of these elements, of which we have in fact introduced two in the norm, and the inner product functions. More generally, a function K may have as argument one or more elements in a space X and produce as result one or more elements of a space Y. The set of permissible arguments, for which the function is defined, is the domain D(K); the set of results which can be produced is the range R(K). These may be the whole spaces X and Y, or subsets of X and Y, respectively.

We consider here mainly linear, homogeneous functions of one variable; these are usually referred to as linear operators.

Definition 12 (see T. p.162)

The (not necessarily linear) operator K from X to Y is bounded if for some C

$$\sup_{0 \neq x \in X} \left\{ ||Kx|| / ||x|| \right\} = C < \infty \qquad (4.1)$$

We then define the norm of K to be $||K|| = C$

Note that in (4.1), x is an element of the normed space Y so that $||Kx||$ is taken in the metric of Y, while $||x||$ is in the metric of X. If Y=X, we say that $K \in [X]$, the space of linear operators from X to X.

We see also that if I is the unit operator $Ix=x \ \forall x$, then from (4.1)

$$||I|| = 1 \qquad (4.2)$$

Example:

Fredholm integral operators with L^2 kernels or with continuous kernels on a compact set, are bounded. For we have in either case

$$||Kx||^2 = \int_a^b \left| \int_a^b K(s,t)\, x(t)dt \right|^2 ds$$

$$\leq \left\{ \int_a^b \int_a^b |K(s,t)|^2 ds\, dt \right\} \int_a^b |x(t)|^2 dt$$

$$\leq C^2 \, ||x||^2 \quad \text{say.}$$

Theorem 4 (T. p.163)

If A,B are bounded operators in [X] then AB is defined and in [X], and

$$||AB|| \leq ||A|| \, ||B||$$

As with functions on a scalar space, the concept of continuity is important.

Definition 13 (T. p.70)

The (not necessarily linear) operator K from X to Y is continuous at the point x_0 if, for every $\delta>0$ there exists an $\epsilon>0$ such that for every member of the set x: $||x-x_0|| < \epsilon$ we have $||K(x)-K(x_0)|| \leq \delta$.

Comment Every bounded linear operator is continuous; for we have $||K(x)-K(x_0)|| \leq ||K|| \, ||x-x_0|| \leq ||K|| \, \epsilon$ so that given δ we may choose $\epsilon = \frac{\delta}{||K||}$ and satisfy the conditions of the definition.

The following two theorems have obvious relevance for the solution of operator equations:-

Theorem 5 (Neumann expansion) (T. p.164)

If $K \in [X]$ and $||K|| < 1$, then $(I-K)^{-1}$ exists and $\in [X]$ and

$$(I-K)^{-1} = I + K + K^2 + \ldots\ldots + K^n +$$

Further

$$||(I-K)^{-1}|| \leq \frac{1}{1-||K||}$$

Theorem 6

Suppose K and K_0 are operators in [X] and K^{-1}, K_0^{-1} exist and are continuous.

hen $$||K^{-1} - K_0^{-1}|| \leq \frac{||K_0^{-1}||\ ||I-K_0^{-1}K||}{I - ||I-K_0^{-1}K||}$$

his theorem K is relevant if K_0^{-1} is a known approximation to K^{-1}, for xample [note that we do not need to know K_0 to use it].

For many purposes the fact that K is bounded (and continuous) llows us to analyse the convergence of numerical methods in ufficient detail. In some cases however, and particularly in the igenvalue problem, more restricted classes of operators have mportance. We introduce two further properties. The first of hese is a simple symmetry property:

efinition 14 (T. p.324)

The operator $K\in[K]$ is Hermitian (or symmetric) if for every x nd y in its domain

$$(y,Kx) = (Kx,y)$$

efinition 15

The operator K* adjoint to K is an operator whose domain is the ange of K such that $(y,Kx)=(K^*y,x)$ whenever x,y belong to D(K) and (K) respectively. Clearly, if D(K)=R(K), and K is Hermitian, K*=K. e then say that K is self-adjoint.

heorem 7 (T. p.314)

The eigenvalues of an Hermitian operator are real; eigenvectors orresponding to different discrete eigenvalues are orthogonal.

It is necessary to make the restriction to the discrete spectrum f the operator, because Hermiticity, and even boundedness, does not uarantee a discrete eigenvalue spectrum. The Fredholm integral perators with continuous or $\mathcal{L}^2$ kernels, however, do have a discrete pectrum although since they may not be Hermitian, the spectrum may ot be real. They are examples of a class of operators known as ompact or completely continuous operators.

efinition 16 (T. p.274)

Let K be an operator from X to Y. K is compact, or completely ontinuous if, for each bounded sequence $\{x_n\}$ in X, the sequence

$\{Kx_n\}$ in Y contains a convergent subsequence.

Comment The drastic "smoothing" effect of such operators can be seen by considering an orthonormal set $\{x_n\}$ in X. Such a sequence clearly has no limit, and nor does any subsequence, since for each n, $m \neq n$, x_m is orthogonal to x_n and $||x_m - x_n||^2 = 2$. However, if K is compact, the sequence $\{Kx_n\}$ is either convergent, or contains a convergent subsequence (the need for the latter qualifying phrase can be seen if we consider shuffling the sequence $\{x_n\}$ so that some of its early members appear arbitrarily late).

Examples:

The Fredholm integral operators with L^2 (or continuous) kernel in L^2 (or C) [a,b] are completely continuous; the unit operator I is not completely continuous. Perhaps the two most important properties of compact operators are summed up in the following theorems:

Theorem 8

The spectrum $\{\lambda\}$ of a compact operator K contains at most a countable set of points, and no point of accumulation except possibly $\lambda = 0$.

Theorem 9

If K is compact then $(\lambda - K)^{-1}$ exists unless λ is an eigenvalue of K, and is continuous if it exists.

Finally, we make a brief if tenuous connection with numerical methods. If a solution x to an integral (or other) equation has the expansion

$$x = \sum_{i=1}^{\infty} b_i h_i \tag{4.3}$$

many of the methods discussed in this book take a finite approximation x_N of the form

$$x_N = \sum_{i=1}^{N} a_i^{(N)} x_i \qquad (4.4)$$

The convergence of (4.4) to (4.3) can be discussed in terms of projection generators.

Definition 17

A linear operator $P \in [X]$ is a projection operator if $D(P)=X$ and $P^2=P$. It is said to project onto the subset M where $M=R(P)$.

Comment Projection operators are most simply constructed in terms of an orthogonal expansion such as (4.3). If the set $\{x_i\}$ is a c.o.n.s. in X, any x may be written in the form (4.3). Then a projection onto the space spanned by $\{x_1 \text{ ----} x_N\}$ is given directly by

$$P_N x = \sum_{i=1}^{N} b_i \; x_i$$

Such a projection is called orthogonal, because it splits X into two orthogonal parts: for every $x \in X$

$$(P_N x, \; (I-P_N)x) = 0 \qquad (4.5)$$

P_N is also a symmetric operator; that is,

$$(P_N x, y) = (x, P_N y) \qquad (4.6)$$

Equ. (4.6) is a necessary and sufficient condition for (4.5).

References

TAYLOR, A.E. (1957) Introduction to Functional Analysis, Wiley & Sons.

CHAPTER 5 BASIC APPROXIMATION THEORY

I. Barrodale
University of Victoria, B.C.

1. Introduction

Approximation theory is the study of how given functions can be approximated by polynomials, spline functions, rational functions, and other useful approximating functions. It is a subject which has expanded very rapidly during the computer era, and which now consists of practical techniques for obtaining approximations as well as many elegant theorems concerning their properties. Applications of approximation theory to the numerical solution of integral equations arise in finite element methods, variational methods, residual minimizing techniques, and other expansion methods.

The central problem of this chapter is the approximation of a given function y, which is defined on a domain X, by an approximating function whose general form f_n depends on n parameters $\alpha_1, \alpha_2, \ldots, \alpha_n$. By choosing values α_j^* for these parameters according to some appropriate approximation criterion, we obtain a particular approximating function f_n^*. For simplicity, we shall consider here only real-valued functions defined on subsets of the real line. (Recommended texts in approximation theory include those of Cheney (1966), Davis (1963), Hayes (1970), Rice (1964, 1969), Rivlin (1969) and Talbot (1970).)

2. Types of Approximating Functions

The key to effective approximation lies in the selection of an appropriate general form f_n for the approximating function. Although nonlinear approximating functions such as

$$\alpha_1 e^{\alpha_3 x} + \alpha_2 e^{\alpha_4 x} \quad \text{and} \quad (\alpha_1 + \alpha_2 x)/(1 + \alpha_3 x + \alpha_4 x^2) \qquad (2.1)$$

are sometimes desirable, in practice we often restrict f_n to be a linear approximating function of the form

$$f_n(x) = \sum_{j=1}^{n} \alpha_j \phi_j(x). \qquad (2.2)$$

Notice that the approximating functions in (2.1) become linear if the parameters α_3 and α_4 are both replaced by known constants.

For most of this chapter we shall concentrate on linear approximation since the practical techniques for determining the parameter values α_j^* are well developed in this case. Some of the most useful types of linear approximating functions are polynomials, trigonometric functions, and spline functions with fixed knots. Consequently, typical choices for $\phi_j(x)$ in (2.2) include

$$x^{j-1},\ x^{2j-1},\ x^{2j-2},\ \sin jx,\ \cos(j-1)x,\ \text{and } (x-t_j)_+^m, \tag{2.3}$$

$$\text{where} \quad (x-t_j)_+^m = \begin{cases} (x-t_j)^m, & x \geq t_j, \\ 0, & x < t_j. \end{cases} \tag{2.4}$$

(Briefly, a spline function s_n of degree m with k prescribed knots $t_1 < t_2 < \ldots < t_k$ is a polynomial of degree m in each of the $k+1$ subintervals determined by the t_j's, and s_n and its first $m-1$ derivatives are continuous at these knots. Using (2.4) we can write

$$s_n(x) = \sum_{j=1}^{k} \alpha_j (x-t_j)_+^m + \sum_{i=1}^{m+1} \alpha_{k+i}\, x^{i-1},$$

which is in the form (2.2) with $n = k + m + 1$). However, for computational purposes s_n is often expressed in terms of B-splines (see Cox, 1972 or De Boor, 1972), which are obtained by taking divided differences of (2.4) with respect to the knots.)

One class of methods for solving integral equations involves substitution of a suitable approximating function f_n in place of the unknown solution. The values α_j^* are then determined using the transformation of f_n which arises from substituting this function in the integral equation (as shown in Chapter 8). It is therefore essential that the techniques used to calculate these parameters are applicable to general linear approximating functions, rather than to just the standard types listed in (2.3).

3. Interpolation (Collocation)

A simple technique for obtaining a linear approximation f_n^* consists of selecting n distinct points $x_i \in X$ and solving the $n \times n$ system of linear equations which results from setting

$$\sum_{j=1}^{n} \alpha_j\, \phi_j(x_i) = y(x_i), \quad \text{for } i = 1, 2, \ldots, n. \tag{3.1}$$

Provided that the matrix of coefficients

$$\begin{bmatrix} \phi_1(x_1) & \phi_2(x_1) & \cdots & \phi_n(x_1) \\ \phi_1(x_2) & \phi_2(x_2) & \cdots & \phi_n(x_2) \\ \vdots & \vdots & & \vdots \\ \phi_1(x_n) & \phi_2(x_n) & \cdots & \phi_n(x_n) \end{bmatrix}$$

is nonsingular, it is well known that the <u>interpolation problem</u> (3.1) has a unique solution. Reliable techniques based on Gaussian elimination can then be used to determine the solution values α_j^* which satisfy (3.1).

To illustrate the interpolation problem consider the case where $n = 2$, $\phi_1(x) = 1$, $\phi_2(x) = x^2$, and $X = [-1, 1]$. Here, problem (3.1) has

(i) <u>no</u> solution if $x_1 = -x_2$ and $y(x_1) \neq y(x_2)$,

(ii) an <u>infinity</u> of solutions if $x_1 = -x_2$ and $y(x_1) = y(x_2)$,

or (iii) a <u>unique</u> solution if $x_1 \neq -x_2$.

If the functions $\phi_1, \phi_2, \ldots, \phi_n$ are <u>linearly independent</u> on X, in practice problem (3.1) usually has a unique solution for any reasonable choice of n distinct points $x_i \in X$. A function f_n which <u>always</u> leads to a unique solution of (3.1) is said to satisfy the <u>Haar condition</u> on X. This is a more stringent requirement on f_n than linear independence of the functions $\phi_1, \phi_2, \ldots, \phi_n$, and only a few standard linear approximating functions satisfy the Haar condition on general domains X. (Functions of more than one independent variable never satisfy the Haar condition, and here the interpolation problem is less likely to be nonsingular.)

Of course, problem (3.1) is uniquely solvable whenever f_n is a polynomial of degree $n-1$. Newton, Lagrange, and others have provided special polynomial <u>interpolation formulae</u> for obtaining f_n^* more conveniently than by using Gaussian elimination on (3.1). Also, when f_n is a polynomial the n points x_i can be selected in a manner which ensures that f_n^* provides an effective approximation to y over the whole domain X (see Powell, 1967).

Within the class of methods referred to at the end of §2, the <u>collocation method</u> for linear integral equations is equivalent to solving the general interpolation problem (3.1).

4. Best Approximations

One way of improving on the technique of the previous section is to select m ($>n$) points $x_i \in X$ and "solve" the overdetermined system of linear equations

$$\sum_{j=1}^{n} \alpha_j \phi_j(x_i) = y(x_i), \quad \text{for } i = 1, 2, \ldots, m. \tag{4.1}$$

This allows us to represent the function y on more than n points of X when determining f_n^*. Furthermore, for problem (4.1) a "solution" always exists in any one of the following senses.

Let the residuals of the equations (4.1) be defined as

$$\delta_n(x_i) = y(x_i) - \sum_{j=1}^{n} \alpha_j \phi_j(x_i), \quad \text{for } i = 1, 2, \ldots, m, \tag{4.2}$$

and consider the problem of determining

$$\|\delta_n^*\|_p = \underset{\alpha_1,\alpha_2,\ldots,\alpha_n}{\text{minimum}} \|\delta_n\|_p, \text{ for } p = 1, 2, \text{ or } \infty, \tag{4.3}$$

where

$$\|\delta_n\|_1 = \sum_{i=1}^{m} |\delta_n(x_i)|, \tag{4.4}$$

$$\|\delta_n\|_2 = \left\{\sum_{i=1}^{m} |\delta_n(x_i)|^2\right\}^{\frac{1}{2}}, \tag{4.5}$$

$$\|\delta_n\|_\infty = \underset{1\leq i\leq m}{\text{maximum}} |\delta_n(x_i)|. \tag{4.6}$$

The quantity $\|\delta_n^*\|_p$ in (4.3) always exists, and the optimal values α_j^* yield

(i) an l_1 (or least-first-power) solution to (4.1) in the case of (4.4),

(ii) an l_2 (or least-squares) solution to (4.1) in the case of (4.5),

or (iii) an l_∞ (or Chebyshev, minimax) solution to (4.1) in the case of (4.6).

The corresponding approximating function $f_n^* = \sum_{j=1}^{n} \alpha_j^* \phi_j$ is called a <u>best l_p approximation</u>. In general f_n^* changes with p, and even for a fixed value of p there may be more than one best approximation (although this rarely happens in practice). Practical techniques for determining best l_p approximations are available for p = 1, 2, and ∞, and in §5 we discuss some of these techniques briefly.

When the given function y is defined on an interval $X = [a,b]$, then instead of selecting m points x_i on which to determine a best approximation we sometimes use all the points of X. This causes us to replace (4.4) - (4.6) in problem (4.3) by

$$||\delta_n||_p = \left\{\int_a^b |\delta_n(x)|^p \, dx\right\}^{1/p} \quad \text{for } p = 1 \text{ and } 2,$$

and $\|\delta_n\|_\infty = \underset{a \leq x \leq b}{\text{maximum}} |\delta_n(x)|$.

The quantity $\|\delta_n^*\|_p$ in (4.3) is still guaranteed to exist, and the corresponding approximating function f_n^* is now called a <u>best L_p approximation</u>.

The following example shows that best approximations for different values of p sometimes differ quite considerably. For $y(x) = x^{1/3}$, $X = [0,1]$, and $f_n(x) = \alpha_1 + \alpha_2 x$, the unique best L_p approximations, are given (to five decimal places) by

$$\left.\begin{aligned} L_1&: f_n^*(x) = 0.49066 + 0.55720\,x\,, \\ L_2&: f_n^*(x) = 0.42857 + 0.64286\,x\,, \\ L_\infty&: f_n^*(x) = 0.19245 + x\,. \end{aligned}\right\} \qquad (4.7)$$

(The reader should not be misled by this example: one interpretation of (4.7) is simply that a straight line cannot provide a good enough approximation to $x^{1/3}$ on $[0,1]$.)

Computational techniques for determining best L_p approximations vary greatly in complexity. Fairly simple techniques are available for polynomials: this is partly because they are easy to manipulate, but also because we can take advantage of certain theorems which characterize best approximations by polynomials. (For instance, we obtained the best L_1 approximation in (4.7) simply by determining the line which interpolates y at $x = \frac{1}{4}$ and $x = \frac{3}{4}$: see Barrodale, 1970.)

The following <u>alternation property</u> for best L_∞ polynomial approximation can be used to reduce considerably the computational complexity of practical techniques for solving this problem.

Theorem The polynomial $f_n^*(x) = \sum_{j=1}^{n} \alpha_j x^{j-1}$ is a best L_∞ approximation to y on $X = [a,b]$ if and only if there exist at least $n+1$ points $x_1 < x_2 < \ldots < x_{n+1}$ of X at which

$$\delta_n^*(x_i) = -\delta_n^*(x_{i+1}) = \pm \|\delta_n^*\|_\infty .$$

Here $\delta_n^*(x) = y(x) - f_n^*(x)$ and $\|\delta_n^*\|_\infty = \underset{a \leq x \leq b}{\text{maximum}} |\delta_n^*(x)|$.

(This theorem also accounts for the important role that Chebyshev polynomials of the first kind occupy in L_∞ approximation by polynomials: see Fox and Parker, 1968, and Mason, 1970.)

However, determining a best L_p approximation is considerably more complicated when f_n is a general linear approximating function. One simple method of avoiding this difficulty consists of solving the corresponding discrete problem instead, i.e. we select a "large" number m ($\gg n$) of points from X and compute a best l_p approximation. Experience suggests that this procedure is quite satisfactory in practice.

Least-squares methods for solving integral equations are well established: see Lonseth (1954) and Berezin and Zhidhov (1965). In Chapter 8 we present several numerical examples of integral equations solved by a Chebyshev approximation method.

5. Approximation Algorithms

The computational aspects of approximation theory have received much attention during the past ten to fifteen years, and there are now many algorithms available for constructing approximations with the aid of a computer. Most of these apply to standard linear approximating functions such as polynomials and splines, whereas in the context of solving integral equations we are more interested in general algorithms.

The interpolation problem (3.1) requires us to solve n linear equations for the n unknowns $\alpha_1, \alpha_2, \ldots, \alpha_n$. This general problem is best solved on a computer by Gaussian elimination, using row interchanges to control the growth of the inevitable rounding errors that occur in floating-point arithmetic: see Forsythe and Moler (1967), Wilkinson and

Reinsch (1971).

Techniques for determining l_2 (or L_2) solutions to problem (4.1) must allow for the fact that best l_2 approximations are characterized by an $n \times n$ system of linear equations which tends to be badly ill-conditioned. It is inadvisable to solve these normal equations directly by Gaussian elimination, but an equivalent system can be solved accurately using orthogonalization methods. These stable algorithms are based either on Householder transformations or on a modification of the Gram-Schmidt process of linear algebra (see Golub and Saunders, 1970, and Wilkinson and Reinsch, 1971).

There exists a general technique (see Barrodale and Roberts, 1973) for determining best l_1 approximations which is based on the simplex method of linear programming. A best l_1 (or L_1) approximation f_n^* usually interpolates y at some n points $x_i \in X$.

Chebyshev approximation is perhaps the most appealing type of best approximation criterion when solving functional equations. A best L_∞ (or l_∞) approximation usually possesses an alternation property, which distributes the residuals of largest magnitude throughout X while making them as small as possible. (Similar alternation properties characterize Chebyshev approximations by some nonlinear approximating functions, such as rational functions and splines with variable knots.) A general technique for determining l_∞ solutions to problem (4.1) is available which is based on solving the equivalent problem

$$\text{minimize} \quad w$$

$$\text{subject to} \quad \begin{cases} w + \sum_{j=1}^{n} \alpha_j \, \phi_j(x_i) \geq y(x_i) \\ \\ w - \sum_{j=1}^{n} \alpha_j \, \phi_j(x_i) \geq -y(x_i) \end{cases} \quad \text{for } i = 1, 2, \ldots, m$$

by applying the simplex method to the dual linear programming problem

$$\text{maximize} \quad \sum_{i=1}^{m} (s_i - t_i)\, y(x_i)$$

$$\text{subject to} \begin{cases} \sum_{i=1}^{m} (s_i + t_i) \leqslant 1 \\ \\ \sum_{i=1}^{m} (s_i - t_i)\, \phi_j(x_i) = 0, \text{ for } j = 1, 2, \ldots, n, \end{cases} \tag{5.1}$$

$$\text{and} \qquad s_i \geqslant 0,\ t_i \geqslant 0, \qquad \text{for } i = 1, 2, \ldots, m.$$

There are certain algorithmic modifications that can be employed to increase the efficiency of the simplex method when it is applied to problem (5.1): see Barrodale and Phillips (1973). Additional linear constraints which might occur in problem (4.1) are quite easily accommodated when using the simplex method.

Finally, techniques for best approximation by nonlinear functions are available only in certain cases. For example, it is more difficult to obtain best L_p (or l_p) approximations by rational functions when $p=1$ or 2 than when $p=\infty$: see Barrodale (1973). On the other hand local-best l_2 approximations can often be determined for general nonlinear functions by using optimization routines such as Marquardt's method: see Fletcher (1971).

References

BARRODALE, I. (1970) On computing best L_1 approximations, Talbot (1970), pp. 205-215.

BARRODALE, I. (1973) Best rational approximation and strict quasi-convexity. SIAM J. numer. Anal. Vol. 10, pp. 8-12.

BARRODALE, I. and PHILLIPS, C. (1973) Solution of an overdetermined system of linear equations in the Chebyshev norm. Preprint, Dept. of Computational and Statistical Science, University of Liverpool.

BARRODALE, I. and ROBERTS, F.D.K. (1973) An improved algorithm for discrete l_1 approximation. SIAM J. numer. Anal. Vol. 10 (to appear).

BEREZIN, I.S. and ZHIDKOV, N.P. (1965) Computing Methods, Vol. II Pergamon.

CHENEY, E.W. (1966) Introduction to Approximation Theory McGraw-Hill.

COX, M.G. (1972) The numerical evaluation of B-splines. J. Inst. Math. and its Appl. Vol. 10, pp. 134-149.

DAVIS, P.J. (1963) Interpolation and Approximation. Blaisdell.

DE BOOR, C. (1972) On calculating with B-splines. J. Approx. Theory Vol. 6, pp. 50-62.

FLETCHER, R. (1971) A modified Marquardt subroutine for nonlinear least squares. AERE Report R6799, Harwell, Berks.

FORSYTHE, G.E. and MOLER, C.B. (1967) Computer Solution of Linear Algebraic Systems Prentice-Hall.

FOX, L. and PARKER, I.B. (1968) Chebyshev Polynomials in Numerical Analysis. Oxford University Press.

GOLUB, G.H. and SAUNDERS, M.A. (1970) Linear least squares and quadratic programming, from Integer and Nonlinear Programming, ed. J. Abadie, North-Holland. pp. 229-256.

HAYES, J.G. (ed.) (1970) Numerical Approximation to Functions and Data Athlone Press.

LONSETH, A.T. (1954) Approximate solutions of Fredholm-type integral equations. Bull. Am. math. Soc. Vol. 60, pp. 415-430.

MASON, J.C. (1970) Orthogonal polynomial approximation methods in numerical analysis, Talbot (1970), pp. 7-33.

POWELL, M.J.D. (1967) On the maximum errors of polynomial approximations defined by interpolation and by least squares criteria. Comput. J. Vol. 9, pp. 404-407.

RICE, J.R. (1964, 1969) The Approximation of Functions, Vols. I and II Addison-Wesley.

RIVLIN, T.J. (1969) An Introduction to the Approximation of Functions Blaisdell.

TALBOT, A. (ed.) (1970) Approximation Theory Academic Press.

WILKINSON, J.H. and REINSCH, C. (1971) Handbook for Automatic Computation, Vol. II Linear Algebra Springer-Verlag.

PART 2

GENERAL NUMERICAL METHODS

CHAPTER 6 QUADRATURE METHODS FOR FREDHOLM EQUATIONS OF THE SECOND KIND

D.F. Mayers
University of Oxford

1. Introduction

Fredholm equations of the second kind can be approximated in a straightforward way by means of quadrature formulae. More detailed discussion of these methods, with practical examples, may be found in Fox (1962, ch.11) and Fox and Goodwin (1953). We start with a simple example. Suppose we require to find an approximate solution of the integral equation

$$\int_0^1 K(x, y)\, f(y)\, dy + g(x) = f(x). \tag{1.1}$$

We use Simpson's rule to approximate the integral, in the form

$$\int_0^1 u(y)\, dy = (1/6)\, \{u(0) + 4\, u(\tfrac{1}{2}) + u(1)\} + E_1,$$

where E_1 represents the error or remainder term. Neglecting this remainder for the moment, we obtain the relation

$$(1/6)\, \{K(x, 0)\, \tilde{f}(0) + 4\, K(x, \tfrac{1}{2})\, \tilde{f}(\tfrac{1}{2}) + K(x, 1)\, \tilde{f}(1)\} + g(x) = \tilde{f}(x).$$

To indicate that this is only an approximation to the integral equation $f(x)$ has been replaced by $\tilde{f}(x)$.

In this equation write $x = 0$, $\frac{1}{2}$ and 1 successively, and we obtain

$$(1/6)\, \{K(0, 0)\, \tilde{f}(0) + 4\, K(0, \tfrac{1}{2})\, \tilde{f}(\tfrac{1}{2}) + K(0, 1)\, \tilde{f}(1)\} + g(0) = \tilde{f}(0),$$

$$(1/6)\, \{K(\tfrac{1}{2}, 0)\, \tilde{f}(0) + 4\, K(\tfrac{1}{2}, \tfrac{1}{2})\, \tilde{f}(\tfrac{1}{2}) + K(\tfrac{1}{2}, 1)\, \tilde{f}(1)\} + g(\tfrac{1}{2}) = \tilde{f}(\tfrac{1}{2}),$$

$$(1/6)\, \{K(1, 0)\, \tilde{f}(0) + 4\, K(1, \tfrac{1}{2})\, \tilde{f}(\tfrac{1}{2}) + K(1, 1)\, \tilde{f}(1)\} + g(1) = \tilde{f}(1).$$

These constitute a set of three linear algebraic equations for the three unknowns $\tilde{f}(0)$, $\tilde{f}(\frac{1}{2})$ and $\tilde{f}(1)$. It is a simple matter to solve the equations, and we then have approximations to the values of $f(x)$ at the three points $x = 0$, $\frac{1}{2}$ and 1.

If we write $\underline{\tilde{f}}$ and $\underline{g}$ for the vectors with elements $(\tilde{f}(0), \tilde{f}(\frac{1}{2}), \tilde{f}(1))$ and $(g(0), g(\frac{1}{2}), g(1))$ respectively, we may write the set of equations in the matrix form

$$M\, \underline{\tilde{f}} + \underline{g} = \underline{\tilde{f}},$$

where M is a 3×3 matrix with elements such as $(1/6)\ K(0, 0)$. The elements of M are in fact the values of the kernel $K(x, y)$, where x and y take the values 0, $\frac{1}{2}$ and 1, but with the three columns multiplied respectively by 1/6, 4/6 and 1/6.

The equations may thus be written as

$$(I - KD)\ \underline{\tilde{f}} = \underline{g}$$

where $M = KD$, K is the matrix whose elements are the values of $K(x, y)$ and D is the diagonal matrix

$$D = \begin{bmatrix} 1/6 & 0 & 0 \\ 0 & 4/6 & 0 \\ 0 & 0 & 1/6 \end{bmatrix}.$$

As an alternative to Simpson's rule we could, for example, use the 4-point Gauss formula

$$\int_0^1 u(y)\ dy = w_1\ u(y_1) + w_2\ u(y_2) + w_3\ u(y_3) + w_4\ u(y_4) + E_2,$$

leading to

$$w_1\ K(x,y_1)\ \tilde{f}(y_1) + w_2 K(x,y_2)\ \tilde{f}(y_2) + w_3 K(x,y_3)\ \tilde{f}(y_3) + w_4 K(x,y_4) \tilde{f}(y_4)$$
$$+ g(x) = \tilde{f}(x).$$

Writing $x = y_1, y_2, y_3, y_4,$ we obtain a set of 4 equations which may be written

$$(I - KD)\ \underline{\tilde{f}} = \underline{g}\ ,$$

as above. The elements of the matrix K are now the values of $K(y_i, y_j)$, and D is the diagonal matrix with elements w_1, w_2, w_3 and w_4.

2. Formulation of Discrete Equations

The general case is now clear. We choose any convenient quadrature formula

$$\int_a^b u(y)\ dy = \sum_1^n w_j\ u(y_j)$$

involving the n points y_j and the corresponding weights w_j. The general Fredholm integral equation of the second kind

$$\int_a^b K(x, y)\ f(y)\ dy + g(x) = f(x)$$

is then replaced by the system of n linear algebraic equations for the unknowns $\tilde{f}_j$, written in the matrix form $(I - KD)\underline{\tilde{f}} = \underline{g}$, where the matrix K has the elements $K_{ij} = K(y_i, y_j)$ and D has the diagonal elements w_j. The solution of these equations, $\tilde{f}_j$, represents the approximate values of $f(x)$ at the points $x = y_j$.

Those familiar with boundary value problems in differential equations will recognise an analogy with the finite difference method for the solution of the problem

$$y'' = p(x)\ y + q(x),\ y(0) = A,\ y(1) = B.$$

In a similar way, if we use the approximation

$$\{y(x + h) - 2y(x) + y(x - h)\}/h^2$$

to the second derivative y'', we obtain a system of linear algebraic equations for the approximate values of $y(x)$ at the points $h, 2h, \ldots, (n-1)h$. There are, however, two important differences between the integral equation problem and the differential equation problem, one practical and one theoretical.

The differential equation leads to a linear system in which the matrix has a very simple band structure. In each row there are only three non-zero elements, symmetrically placed about the diagonal. For such matrices there is a very simple and efficient method of numerical solution. The matrix $(I - KD)$ derived from the integral equation has no such simple structure, and in general there is no particular reason why any of the elements should be zero. It is therefore necessary to solve the equations by a general method. There is, of course, one important special case; for a Volterra equation the matrix will be essentially triangular, but this is a special type of equation which will be considered separately in Chapter 11.

The result of the computation is in each case a set of discrete values, giving approximate values of the solution at certain points. In the differential equation problem intermediate values can be found by interpolation as required, but there is no obvious choice of interpolation procedure related to the differential equation being considered. For the Fredholm integral equation, however, we may write the approximate equation in the form

$$\tilde{f}(x) = g(x) + \sum_1^n w_j\ K(x, y_j)\ \tilde{f}(y_j).$$

Having computed the values of $\tilde{f}(y_j)$ this defines the function $\tilde{f}(x)$ for all values of x in a very natural way, and of course this function, by construction, takes the values $\tilde{f}(y_j)$ at the points y_j.

The computational procedure is straightforward. Having decided which quadrature formula to use, we compute the values of the kernel $K(y_i, y_j)$. Taking this to form the matrix K, we multiply the columns by the respective weights (with negative sign), $-w_j$, and add unity to the diagonal elements, so giving the matrix $(I - KD)$. Then we compute the values of the function $g(y_j)$, and solve the system of n linear algebraic equations. This last is a straightforward matter, and the use of Gaussian elimination is convenient and efficient. At this stage it is worth pointing out that the labour involved in constructing the matrix is proportional to n^2; for the solution of the equations we may be more precise and say that this involves $n^3/3$ multiplications. It is therefore clear that we should not use a value of n which is larger than is necessary to achieve the required accuracy.

3. Choice of Quadrature Formula

The computational procedure outlined in the previous section is, of course, incomplete, and there is another stage to be considered. The result of the computation is an approximation to the solution of the integral equation, and we must now estimate the error of the approximation, and study means by which the error may be reduced. This problem is naturally linked with the original choice of quadrature formula. Our choice of Simpson's rule in §1 was made for simplicity. It is well known that the error in approximating the integral of $u(y)$ over $(0, 1)$ by Simpson's rule may be written $\dfrac{u^{iv}(\theta)}{2800}$, under appropriate conditions of differentiability, where θ is some value between 0 and 1. This means that our approximate solution satisfies exactly the integral equation

$$\int_0^1 K(x, y)\, \tilde{f}(y)\, dy + g(x) = \tilde{f}(x) + u^{iv}(\theta(x))/2800 ,$$

where u^{iv} denotes the 4th partial derivative with respect to y of the function $u(y) = K(x, y)\, \tilde{f}(y)$, and the notation is intended to indicate that $\theta(x)$ is a function of x. There are clearly two stages to the discussion of the error in the solution; we must first estimate the size of u^{iv}, and then deduce the resulting size of the error in the result, $\tilde{f}(x) - f(x)$. We shall return to a fuller discussion of these points in §11.

There are some simple features of the error term which are common to all methods of this type, and are worth mentioning here. First of all, the function $u(y)$ depends on the solution $\tilde{f}(y)$. It will obviously require some elaborate manipulation to obtain any sort of expression for u^{iv}, and it is extremely unlikely that any close estimate of its magnitude could be obtained by analytical means, in advance of the numerical computation. Use of other quadrature formulae in place of Simpson's rule may involve an error term containing a derivative of even higher order, whose estimation will be even more difficult. Investigation of the error term will therefore require either estimation of the derivative by use of finite differences, or some other expression requiring a second computation using different points y_j.

4. Use of Finite Differences

An improvement on Simpson's rule can be obtained, for example, by use of Gregory's formula for the integration, in the form

$$\int_a^b u(y)\,dy = h\{\tfrac{1}{2}u_0 + u_1 + \ldots + u_{m-1} + \tfrac{1}{2}u_m\}$$

$$- (h/12)\{\nabla u_m - \Delta u_0\} - (h/24)\{\nabla^2 u_m + \Delta^2 u_0\} - \ldots$$

In the usual notation $h = (b - a)/n$ and $u_j = u(a + jh)$. Notice that this formula will lead to a system of $m + 1$ linear equations. Once we have decided how many terms to use in this series, we obtain a quadrature formula to which the general method applies. For example, truncating after the terms involving first differences, we shall of course use the points $y_j = a + jh$, and the weights are now

$$(7h/12,\ 11h/12,\ h,\ h,\ \ldots,\ h,\ 11h/12,\ 7h/12).$$

To make use of Gregory's formula, we first choose an interval h, and then decide how many differences to use. Since the remainder term will involve the unknown function $\tilde{f}(x)$ it is almost certainly impossible to predict in advance how many differences should be included, though on general grounds we may be able to make a sensible choice of interval size which will ensure that the differences converge. The method of "deferred correction" avoids this difficulty.

We begin by finding an approximate solution, neglecting the difference terms altogether; this is of course equivalent to the simple use of the composite trapezium rule. Having obtained this solution, we form tables of the function $K(x, y)\tilde{f}(y)$ and its differences. Using Gregory's formula we then write an approximation to the integral equation in the form

$$h\{\tfrac{1}{2}K(x, y_0)\tilde{f}_0 + K(x, y_1)\tilde{f}_1 + \ldots + \tfrac{1}{2}K(x, y_m)\tilde{f}_m\} + g(x)$$
$$= \tilde{f}(x) + (h/12)\{\nabla u_m - \Delta u_0\} + \ldots$$

The correction terms on the right-hand side involve differences of the function $u(y) = K(x, y)\tilde{f}(y)$. We compute this term, using differences of the function formed from our first approximation $\tilde{f}$, and include as many differences as are necessary. At this stage, if the differences do not converge, we must begin the whole computation again with a smaller interval size.

We can now solve a new set of linear equations. This system has the same matrix as before, but the vector g_j on the right hand side is modified by the inclusion of the correction terms computed from the first approximation. This first correction gives a second approximation to $\tilde{f}(x)$. The difference correction can then be re-computed in terms of the new approximation, and the process repeated. This leads to an iterative procedure, which can be represented symbolically in the form

$$(I - KD)\underline{\tilde{f}}^{(k+1)} = \underline{g} - C(\underline{\tilde{f}}^{(k)}),$$

where $C(\underline{\tilde{f}})$ represents a vector of correction terms. The first iteration corresponds to $\underline{\tilde{f}}^{(0)} = 0$, so that we compute $\underline{\tilde{f}}^{(1)}$ with neglect of the correction. In practice this iteration should converge rapidly; it is unlikely that more than 2 iterations will be necessary.

The important point about the iteration procedure is that we solve a system of linear equations at each step; the matrix $I - KD$ is the same each time, and only the right-hand side is altered. If we use the normal Gauss elimination process to represent the matrix in the product form LU, where L and U are triangular matrices, the elimination only has to be done once. The first step thus requires of order n^3 operations, but the other steps require only of order n^2.

This type of difference correction has long been advocated in the solution of ordinary differential equations, but its use in the integral equation problem is distinctly more laborious. The reason is that we must tabulate and examine the differences of the function $K(x, y)\tilde{f}(y)$ for each value of x; the difference correction term must be computed $m + 1$ times in each step, giving the elements of the correction vector $C(\underline{\tilde{f}})$. The number of differences to be included may of course not be the same for all the elements.

5. Deferred Approach to the Limit

An alternative way of estimating, and ultimately eliminating, the error in the quadrature formula is to repeat the computation with a different interval size. For example, if we use the composite trapezium rule just described, neglecting the correction terms in the Gregory formula, we expect that under suitable conditions of differentiability,

$$\int_a^b u(y)\, dy = h\{\tfrac{1}{2}u_0 + u_1 + \ldots + u_{m-1} + \tfrac{1}{2}u_m\}$$
$$+ A_2h^2 + A_4h^4 + \ldots,$$

where the coefficients A_2, A_4, etc., are in general unknown. This leads to the result that the corresponding approximate solution of the integral equation can be expressed in the form

$$\tilde{f}(x, h) = f(x) + B_2(x)\, h^2 + B_4(x)h^4 + \ldots$$

Here we have written $\tilde{f}(x, h)$ to emphasise the fact that the approximate solution obviously depends on the interval size used, and clearly the coefficients B_2, B_4 depend on x.

These coefficients will be unknown, but it is a simple matter to eliminate one or more of them by computing $\tilde{f}(x, h)$ for different values of h. Choosing h_1 and h_2 it is easy to see that

$$\tilde{f}(x, h_1, h_2) = f(x) + C_4(x)\, h^4 + O(h^6),$$

where

$$\tilde{f}(x, h_1, h_2) = \frac{h_2{}^2\tilde{f}(x, h_1) - h_1{}^2\tilde{f}(x, h_2)}{h_2{}^2 - h_1{}^2},$$

and in a similar way we might eliminate the term in h^4. The same idea may be applied to the use of the composite Simpson's rule, but here the term in h^2 does not appear, and we should begin by eliminating the

h^4 term.

When using a method of this kind we must solve two systems of linear equations to determine $\tilde{f}(x, h_1)$ and $\tilde{f}(x, h_2)$; the systems do not, of course, have the same matrix, as they are obviously of different sizes. They are best regarded as completely separate computations, except that a rather natural choice $h_2 = \frac{1}{2}h_1$ will mean that some of the values of the kernel $K(y_i, y_j)$ will already be available and will not have to be computed again.

Examination of the values $\tilde{f}(x, h_1)$ and $\tilde{f}(x, h_2)$ and the extrapolated value $\tilde{f}(x, h_1, h_2)$, for various values of x, will give a good estimate of the accuracy of the result, and will indicate whether a further computation of $\tilde{f}(x, h_3)$ is necessary. In computing integrals by this type of method there is some danger of accidental agreement; here the coefficient A_2 might be accidentally small, so that $\tilde{f}(x, h_1)$ and $\tilde{f}(x, h_2)$ agree together more closely than they agree with $f(x)$. In solving integral equations this may easily occur for a particular value of x, but it is rather unlikely that such spurious agreement will happen at several different points. A serious underestimate of the error in the result is therefore unlikely.

The suggestions above for improving the accuracy of quadrature formulae turn out in practice to be rather academic. They both require certain conditions of differentiability on the function $u(y)$, and it is an unfortunate fact that almost all integral equations which arise in practice are singular, in a rather loose sense. Usually the kernel, or some of its derivatives, is infinite at some point, or sometimes the interval of integration is infinite. In such cases special quadrature formulae are used; these are discussed in §§7-10.

6. Non-linear Equations

If the integral equation is not linear, it is still possible to approximate the integral by a quadrature formula. Returning to the simple use of Simpson's rule, we might replace

$$\int_0^1 f(y)^2 \, dy + g(x) = f(x)$$

by the approximation

$$(1/6)\ \{\tilde{f}(0)^2 + 4\,\tilde{f}(\tfrac{1}{2})^2 + \tilde{f}(1)^2\} + g(x) = \tilde{f}(x).$$

Now writing $x = 0$, $\frac{1}{2}$ and 1 gives a system of 3 non-linear equations to solve for the three unknown values $\tilde{f}(0)$, $\tilde{f}(\frac{1}{2})$ and $\tilde{f}(1)$. In general we should obtain a system of n simultaneous non-linear equations to solve; this is a much more difficult problem than the linear case, and will be discussed in Chapter 16.

7. Singular Integral Equations

As we mentioned in §5, most integral equations which arise in practical problems are, loosely speaking, "singular". We shall now discuss some methods which may be used for such equations.

First, it is useful to think of "mild" and "severe" singularities, but without giving any precise definitions. Suppose we wish to integrate, over the interval $(0, 1)$, the function

$$u(y) = e^y \log(|x - y|),$$

and we use Simpson's rule with the three points $y = 0, \frac{1}{2}$ and 1. This will obviously fail at $x = \frac{1}{2}$, for example, for the value of the integrand at $y = \frac{1}{2}$ is infinite, and Simpson's rule will give an infinite result. This is quite a common type of problem, as the logarithmic kernel $\log(|x - y|)$ often appears in integral equations derived from two-dimensional potential problems. This is an example of a severe singularity, where the quadrature method breaks down completely.

If, however, x has the value $\frac{1}{4}$, Simpson's rule will give a numerical result which will be finite. It is likely to be much less accurate than the corresponding result for a non-singular integrand, and in addition it is difficult to estimate how accurate it is. The standard form for the remainder in Simpson's rule involves the factor $u^{(iv)}(\theta)$, the fourth derivative of the integrand at an unknown point. If a bound on the fourth derivative is known, or if it can be estimated by examining the finite differences of the function, this error term can also be estimated. But the fourth derivative of our integrand is certainly infinite at the point $y = x = \frac{1}{4}$; to obtain a useful error bound we must either locate the point θ more closely, or find some other expression for the error in a different form. This is an example of a mild singularity.

A simple and effective method of dealing with a singularity in numerical integration is to remove it by a change of variable, where this is possible. In our example we could write

$$y = x + e^s$$

and evaluate the new integral in the variable s. Our problem is, however, rather different. In solving an integral equation we do not wish to evaluate the integral, but rather to express it as a linear combination of values of the integrand at certain points. The simple change of variable does not help, since the new variable depends also on the value of x.

8. Removal of the Diagonal Term

A common type of singular integral equation has a kernel which becomes infinite when $x = y$. The example given in the previous section corresponded to

$$K(x, y) = \log(|x - y|),$$

which is just of this form. Any quadrature formula of the types considered in §§1-3 will fail in this case. The matrix there denoted by K has the elements $K_{ij} = K(y_i, y_j)$, so that each of its diagonal elements is infinite. It is a simple matter to transform this "severe" singularity into a "mild" one, however.

We write

$$\int_a^b K(x, y)\ f(y)\ dy = \int_a^b K(x, y)\{f(y) - f(x)\}dy + \int_a^b K(x, y)\ f(x)\ dy$$

$$= \int_a^b K(x, y)\{f(y) - f(x)\}dy + f(x)\ H(x)\ ,$$

where $H(x) = \int_a^b K(x, y)\ dy$ is assumed to be a known function. In this form the integrand has only a mild singularity; although $K(y, y)$ is infinite, the other factor $f(y) - f(x)$ vanishes when $y = x$, and the product must also vanish, or the singularity would be so severe that the integral would not exist at all.

We may therefore replace the integral equation by the modified approximate form

$$\tilde{f}(x) = g(x) + \sum_{j=1}^{n} w_j\ K(x, y_j)\ \{\tilde{f}(y_j) - \tilde{f}(x)\} + \tilde{f}(x)H(x),$$

where the points y_j and the weights w_j correspond to any convenient quadrature formula, and in the sum the term $y_j = x$ is taken to be zero. The integral equation is again approximated by a system of linear equations, which may now be written

$$(I - M)\ \tilde{\underline{f}} = \underline{g}$$

in the notation introduced in §1. The off-diagonal elements of the matrix M are $M_{ij} = K_{ij} w_j$ just as in the non-singular case, but the diagonal elements are more complicated. We find that

$$M_{ii} = H_i - \sum K_{ik} w_k,$$

where the sum is taken over those values of k not equal to i.

The computational procedure is thus very little more complicated than in the non-singular case, but it is important to notice that we have weakened the singularity, and not eliminated it. The original integrand $K(x, y)\ f(y)$ was infinite when $y = x$; the new integrand $K(x, y)\{f(y) - f(x)\}$ is everywhere finite, but it clearly has an infinite derivative with respect to y at this point.

9. Use of Product Integration Methods

Methods involving product integration formulae can be devised to remove the singularity entirely. There are many variants, but we shall introduce only the basic idea. Having chosen the points y_j, which might for example be equally spaced, the Newton-Cotes type formulae like

$$\int_a^b u(y)\ dy = \sum_1^n w_j\ u(y_j) + E$$

are designed to give the exact result, with $E = 0$, when $u(y)$ is a polynomial of a certain degree. In a similar way, a product integration formula will give

$$\int_a^b K(x, y)\ f(y)\ dy = \sum_1^n z_j\ f(y_j) + E_z$$

where now the result is exact, with zero remainder $E_z = 0$, when $f(y)$, rather than the whole integrand, is a polynomial of the appropriate degree. In the usual way, by writing successively $f(y) = 1, y, y^2, \ldots, y^{n-1}$ in this formula we obtain a system of n linear equations for the weights z_j; the result will now be exact for all polynomials of degree $n - 1$ or less, and we should expect to express the error term E_z in terms of the nth derivative of $f(y)$.

If the required solution $f(y)$ is sufficiently smooth, this method will avoid the singularity in the kernel completely. However there is a heavy price to pay. If we look more closely at the process just described, we notice that the weights z_j depend on the value of x. Thus for each point x we must evaluate each of the integrals

$$\int_a^b K(x, y)\, y^r\, dy, \quad r = 0, 1, \ldots, n-1,$$

and solve a system of n linear equations for the weights $z_j(x)$. The main part of the numerical procedure is then the same as before, with the minor difference that in the final matrix the weights are not the same all the way down each column. But note that instead of solving one system of n simultaneous equations as in the non-singular case, we now have to solve n + 1 such systems, which represents an enormous increase in labour.

Various forms of composite formulae have been suggested to avoid this difficulty. For example, in the non-singular case we might use the composite trapezium rule, without the difference correction, as described in §4. If the kernel is singular at t = x, we might expect to use the trapezium rule safely over the intervals (a, x-h) and (x+h, b), with a special product formula over the danger zone (x-h, x+h) only. This special formula will involve only 3 points, and will be exact when f(y) is any quadratic polynomial. To construct the weights for such a formula is relatively trivial, and involves only 3 simultaneous equations. Moreover, in the important special case of a convolution kernel of the form

$$K(x, y) = k(x - y),$$

we see that the weights in the approximation

$$\int_{x-h}^{x+h} k(x - y) f(y)\, dy = \sum_{j=1}^{n} z_j\, f(y_j) + E_z$$

do not depend on x, and therefore only have to be computed once.

These two particular methods are extreme forms. In the one case we have a different n-term formula for each value of x. In the other we use the trapezium rule, except for a three-term formula across the point y = x. There are of course many other possibilities; in particular, elaborate methods have been devised for deciding automatically how wide a region about the point y = x should be treated specially. It may also be convenient to write the kernel itself in the form of a product

$$K(x, y) = K_1(x, y)\, K_2(x, y),$$

where K_1 is singular, but K_2 is not. Only the part K_1 now needs special treatment. This may be useful if evaluation of the integrals $\int_a^b K(x, y)y^r\,dy$ is difficult, while those involving K_1 only may be much simpler.

10. Singularity in the Solution

In the methods just described we have taken account of the singularity in the kernel, but have assumed sufficient differentiability in the solution $f(y)$. Unfortunately, a singular integral equation often has a solution with mild singularities at the end-points a and b. Simple examples are difficult to construct, but plausible practical cases are clear enough. Integral equations often arise from potential problems, for example in electrostatics. The unknown function $f(y)$ will then be a charge density. Considering the charge density on a conducting strip, it is intuitively obvious that there will be singularities of some sort at the edges of the strip, and even if the charge density is finite there an infinite derivative of some order is very likely.

When constructing examples for testing a numerical procedure, it is very easy to begin with a specified kernel, choose a simple solution $f(y)$, and determine the right-hand side $g(x)$ so that the integral equation is satisfied. This may give a false impression of the accuracy of the process when applied to a practical case where $f(x)$ is not smooth.

To deal completely with this problem it would be necessary first to determine exactly what type of singularity $f(x)$ has at the ends, if any. Then we should have to construct quadrature formulae to deal with integrands which have singularities at $y = x$, and at the end-points. This is obviously very difficult, and little seems to be known about the sort of singularities to be expected in the solution. For this reason it seems unwise to use product type quadrature formulae of high order, which have error terms involving high derivatives of $f(x)$. Such formulae might give a false impression of the accuracy of the result.

11. Error Analysis of Quadrature Methods

We now return to the problem of estimating the accuracy of the result of the calculation, for any of the quadrature methods discussed. The function $f(x)$ is required to satisfy the integral equation. The computed function $\tilde{f}(x)$ actually satisfies exactly a slightly different equation, of the form

$$\int_a^b K(x, y)\, \tilde{f}(y)\, dy + g(x) = \tilde{f}(x) - E(x). \qquad (11.1)$$

The term $E(x)$ denotes the error term in the quadrature formula. It will usually be expressed in terms of a derivative, either of the complete integrand in the nonsingular case, or perhaps of $\tilde{f}(x)$ only in a singular case. If we write $e(x) = f(x) - \tilde{f}(x)$ for the error in our result, we easily find by subtraction that

$$\int_a^b K(x, y)\, e(y)\, dy = e(x) + E(x).$$

In the usual way, writing K for the integral operator

$$K(f) = \int_a^b K(x, y)\, f(y)\, dy,$$

the error is given by

$$e(x) = -(1 - K)^{-1}\, E(x),$$

and so

$$\|e\| \leqslant \|(1 - K)^{-1}\|\, \|E\|.$$

Now $\|E\|$ can easily be estimated. Under suitable assumptions about the smoothness of the various terms we can estimate the magnitudes of the derivatives by examining finite differences. It therefore remains to estimate $\|(1 - K)^{-1}\|$. A well known result states that

$$\|(1 - K)^{-1}\| \leqslant 1/(1 - \|K\|),$$

where K is any bounded linear operator in a Banach space, provided that $\|K\| < 1$. We usually choose the norm $\|f\| = \max|f|$, over the interval $[a, b]$. Then the norm of K is

$$\|K\| = \max_{a \leqslant x \leqslant b} \int_a^b |K(x, y)|\, dy.$$

Provided this number is less than unity we therefore have a rigorous bound for the error in the result.

An alternative approach is often useful, as it may happen that although this condition is not satisfied, the error is still acceptably small. Instead of working with the two integral equations (1.1) and (11.1), we can consider the matrix system

$$(I - M)\underline{\tilde{f}} = \underline{g} .$$

In the nonsingular case M will be the matrix KD of §1, but this generalisation allows for the singular case as well. The vector $\underline{\tilde{f}}$ satisfies this equation exactly. If we apply the approximate quadrature formula to the function f we should obtain

$$(I - M)\underline{f} = \underline{g} + \underline{E}_f$$

where $\underline{E}_f$ is of the same form as E above, but involves derivatives of the function f instead of $\tilde{f}$. By subtraction we now find that

$$(I - M)\ \underline{e} = \underline{E}_f,$$

so that

$$||\underline{e}|| \leqslant ||(I - M)^{-1}||\ ||\underline{E}_f|| .$$

This result looks very much the same as the previous one, but it is rather different in practice. The term I - M is a matrix, not an integral operator. It is therefore quite possible to determine $||(I-M)^{-1}||$ explicitly, at the expense of inverting an $n \times n$ matrix and forming the row sums. Thus one of the two terms is much easier to compute in this case. However, the other term $||\underline{E}_f||$ is more difficult than before, as it involves derivatives of the unknown function f.

It remains to combine the two processes. We can summarise a great deal of complicated analysis by saying that, under rather general conditions of smoothness in the original problem, and under certain conditions on the quadrature formulae used, the norm of $(I - M)^{-1}$ converges to the norm of $(I - K)^{-1}$ as the number of quadrature points is increased. In other words, if a reasonably large number of points is used with a suitable quadrature formula, we may replace f by $\tilde{f}$ in computing the error term $\underline{E}_f$, so giving the easily computed error bound

$$||\underline{f} - \underline{\tilde{f}}|| \leqslant ||(I - M)^{-1}||\ ||\underline{E}_f(\tilde{f})|| ,$$

where M is the $n \times n$ matrix, and $\underline{E}_f(\tilde{f})$ is a vector of error terms in the quadrature formula. This error estimate is asymptotically exact for large values of n.

A full derivation of these results, for both the singular and the non-singular case, is given by Atkinson (1967, 1972).

References

ATKINSON, K.E. (1967) The numerical solution of Fredholm integral equations of the second kind. SIAM J. numer. Anal. Vol. 4, pp.337-348.

ATKINSON, K.E. (1971) A survey of numerical methods for the solution of Fredholm integral equations of the second kind. Dept. of Mathematics, Indiana University. (To appear in Proceedings of SIAM National Meeting, Fall 1971.)

FOX, L. (ed.)(1962) Numerical Solution of Ordinary and Partial Differential Equations. Pergamon.

FOX, L. and GOODWIN, E.T. (1953) The numerical solution of nonsingular linear integral equations. Phil. Trans. Roy. Soc. A Vol. 245, pp. 501-534.

CHAPTER 7 EXPANSION METHODS

C.T.H. Baker
University of Manchester

We describe methods of approximating the solution of a Fredholm integral equation of the second kind by expansions of the form

$$\tilde{f}(x) = \sum_{i=0}^{n} \tilde{a}_i \, \phi_i(x) \quad \text{where} \quad \phi_0(x), \phi_1(x), \ldots, \phi_n(x)$$

are suitably chosen functions. The techniques described here can also be applied to Volterra equations of the second kind, though such equations are usually treated by step-by-step or block-by-block methods (see Chapter 11). For further details of many aspects covered here see Baker (to appear).

1. Nature of Approximating Function

If we consider the Nyström method (Chapter 6) for a Fredholm equation

$$f(x) - \lambda \int_a^b K(x,y)\, f(y)dy = g(x)$$

we obtain an approximation of the form

$$\tilde{f}(x) = g(x) + \lambda \sum_{j=0}^{n} w_j \, K(x,y_j)\, \tilde{f}(y_j) \tag{1.1}$$

and the approximation $\tilde{f}(x)$ is a linear combination (determined by the weights and abscissae of a quadrature rule) of $g(x)$ and the functions $K(x,y_j)$.

It is sometimes more convenient to seek an approximate solution $\tilde{f}(x)$ in terms of prescribed functions $\phi_0(x)$, $\phi_1(x), \ldots, \phi_n(x)$ (which may depend on n) which are generally independent of $K(x,y)$. Natural choices of functions $\phi_i(x)$ are those which are commonly used to approximate a known function. In particular, we may choose polynomial approximations, orthogonal functions such as Chebyshev polynomials, trigonometric functions, or splines, all of which are mentioned in Chapter 5.

It is natural to choose functions $\phi_i(x)$ $(i=0,1,\ldots,n)$ which are linearly independent. This we shall assume; a function

$$\tilde{f}(x) = \sum_{r=0}^{n} \tilde{a}_r \, \phi_r(x)$$

which is a linear combination of $\phi_0(x), \phi_1(x), \dots, \phi_n(x)$ then determines uniquely its expansion coefficients $\tilde{a}_0, \tilde{a}_1, \dots, \tilde{a}_n$. In theory, two different choices of $\left\{\phi_0(x), \phi_1(x), \dots, \phi_n(x)\right\}$ are equivalent if the function spaces spanned by each set are the same. (Thus, for example, the choices (i) $\phi_r(x) = x^r$ (ii) $\phi_r(x) = T_r(x)$ are equivalent since any polynomial of degree n can be expressed uniquely in terms either of the powers of x or of the Chebyshev polynomials.) In practice, the form of the functions $\phi_i(x)$ is of some importance and can govern the numerical accuracy obtainable. The theory here is incomplete, though some results are known and will be summarised later.

2. Criteria for Determining the Approximation

Suppose that the equation to be solved has the form

$$f(x) = \lambda \int_a^b K(x,y) \, f(y)dy + g(x) \quad (a \leq x \leq b) \tag{2.1}$$

where λ is not a characteristic value of $K(x,y)$, or (to consider a formal extension to non-linear equations)

$$F_0(x; f(x)) - \int_a^b F_1(x,y; f(x), f(y))dy = 0 \quad (a \leq x \leq b). \tag{2.2}$$

To find an approximate solution

$$\tilde{f}(x) = \sum_{i=0}^{n} \tilde{a}_i \, \phi_i(x)$$

in terms of predetermined functions $\phi_i(x)$, we seek a prescription for choosing the coefficients $\tilde{a}_0, \tilde{a}_1, \dots, \tilde{a}_n$. Now for any such choice, substitution of $\tilde{f}(x)$ into the appropriate integral equation will give a residual

$$\eta(x) \; (\equiv \eta(x; \tilde{a}_0, \tilde{a}_1, \dots, \tilde{a}_n))$$

where

$$\eta(x) = \tilde{f}(x) - \lambda \int_a^b K(x,y) \, \tilde{f}(y)dy - g(x), \tag{2.3}$$

or

$$\eta(x) = F_0(x; \tilde{f}(x)) - \int_a^b F_1(x,y; \tilde{f}(x), \tilde{f}(y))dy, \tag{2.4}$$

respectively. Unless the true solution is a linear combination of $\phi_0(x),\phi_1(x),\ldots,\phi_n(x)$, we cannot choose $\tilde{a}_0,\tilde{a}_1,\ldots,\tilde{a}_n$ to make $\eta(x)$ vanish identically. However, suitable constraints can be imposed on the choice of $\tilde{a}_0,\tilde{a}_1,\ldots,\tilde{a}_n$ which ensure that $\eta(x)$ is in some sense "small".

To motivate such a method, suppose that we choose $z_0^{(n)},z_1^{(n)},\ldots,z_N^{(n)}$ as distinct points lying in $[a,b]$. For convenience we write $z_i^{(n)} = z_i$. A (polynomial) interpolant to $\eta(x)$ at $z_0,z_1,\ldots,z_N$ (of degree N) will be zero if $\eta(z_i) = 0$ for $i=0,1,\ldots,N$ and this suggests that $\eta(x)$ will be in some sense small if we choose $\tilde{a}_0,\tilde{a}_1,\ldots,\tilde{a}_n$ to satisfy these conditions. (There is no need to construct the interpolant itself.)

In the linear case (2.1), the condition that $\eta(z_i) = 0$ is

$$\sum_{j=0}^{n} \tilde{a}_j\, \phi_j(z_i) - \lambda \sum_{j=0}^{n} \tilde{a}_j \int_a^b K(z_i,y)\, \phi_j(y)dy = g(z_i), \tag{2.5}$$

and with $i=0,1,\ldots,N$ we obtain a system of linear equations governing the choice of $\tilde{a}_0,\tilde{a}_1,\ldots,a_n$. A similar non-linear system of equations is produced for equation (2.2). Thus, for the Urysohn equation

$$f(x) - \int_a^b F(x,y;\, f(y))dy = g(x), \tag{2.6}$$

we have the non-linear equations

$$\sum_{j=0}^{n} \tilde{a}_j\, \phi_j(z_i) - \int_a^b F(z_i,y;\, \sum_{k=0}^{n} \tilde{a}_k\, \phi_k(y))dy = g(z_i). \tag{2.7}$$

Evaluation of the integral as a function of $\tilde{a}_0,\tilde{a}_1,\ldots,\tilde{a}_n$ may well be difficult, in the last equation.

In setting up the equations (2.5) we obtain

$$\eta(z_i) = 0 \quad (i=0,1,\ldots,N) \tag{2.8}$$

where

$$\eta(z_i) = \sum_{j=0}^{n} A_{ij}\, \tilde{a}_j - \lambda \sum_{j=0}^{n} B_{ij}\, \tilde{a}_j - g(z_i) \tag{2.9}$$

and

$$A_{ij} = \phi_j(z_i) \quad \text{and} \quad B_{ij} = \psi_j(z_i)$$

with

$$\psi_j(x) = \int_a^b K(x,y)\, \phi_j(y)dy. \tag{2.10}$$

The system of equations (2.8) may be over- or under-determined; this depends both upon the choice of $z_0, z_1, \ldots, z_N$ and the choice of the functions $\phi_i(x)$ $(i=0,1,\ldots,n)$. We obtain the usual method of collocation if we choose $n=N$ and seek a solution of (2.8); a unique solution then exists if $\det(A-\lambda B) \neq 0$.

Since there may be no unique solution of (2.8) we may instead seek a vector $\tilde{\underset{\sim}{a}}=[\tilde{a}_0,\tilde{a}_1,\ldots,\tilde{a}_n]^T$ which minimizes

$$\max_{i=0,1,\ldots,N} |\eta(z_i)|$$

or

$$\sum_{j=0}^{N} |\eta(z_j)|$$

or

$$\sum_{j=0}^{N} \Omega_j |\eta(z_i)|^2 \quad (\text{where } \Omega_j > 0 \text{ for } j=0,1,\ldots,N).$$

Such approaches lead to linear programming (in the real case)(Barrodale and Young (1970)) or to least-squares methods for (2.5). In the non-linear case there are complications, and it is probably simpler to treat the case of collocation with $N=n$, endeavouring to solve the non-linear equations for $\tilde{a}_0,\tilde{a}_1,\ldots,\tilde{a}_n$. (However, non-linear discrete least-squares methods are available (see Osborne (1972)).

If we restrict our attention to the collocation method we see in the linear case that the functions $\{\phi_i(x)\}$ and the points $\{z_i\}$ should be matched so that the matrix $A-\lambda B$ is non-singular, and well-conditioned. It is impossible to choose a fixed scheme which will achieve this for all $\lambda K(x,y)$ since B depends on the kernel. For the non-linear case the collocation equations will in general be solved by an iterative technique in which a set of linear equations is solved at each stage, and it is desirable that these systems should be well-conditioned. Thus for (2.7) the Newton method corresponds to an iterative solution of systems of the form (2.5) in which $K(x,y)$ changes at each step. The inherent difficulties are clear.

In the classical Galerkin method we use an apparently different technique. Recall that our aim is to ensure that $\eta(x)$ (see (2.3) and (2.4)) is, in some sense, small, and we consider that we achieve this aim if $\eta(x)$ is chosen so that

$$(\eta,\phi_i) = \int_a^b \eta(x)\,\overline{\phi_i(x)}\,dx$$

vanishes for i=0,1,...,n. The method is simplified if $(\phi_i,\phi_j) = \delta_{ij}$; we then see that if $(\eta,\phi_i) = 0$ the generalized Fourier sum

$$\sum_{i=0}^{n} (\eta,\phi_i)\,\phi_i(x) = \hat{\eta}(x)$$

vanishes, and since $\hat{\eta}(x) \simeq \eta(x)$, $\eta(x)$ may then be considered "small".

The condition $(\eta,\phi_i) = 0$ produces for (2.3) the equation

$$\sum_{j=0}^{n} C_{ij}\,\tilde{a}_j - \lambda \sum_{j=0}^{n} D_{ij}\,\tilde{a}_j = \gamma_i \quad (i=0,1,\dots,n) \tag{2.11}$$

where

$$C_{ij} = (\phi_j,\phi_i),$$
$$D_{ij} = (\psi_j,\phi_i)$$

and

$$\gamma_i = (g,\phi_i).$$

Thus

$$C_{ij} = \int_a^b \phi_j(x)\,\overline{\phi_i(x)}\,dx$$

$$\gamma_i = \int_a^b g(x)\,\overline{\phi_i(x)}\,dx, \tag{2.12}$$

and

$$D_{ij} = \int_a^b \int_a^b K(x,y)\,\phi_j(y)\,\overline{\phi_i(x)}\,dx\,dy.$$

The matrix C is a Gram matrix and is symmetric and positive-definite, (see Davis, (1965) p 176) and it reduces to the identity matrix when the functions $\phi_i(x)$ are orthonormal. If λ is not a characteristic value of $K(x,y)$ then $C-\lambda D$ is non-singular if $\phi_0(x),\phi_1(x),\dots,\phi_n(x)$ form part of a complete system and n is sufficiently large; however, the behaviour of the condition number of the matrix $C-\lambda D$ depends on the choice of the functions $\phi_i(x)$. The Galerkin method reduces to the Rayleigh-Ritz method if $K(x,y) = \overline{K(y,x)}$.

The implementation of the Galerkin method is complicated by the need to calculate the elements C_{ij}, D_{ij} and γ_i in (2.12), and this involves evaluating single or double integrals. In practice it is often ecessary to approximate these integrals numerically. Whereas D_{ij} can e calculated by a method of approximate cubature, there is some ractical motivation for calculating values of $\psi_i(x)$ and storing these or future use (see below). (This does not apply in the non-linear case here the Galerkin equations are more complicated.)

The classical Galerkin method can be extended to the "method of oments" if we choose linearly independent functions $\chi_0(x), \chi_1(x), \ldots, \chi_n(x)$ nd choose

$$\tilde{f}(x) = \sum_{i=0}^{n} \tilde{a}_i \, \phi_i(x)$$

o that (η, χ_i) vanishes for $i=0,1,\ldots,n$. In particular we may choose a unction $w(x)$ which is positive on $[a,b]$ and set

$$\chi_i(x) = w(x) \, \phi_i(x).$$

e call this method the weighted Galerkin method and we obtain, in the inear case, an equation of the form (2.11) with

$$C_{ij} = \int_a^b w(x) \, \phi_j(x) \, \overline{\phi_i(x)} \, dx$$

$$D_{ij} = \int_a^b w(x) \, \psi_j(x) \, \overline{\phi_i(x)} \, dx$$

nd

$$\gamma_i = \int_a^b w(x) \, g(x) \, \overline{\phi_i(x)} \, dx.$$

As an example where this method is computationally convenient consider he case $[a,b] = [-1,1]$, $\phi_0(x) = \frac{1}{2}$, $\phi_r(x) = T_r(x)$ $(r \geq 1)$ with $w(x) = (1-x^2)^{-\frac{1}{2}}$.)

Suppose, in the linear case, that we set

$$\chi_i(x) = \phi_i(x) - \lambda \int_a^b K(x,y) \, \phi_i(y) \, dy$$

nd then choose $\tilde{a}_0, \tilde{a}_1, \ldots, \tilde{a}_n$ to ensure that $(\eta, \chi_i) = 0$ for $i=0,1,\ldots,n$. his is equivalent to choosing $\tilde{a}_0, \tilde{a}_1, \ldots, \tilde{a}_n$ to minimize

$$\int_a^b |\eta(x)|^2 \, dx$$

(a continuous least-squares method). It is of some interest to note that this is also equivalent to applying the (Rayleigh-Ritz) Galerkin method to the symmetrised equation

$$f(x) - \int_a^b \left\{ \lambda K(x,y) + \overline{\lambda K(y,x)} - |\lambda|^2 \; K^*K(x,y) \right\} f(y)$$

$$= g(x) - \overline{\lambda} \int_a^b \overline{K(y,x)} \, g(y) \, dy \qquad (2.13)$$

where

$$K^*K(x,y) = \int_a^b \overline{K(z,x)} \, K(z,y) \, dz.$$

Because the conditioning of this integral equation is in general worse than the conditioning of equation (2.1) we do not recommend this choice of functions $\chi_i(x)$ or the equivalent least-squares method. (Compare (2.13) with the normal equations for least-squares problems in linear algebra.)

We observe (Collatz (1960)) that in variational methods for (linear and) non-linear integral equations we frequently produce a functional $J[\phi]$ which assumes a stationary value when $\phi(x)$ is $f(x)$, the solution of the equation. We then choose

$$\tilde{f}(x) = \sum_{i=0}^{n} \tilde{a}_i \, \phi_i(x)$$

so that

$$\left(\frac{\partial}{\partial \tilde{a}_i}\right) J[\tilde{f}] = 0$$

for $i=0,1,\ldots,n$. The process is always equivalent to applying the classical Galerkin method to some reformulation of the integral equation and this may suggest useful formulations of the Galerkin method, to which the theory of Mikhlin on stability (§4) can be applied.

Before considering further practical details we remark on a feature of our methods applied to the linear equation (2.1). In all the methods we obtain an approximation of the form

$$\tilde{f}(x) = \sum_{i=0}^{n} \tilde{a}_i \, \phi_i(x)$$

and in the process of the computation we require certain values of $\psi_i(x)$. A further approximation is obtained if we substitute $\tilde{f}(x)$ in the right-hand side of (2.1) to obtain

$$\overset{o}{f}(x) = \lambda \sum_{i=0}^{n} \tilde{a}_i \, \psi_i(x) + g(x). \tag{2.14}$$

hat is, we carry out one stage of Neumann iteration from the "point" $\tilde{f}$. t frequently happens that $\overset{o}{f}(x)$ provides a better approximation than (x) (though we observe that for the collocation method $\tilde{f}(z_i) = \overset{o}{f}(z_i)$ or $i=0,1,\ldots,N=n$, and only the values $\psi_j(z_i)$ are naturally available).

. Choice of $\phi_0(x),\phi_1(x),\ldots,\phi_n(x)$

As we have indicated, the functions $\phi_i(x)$ $(i=0,1,\ldots,n)$ employed in he methods of §2 in general depend upon n, that is $\phi_i(x) \equiv \phi_{i,n}(x)$. If e use the convergence theory as motivation for a practical choice with inite n, it appears that the functions

$$\begin{array}{l} \phi_{0,0}(x); \\ \phi_{0,1}(x),\ \phi_{1,1}(x); \\ \cdot\quad\cdot\quad\cdot\quad\cdot\quad\cdot\quad\cdot \\ \phi_{0,n}(x),\ \phi_{1,n}(x),\ldots,\phi_{n,n}(x) \end{array} \tag{3.1}$$

hould be chosen in such a way that

$$\lim_{n \to \infty} \inf_{a_{i,n}} \left\| f(x) - \sum_{i=0}^{n} a_{i,n} \, \phi_{i,n}(x) \right\| = 0 \tag{3.2}$$

where $\|\epsilon(x)\|$ is a norm measuring the 'size' of $\epsilon(x)$. (In general we an employ the mean-square norm or the uniform norm.) Since f(x) is nknown we in general make a choice which ensures that (3.2) is satisfied when f(x) is replaced by any function $\phi(x)$ in a class known to contain f(x), for example $\phi(x) \in C[a,b]$ or $\phi(x) \in C^1[a,b]$.

Let us consider an example. We suppose that $[a,b] = [-1,1]$ and seek polynomial approximation of degree n with an appropriate choice of polynomials $\phi_{0,n}(x),\phi_{1,n}(x),\ldots,\phi_{n,n}(x)$. We may set, for example,

(i) $\phi_{r,n}(x) = x^r$

(ii) $\phi_{r,n}(x) = \sqrt{(2r+1)/2}\, P_r(x)$

(iii) $\phi_{r,n}(x) = T_r(x)$.

The choice (i) is unwise since for large n it usually leads to ill-conditioning. (However, we should comment that this ill-conditioning does not appear to affect the numerical examples given in Chapter 8.) The choice (ii) seems obvious for the classical Galerkin method (since C=I) and for collocation using points z_i with $P_{n+1}(z_i) = 0$.

In the latter case

$$A_{ij} = [(2j+1)/2]^{\frac{1}{2}} P_j(z_i)$$

and (by the theory of Gaussian quadrature) $A^*DA = I$, where I is the identity and D is the diagonal matrix whose entries are the weights of the Gauss-Legendre rule with abscissae $z_0, z_1, \ldots, z_n$. Thus $A^{-1} = A^*D$ is readily obtained, and we can easily set up the equations $(I - \lambda B A^{-1})\tilde{f} = g$ for the vector $\tilde{f} = [\tilde{f}(z_0), \tilde{f}(z_1), \ldots, \tilde{f}(z_n)]^T$, where the components of g are $g(z_i)$, $i=0,1,\ldots,n$. On the other hand, the use of the quadrature method with the Gauss-Legendre rule would yield a vector of function values from which an interpolating polynomial could be easily derived as a sum of Legendre polynomials.

We favour the choice (iii) using the weighted Galerkin method with a weight function $w(x) = (1-x^2)^{-\frac{1}{2}}$ or collocation using either the zeros of $T_{n+1}(x)$ or the points of extrema of $T_n(x)$ as the points z_i (or similar discrete least-squares methods).

Finally we note that for any choice of $z_0, z_1, \ldots, z_n$ in the collocation method, we can employ for $\phi_i(x)$ the cardinal function of Lagrangean interpolation,

$$\phi_i(x) = \prod_{\substack{j \neq i \\ j=0}}^{n} \left\{ (x-z_j)/(z_i-z_j) \right\} .$$

For this choice, if

$$\tilde{f}(x) = \sum_{i=0}^{n} \tilde{a}_i \, \phi_i(x)$$

then $\tilde{f}(z_i) = \tilde{a}_i$.

With the exception of the last choice, the functions $\phi_{i,n}(x)$ are actually independent of n. This is an advantage if we seek to check our results by comparing the calculations resulting with $n=n_1$ and $n=n_2>n_1$ for consistency. For if

$$f_j(x) = \sum_{r=0}^{n} \tilde{a}_r^{(j)} \, \phi_r(x)$$

$(j=1,2)$ and $\phi_r(x)$ does not depend upon n, the coefficients $\tilde{a}_r^{(1)}$ and

$\tilde{a}_r^{(2)}$ (for $r=0,1,\ldots,n_1$) should be comparable when n_1 is large enough and good accuracy has been obtained. (This may not be so for a non-orthogonal set of functions such as x^r; it is likely to be so (Mikhlin (1971)) if the $\phi_i(x)$ are orthonormal with respect to some weight function.) There is also the computational advantage that in these circumstances the matrices C,D for the Galerkin methods can be obtained with $n=n_2$ by bordering the matrices computed with $n=n_1$. A similar device can sometimes be used with the collocation method if

$$z_0^{(n_2)}, z_1^{(n_2)}, \ldots, z_{n_2}^{(n_2)}$$

contains as a subset the points

$$z_0^{(n_1)}, z_1^{(n_1)}, \ldots, z_{n_1}^{(n_1)}$$

In particular the points of extrema of $T_n(x)$ ($z_j^{(n)} = \cos(j\,\pi/n)$) are convenient if $n_2 = 2n_1$.

If we choose for $\{\phi_r(x)\}$ a set of orthogonal polynomials, these polynomials satisfy a recurrence relation which can be used to generate them and can sometimes be used to generate the functions $\psi_r(x)$. For example with $[a,b] = [-1,1]$, $\phi_r(x) = P_r(x)$ we have

$$(r+1)\,P_{r+1}(x) - (2r+1)\,x\,P_r(x) + r\,P_{r-1}(x) = 0$$

and if

$$K(x,y) = \log|x-y|$$

we can deduce that (Phillips, 1969)

$$(r+2)\,\psi_{r+1}(x) - (2r+1)x\,\psi_r(x) + (r-1)\psi_{r-1}(x) = 0 \quad (r>1).$$

For any particular kernel $K(x,y)$ it is necessary to check that the recurrence relation for the functions $\psi_r(x)$ is stable.

It is known that global approximation by polynomials may sometimes be inferior to a piecewise polynomial approximation. Splines (including continuous piecewise-polynomial functions) are currently fashionable and can be used to advantage if $f(x)$ has a discontinuous derivative, of some order, at a known point.

If we construct a piecewise-constant approximation $\tilde{f}(x)$ (or a continuous piecewise-linear $\tilde{f}(x)$) and if this is constructed by collocation at the knots, the process is equivalent to using a modified-quadrature method (consider for example, the use of the generalized trapezium rule). To try to emulate the modified quadrature method using the generalized Simpson's rule, we can employ collocation with cubic splines. (Splines of higher (odd) order have been suggested: see Phillips (1969).)

If $f(x)$ is a cubic spline with knots at $a = \xi_0 < \xi_1 < \ldots < \xi_{n-2} = b$ it may be written in the form

$$\alpha_0 + \alpha_1 x + \alpha_2 x^2 + \sum_{i=0}^{n-3} \beta_i (x-\xi_i)_+^3 \tag{3.3}$$

where $z_+ = z$ if $z > 0$ and 0 if $z \leqslant 0$.

(There is a corresponding form for higher degree splines. We shall consider only splines with fixed knots ξ_i.) In view of the representation above we could take $\phi_0(x) = 1$, $\phi_1(x) = x$, $\phi_2(x) = x^2$ and $\phi_r(x) = (x-\xi_{r-3})_+^3$ for $r=3,4,\ldots,n$. Unfortunately the evaluation of the form (3.3) (and its high-order analogues) is regarded (Rice (1969)) p 159) as ill-conditioned, and the preceding choice of functions $\phi_r(x)$ is not recommended. (The effects of ill-conditioning depend of course on the computer arithmetic, the distribution of the knots, etc.; we incline to the view that for cubic and quintic splines such representation is not usually disastrous, at least on the CDC 7600 (60 bit) computer in use at Manchester University.)

Ahlberg, Nilson and Walsh (1967), p 57) suggest the use of "cardinal splines" (loc. cit. p 52) for the functions $\phi_r(x)$, performing collocation at the knots. (In this way the spline approximation $\tilde{f}(x)$ is forced to satisfy certain derivative conditions and if the collocation conditions are independent they define $\tilde{f}(x)$ uniquely.) In a numerical example, however, Ahlberg et al. (p 58) employ the standard form of a spline (which expresses it as a cubic in each interval between knots) and collocate at the knots ($z_i = \xi_i$), adding derivative conditions which they consider appropriate in order to define $\tilde{f}(x)$ uniquely.

Phillips (1969) chooses the B-splines for a basis $\phi_0(x), \phi_1(x) \ldots, \phi_n(x)$, which seems to be a good choice. The (cubic) B-splines can be defined theoretically in terms of certain divided differences of $(x-t)_+^3$; there has recently been some discussion (Cox (1972); de Boor (1972)) on stable methods of evaluating these functions.

It should be noted that collocation at the knots of the cubic spline approximation (setting $z_i = \xi_i$ for $i=0,1,\ldots,n-2$) is not sufficient to determine the approximation and we must add two further collocation points z_{n-1}, z_n (or impose additional derivative conditions on the spline).

The use of cubic or higher degree splines seems to us to be worth considering when $f(x)$ is known to have badly behaved derivatives at points which can be taken as knots. It is true that spline approximations frequently provide good approximations to the function and its derivatives simultaneously, but if we obtain $\tilde{f}(x)$ using the Nyström method and $g(x)$ and $K(x,y)$ are differentiable we can approximate $f'(x)$ by setting, when appropriate,

$$\tilde{f}'(x) = g'(x) + \lambda \sum_{j=0}^{n} w_j \, K_x(x,y_j) \, \tilde{f}(y_j).$$

The product integration techniques appear, moreover, to provide greater flexibility than the use of splines in collocation methods (with comparable accuracy). Further experience is necessary here.

Whereas the form of our approximation $\tilde{f}(x)$ is of importance in obtaining accuracy, the choice of the basis $\phi_0(x), \phi_1(x), \ldots, \phi_n(x)$ affects the accuracy obtainable in numerical work. This is an important aspect of the topic which we consider is not yet fully understood. An obvious criterion is that, given $\tilde{a}_0, \tilde{a}_1, \ldots, \tilde{a}_n$, we should be able to compute

$$\sum_{r=0}^{n} \tilde{a}_r \, \phi_r(x)$$

accurately; this depends on our ability to generate $\phi_r(x)$ as well as on the stability of the summation process. In terms of the choice $\phi_r(x)$ we note that perturbations δa_r in the coefficients cause a perturbation

$$\epsilon(x) = \sum_{r=0}^{n} \delta a_r \, \phi_r(x)$$

with

$$\|\epsilon(x)\|_\infty \leqslant \max |\delta a_r| \max_{a \leqslant x \leqslant b} \sum_{r=0}^{n} |\phi_r(x)|$$

and it seems desirable to have the last quantity small. Bounds on the errors δa_r resulting from the application of our methods have been analyzed in the linear case by Phillips (1969) Ikebe (1970) Mikhlin (1971, p 76) and Coldrick (1972, §3.2). In particular, a factor in bounding the condition numbers $\kappa(A-\lambda B)$ or $\kappa(C-\lambda D)$ (using the notation we employed above) is shown by Phillips to be the quantity $\kappa(A)$ or $\kappa(C)$ respectively, and it is advantageous to keep the appropriate quantity small.

The theories alluded to above are principally for linear integral equations. Mikhlin (1971) considers variational* methods for linear equations and extends his theory to certain types of non-linear equations. (In particular, Mikhlin (pp 359-361) examines the non-linear integral equation

$$f(x)\ [1 + \int_0^1 K(x,y) \left\{f(y)\right\}^2 dy] = g(x)$$

where $K(x,y)$ satisfies certain conditions, as an example.) To summarise Mikhlin's work, one conclusion is to the effect that a choice of functions $\phi_r(x)$ leads to stable calculations if $\phi_0(x),\phi_1(x),\ldots,\phi_n(x)$ is part of a complete orthonormal system. More generally, the choice $\left\{\phi_r(x)\right\}$ is "stable" for the Galerkin method for a linear equation if the system of functions $\phi_0(x),\phi_1(x),\ldots,\phi_n(x)$ is "strongly minimal" (Mikhlin p 10) in the space $\mathcal{L}^2\ [a,b]$ on which the integral operator is completely continuous (see Mikhlin p 68, p 178) so that there is a uniform bound on $||C^{-1}||_2$, as $n \to \infty$. Further, it is an advantage to choose an "almost orthonormal system" (Mikhlin, p 6) so that $\kappa(C) = ||C||_2\ ||C^{-1}||_2$ is uniformly bounded as $n \to \infty$. (An example of an unsatisfactory system is $\phi_r(x) = x^r$, $(r=0,1,\ldots)$ which is not strongly minimal; almost orthonormal systems are a subset of strongly minimal systems.)

4. Theory of Projection Methods

We shall outline briefly a framework within which the Galerkin and collocation methods can be analyzed theoretically (Phillips (1969), Vainikko (1967)). We shall not go into the details.

Suppose X is a Banach space consisting of functions, and our solution $f(x)$ lies in X. Suppose we seek an approximation $\tilde{f}(x) \in X_n$ where X_n is a Banach space spanned by $\phi_0(x),\phi_1(x),\ldots,\phi_n(x)$, and suppose

---

* In this description, Mikhlin includes the Rayleigh-Ritz and Galerkin methods. The Russian terminology is somewhat different from that of this chapter.

that A_n is a linear approximation operator which assigns to $\phi(x) \in X$ an element $A_n \phi(x) \in X_n$. We suppose that A_n is a projection of X onto X_n. If we consider a non-linear operator T: X → X a projection method for the solution of Tf = 0 consists in solving $A_n T\tilde{f} = 0$ for $\tilde{f} \equiv A_n \tilde{f} \in X_n$.

For example, suppose X is an inner-product space,

$$T\phi(x) = (\phi - \lambda K\phi - g)(x) \equiv \phi(x) - \lambda\int_a^b K(x,y)\,\phi(y)dy - g(x),$$

and

$$(A_n \phi)(x) = \sum_{i=0}^{n} (\phi,\phi_i)\,\phi_i(x)$$

for each $\phi(x) \in X$. (A_n is a projection.) The Galerkin method consists in solving

$$A_n\left\{\tilde{f} - \lambda K\tilde{f} - g\right\}(x) = 0 \qquad (4.1)$$

where

$$\tilde{f}(x) = A_n\tilde{f}(x)$$

and K is the integral operator with kernel K(x,y). (To demonstrate that this is so we write

$$\eta(x) = \tilde{f}(x) - \lambda\, K\tilde{f}(x) - g(x),$$

and in the notation of § 2 $\hat{\eta}(x) = A_n\, \eta(x)$.)

The general framework indicated above can be used to provide convergence results and error estimates, and has been used to analyze the stability of the Galerkin methods. To give the flavour of the convergence results, we state a theorem of the type given by Vainikko (1967).

<u>Theorem</u>. Let X be a Banach space and for n=0,1,... let A_n be a bounded projection of X onto a Banach space $X_n \subseteq X$ such that for any $\phi \in X$,

$$\lim_{n\to\infty} ||\phi - A_n\phi|| = 0.$$

Suppose that f is the solution of the equation f = Sf where S is a (possibly non-linear) operator from X to X which is completely continuous (compact) on an open set containing f. Suppose also that S is continuously Fréchet-differentiable at f and that I - S′(f) is invertible. Then the solution f is unique in some sphere, and for n sufficiently large, the equation $f_n = A_n Sf_n$ has a unique solution in this sphere such that

$$||f - f_n|| = O(||A_n f - f||) \text{ as } n \to \infty.$$

In analyzing (4.1) we would set $Sf = \lambda Kf - g$ for (say) $f \equiv f(x) \in C[a,b]$. This and similar theorems allow us to deduce that the order of accuracy obtainable in a collocation or Galerkin method is the order of accuracy obtainable in approximating the solution by interpolation (in the case of collocation) or a Fourier-type sum (in the Galerkin method). The theory must be applied rigorously, however, or misleading results are obtained.*

5. Other Methods

We do not have an opportunity to discuss all methods here, but it is worth exploring some further possibilities. We consider the linear equation (2.1). Suppose that from the known properties of the solution $f(x)$ we can choose linearly independent functions $\phi_i(x)$ with

$$f(x) = \sum_{i=0}^{\infty} a_i \, \phi_i(x).$$

(The infinite series is presumed to converge in some norm.) If we write

$$\psi_i(x) = \int_a^b K(x,y)\, \phi_i(y)dy$$

and proceed formally, we obtain from (2.1) the identity

$$\sum_{j=0}^{\infty} a_j \left\{ \phi_j(x) - \lambda \, \psi_j(x) \right\} = g(x). \tag{5.1}$$

Using the system of functions $\phi_i(x)$ $(i=0,1,2,\ldots)$ we can obtain from (5.1) the equations

$$\sum_{j=0}^{\infty} (C_{ij} - \lambda \, D_{ij}) \, a_j = \gamma_i \qquad (i=0,1,2,\ldots)$$

where C_{ij}, D_{ij} and γ_i are defined above. We may truncate this infinite system of equations in an infinite number of unknowns (see for example Chapter 24) to obtain the Galerkin system (2.11).

An approach which at first sight is completely different to those of §2 is to approximate the continuous kernel $K(x,y)$ in (2.1) by a degenerate kernel and solve the equation exactly. If we set

---

*Less restrictive versions of the theorem quoted can be found in Vainnikko (1967).

$$g_n(x) = g(x),$$

$$K_n(x,y) = \sum_{i=0}^{n} \phi_i(x) \ \nu_i(y)$$

where the functions $\nu_i(x)$ are chosen appropriately, the equation

$$f_n(x) - \lambda \int_a^b K_n(x,y) \ f_n(y) \ dy = g_n(x)$$

has a solution of the form

$$g_n(x) + \sum_{i=0}^{n} \alpha_i \ \phi_i(x).$$

If K(x,y) is continuous and we set

$$g_n(x) = \sum_{i=0}^{n} \gamma_i \ \phi_i(x)$$

and choose $\gamma_0, \gamma_1, \ldots, \gamma_n$ and $\nu_0(x), \nu_1(x), \ldots, \nu_n(x)$ appropriately, the process can be made equivalent to each one of the methods of §2. See Baker (to appear) for details.

References

AHLBERG, J.H., NILSON, E.N., and WALSH, J.L. (1967) The theory of splines and their applications. Academic Press, New York and London.

ARTHUR, D.W. (1973) The solution of Fredholm integral equations using spline functions. J. Inst. Math and its Appl. vol. 11 pp 121-129.

BAKER, C.T.H. (to appear) Numerical solution of integral equations. Clarendon Press.

BARRODALE, I and YOUNG, A. (1970) see Hayes. (editor) (1970) pp 115-142.

COLDRICK, D.B. (1972) Methods for the numerical solution of integral equations of the second kind. Tech. Rep. 45. Department of Computer Science, University of Toronto (Ph.D. thesis).

COLLATZ, L. (1960) The numerical treatment of differential equations Springer-Verlag, Berlin (3rd edition, 1966).

COX, M.G. (1972) The numerical evaluation of B-splines. J. Inst. Math. and its Appl. vol. 10 pp 134-149

DAVIS, P.J. (1965) Interpolation and approximation (second edition) Blaisdell Publishing Co. New York, Toronto, etc.

de BOOR, C. (1972) On calculating with B-splines. J. Approx. Theory vol. 6 pp 50-62.

HAYES, J.G. (1970) Numerical approximations to functions and data. The Athlone Press, London.

HUNG, HING-SUM (1970) The numerical solution of differential and integral equations by spline functions. MRC Tech. Rep. 1053 Madison, Wisc.

IKEBE, YASUHIKO (1970) The Galerkin method for numerical solution of Fredholm integral equation of the second kind. Report CNA-5, Center for Numerical Analysis, University of Texas at Austin.

MIKHLIN, S.G. (1971) The numerical performance of variational methods. (transl: R.S. Anderssen) Wolters-Noordhoff, Groningen.

OSBORNE, M.R. (1972) Some aspects of non-linear least squares calculations in numerical methods for non-linear optimization. F.A. Lootsman (editor) Academic Press, New York and London.

PHILLIPS, J.L. (1972) The use of collocation as a projection method for solving linear operator equations. SIAM J. Numer. Anal. vol. 9 pp 14-27.

PHILLIPS, J.L. (1969) Collocation as a projection method for solving integral and other operator equations. Ph.D. Thesis, Purdue University (Lafayette, Indiana).

RICE, J.R. (1969) The approximation of functions - vol. 2. - Advanced topics. Addison - Wesley, Reading, Mass. etc.

VAINNIKKO, G.M. (1967) Galerkin's perturbation method and the general theory of approximate methods for non-linear equations (transl:) USSR Comput. Math. and Math. Phys. vol. 7 pp 1-41.

CHAPTER 8 LINEAR PROGRAMMING SOLUTIONS TO INTEGRAL EQUATIONS

I. Barrodale

University of Liverpool/University of Victoria

1. Outline of the Method

A given operator equation

$$Pf = g \tag{1.1}$$

can be solved approximately for f by putting

$$f \approx f_n, \tag{1.2}$$

where f_n is a suitable approximating function depending upon n parameters $\alpha_1, \alpha_2, \ldots, \alpha_n$, substituting (1.2) in (1.1) to give

$$\delta_n = g - Pf_n, \tag{1.3}$$

and then minimizing the residual function δ_n in some sense. Provided that this minimum δ_n^* is small enough, the function f_n^* determined by the optimal values α_j^* of the parameters in (1.3) usually provides a satisfactory approximation to the solution of (1.1).

This type of method can be applied to nonlinear operator equations, but most of our experience with it has been limited to the case of linear operator equations and linear approximating functions (see Barrodale and Young, 1970). In this case we solve

$$Lf = g, \tag{1.4}$$

where L is a linear operator, by substituting

$$f \approx f_n = \sum_{j=1}^{n} \alpha_j \phi_j \tag{1.5}$$

in (1.4) and minimizing

$$\delta_n = g - Lf_n = g - \sum_{j=1}^{n} \alpha_j (L\phi_j) = g - \sum_{j=1}^{n} \alpha_j \psi_j \quad \text{(say)}. \tag{1.6}$$

In this chapter we present several numerical examples of linear integral equations in which the maximum absolute value of the residual δ_n in (1.6) is a minimum, i.e. we determine

$$||\delta_n^*||_\infty = \underset{\alpha_1,\alpha_2,\ldots,\alpha_n}{\text{minimum}} \; ||g - \sum_{j=1}^{n} \alpha_j \psi_j||_\infty . \qquad (1.7)$$

The parameter values α_j^* calculated in (1.7) give an approximate solution

$$f_n^* = \sum_{j=1}^{n} \alpha_j^* \phi_j$$

to (1.4). In practice we do not solve (1.7) exactly, but instead a corresponding discrete problem is solved using the simplex method of linear programming.

2. A Simple Example

Consider the Fredholm integral equation of the second kind

$$f(x) - \frac{1}{4}\int_0^{\pi/2} xy\, f(y)\, dy = \sin x - \frac{1}{4} x, \quad 0 \leqslant x \leqslant \frac{\pi}{2}, \qquad (2.1)$$

for which the exact solution is $f(x) = \sin x$. Suppose that the approximating function is chosen to be

$$f_3(x) = \alpha_1 x + \alpha_2 x^3 + \alpha_3 x^5 . \qquad (2.2)$$

In the notation of (1.5) we thus have $n = 3$, $\phi_1 = x$, $\phi_2 = x^3$, and $\phi_3 = x^5$. Substituting (2.2) in (2.1) yields

$$\delta_3(x) = g(x) - \sum_{j=1}^{3} \alpha_j \psi_j(x) \qquad (2.3)$$

where

$$\left.\begin{aligned} g(x) &= \sin x - \frac{1}{4} x \\ \psi_1(x) &= x - \frac{x}{4}\int_0^{\pi/2} y^2\, dy = x\left(1 - \frac{1}{12}\left(\frac{\pi}{2}\right)^3\right) \\ \psi_2(x) &= x^3 - \frac{x}{4}\int_0^{\pi/2} y^4\, dy = x\left(x^2 - \frac{1}{20}\left(\frac{\pi}{2}\right)^5\right) \\ \psi_3(x) &= x^5 - \frac{x}{4}\int_0^{\pi/2} y^6\, dy = x\left(x^4 - \frac{1}{28}\left(\frac{\pi}{2}\right)^7\right). \end{aligned}\right\} \qquad (2.4)$$

The next task is to minimize the residual $\delta_3(x)$ in (2.3). The criterion (1.7) adopted in this chapter requires us to solve the Chebyshev approximation problem

$$\underset{\alpha_1,\alpha_2,\alpha_3}{\text{minimize}} \quad \underset{0\leq x\leq \frac{\pi}{2}}{\text{maximum}} \quad \left| g(x) - \sum_{j=1}^{3} \alpha_j \, \psi_j(x) \right| \qquad (2.5)$$

where the functions $g(x)$ and $\psi_j(x)$ are defined by (2.4). In practice we replace problem (2.5) by a discrete problem

$$\underset{\alpha_1,\alpha_2,\alpha_3}{\text{minimize}} \quad \underset{1\leq i\leq m}{\text{maximum}} \quad \left| g(x_i) - \sum_{j=1}^{3} \alpha_j \, \psi_j(x_i) \right| \qquad (2.6)$$

in which the $m(>>n)$ points x_i are chosen from $[0,\frac{\pi}{2}]$. In general a discrete problem such as (2.6) can be solved with much less effort than (2.5). Furthermore, for most practical purposes a solution to (2.6) can be regarded as a solution to (2.5), provided that the discretization fairly represents the domain of x. (Typically, we choose $m \approx 5n$ or $10n$ and arrange for the points x_i to be uniformly spaced throughout the domain).

For example, the values α_j^* obtained by solving (2.6) with $m = 33$ points defined by $x_i = .05(i-1)$, for $i = 1,2,\ldots,32$, and $x_{33} = \frac{\pi}{2}$, give the approximate solution

$$f_3^*(x) = .99970\,x - .16567\,x^3 + .00751\,x^5.$$

This agrees with the exact solution of (2.1) to almost four decimal places on $[0, \frac{\pi}{2}]$.

3. Further Details of the Method

The choice of form for the approximating function is the most important factor in applying this method effectively. Of course, given some a priori information about the desired solution, we can sometimes include in f_n functions ϕ_j which are particularly appropriate for the problem at hand. Even or odd functions of x are often suitable choices in this respect. However, in the absence of such information we usually rely on standard linear approximating functions such as algebraic polynomials, trigonometric polynomials, or spline functions with fixed knots.

In contrast with the choice of form for f_n, computational experience suggests that the actual method of determining $f_n{}^*$ is rarely of crucial importance. Although the criterion of minimizing the l_∞ norm of δ_n is the only one considered here, there are other criteria which are frequently used instead. These include the minimization of δ_n in other norms, and Galerkin's method (see Chapters 7 and 9). This is another reason why in practice we do not need to solve (1.7) exactly.

Before determining the parameter values $\alpha_j{}^*$ it is necessary to evaluate the functions ψ_j of (1.6). In (2.4) we were able to obtain these functions by evaluating the integrals involved analytically. This is quite often the case, since we can choose the functions ϕ_j partly with this objective in mind. Otherwise, we resort to numerical integration techniques to estimate the ψ_j's. The use of a quadrature formula here is straightforward for Fredholm equations, but it is less convenient for Volterra equations since the upper limit of integration is variable.

4. Error Bounds

Whenever the approximating function f_n is such that the minimum residual $\delta_n{}^*$ is small, the resulting approximate solution $f_n{}^*$ is then the exact solution to a small perturbation of the original problem. Furthermore, this perturbation is known a posteriori since it is precisely $\delta_n{}^*$ itself. This is a very satisfactory situation, especially when the function g is known to only modest accuracy. However, in many circumstances it is also desirable to know how close $f_n{}^*$ is to the exact solution f.

If we define the <u>error</u> function ϵ_n by

$$f = f_n + \epsilon_n, \tag{4.1}$$

then substituting (4.1) into the linear operator equation $Lf = g$ gives

$$\delta_n = L\epsilon_n \tag{4.2}$$

in view of (1.6). It is clear from (4.2) that minimizing the residual is equivalent to minimizing the image (under L) of the error, and that in order to bound the error we must essentially obtain a bound on the inverse operator L^{-1}. In many problems this latter task is impossible, but at least for some Fredholm equations of the second kind the following upper and lower bounds are available (see Barrodale and Young, 1970).

If the equation

$$f(x) - \int_a^b K(x,y)\, f(y)\, dy = g(x), \quad a \leqslant x \leqslant b, \tag{4.3}$$

is uniquely solvable, the inequality

$$\frac{||\delta_n||_\infty}{1+||K||_\infty} \leqslant ||\epsilon_n||_\infty \leqslant \frac{||\delta_n||_\infty}{1-||K||_\infty} \tag{4.4}$$

is valid provided that

$$||K||_\infty = \underset{a\leqslant x\leqslant b}{\text{maximum}} \int_a^b |K(x,y)|\, dy < 1. \tag{4.5}$$

Determining the quantity $||K||_\infty$ presents no difficulty in practice since it need only be known to low accuracy. The restriction (4.5) sometimes prevents us from applying the bounds (4.4) to a given problem of the type (4.3), but the usefulness of these bounds is demonstrated in the first two numerical examples which follow. Finally, although the bounds (4.4) apply to any residual function δ_n and its corresponding error function ϵ_n defined by (4.2), the best use of these bounds is made when $||\delta_n||_\infty$ is a minimum. This is just what our method sets out to accomplish through criterion (1.7).

5. Numerical Examples

In the numerical examples of this section the problem

$$\underset{\alpha_1,\alpha_2,\ldots,\alpha_n}{\text{minimize}} \quad \underset{1\leqslant i\leqslant m}{\text{maximum}} \quad \left| g(x_i) - \sum_{j=1}^{n} \alpha_j\, \psi_j\, (x_i)\right| \tag{5.1}$$

is solved for various choices of n, m, g, ψ_j, and x_i. For each example the m points x_i arise from discretizing the continuous problem (1.7). In each case the value shown for $||\delta_n^*||_\infty$ refers only to the solution of (5.1), but in order to apply the error bounds (4.4) we assume that this value actually refers to the solution of the corresponding problem (1.7).

There are several computer programs available for solving (5.1), although some of these programs require the functions ψ_j to be of certain types. Our results were obtained using a Fortran program (see Barrodale and Phillips, 1973) which solves (5.1) without imposing any restrictions on the ψ_j's: it is based on the simplex method of linear programming. The numerical examples are such that the ψ_j's can be obtained from the ϕ_j's by evaluating the integrals involved analytically.

Example 1

Solve the Fredholm integral equation of the second kind

$$f(x) + \int_0^1 K(x,y)\, f(y)\, dy = x^3 , \quad 0 \leq x \leq 1, \tag{5.2}$$

where
$$K(x,y) = \begin{cases} x(1-y) & \text{for } 0 \leq x \leq y \leq 1, \\ y(1-x) & \text{for } 0 \leq y \leq x \leq 1. \end{cases}$$

Using the approximating function

$$f_n(x) = \sum_{j=1}^{n} \alpha_j\, x^{j-1} , \qquad \text{for } n = 1,2,\ldots,6,$$

and discretizing $[0,1]$ with $m = 51$ points defined by $x_i = .02(i-1)$ for $i = 1,2,\ldots,51$, the following results are obtained:

n	$\|\|\delta_n^*\|\|_\infty$	n	$\|\|\delta_n^*\|\|_\infty$
1	0.52×10^{0}	4	0.10×10^{-2}
2	0.21×10^{0}	5	0.11×10^{-3}
3	0.36×10^{-1}	6	0.21×10^{-5}

In this example polynomials of increasing degree provide a satisfactory rate of convergence for the sequence of values $||\delta_n^*||_\infty$. The error bounds (4.5) can be used here to estimate $||\epsilon_n^*||_\infty$ since

$$||K||_\infty = \underset{0\leq x\leq 1}{\text{maximum}} \int_0^1 |K(x,y)|\, dy = \frac{1}{8} < 1.$$

Thus, without actually making use of the exact solution $f(x) = \frac{7e}{e^2-1}(e^x - e^{-x}) - 6x$ of equation (5.2), we can state that

$$0.18 \times 10^{-5} \leq \underset{0\leq x\leq 1}{\text{maximum}} |f(x) - f_6^*(x)| \leq 0.24 \times 10^{-5}.$$

Example 2

Solve the Fredholm integral equation of the second kind

$$f(x) + \frac{d}{\pi}\int_{-1}^{1} \frac{f(y)\, dy}{d^2 + (x-y)^2} = 1, \quad -1 \leq x \leq 1, \tag{5.3}$$

for $d = \frac{1}{2}$ and $d = 1.$

Using the approximating function

$$f_n(x) = \sum_{j=1}^{n} \alpha_j \, x^{2j-2} \,, \qquad \text{for } n = 1,2,\ldots,6,$$

and discretizing $[-1,1]$ with $m = 101$ points defined by $x_i = -1 + .02(i-1)$ for $i = 1,2,\ldots,101$ (but, in view of symmetry, solving problem (5.1) only on the 51 points x_i satisfying $0 \leqslant x_i \leqslant 1$), the following results are obtained:

	$d = \frac{1}{2}$	$d = 1$
n	$\|\|\delta_n^*\|\|_\infty$	$\|\|\delta_n^*\|\|_\infty$
1	0.90×10^{-1}	0.52×10^{-1}
2	0.80×10^{-2}	0.11×10^{-2}
3	0.14×10^{-2}	0.23×10^{-3}
4	0.30×10^{-3}	0.20×10^{-4}
5	0.30×10^{-4}	0.10×10^{-5}
6	0.12×10^{-4}	0.24×10^{-6}

Love (1949) proves that equation (5.3) possesses a unique even solution representing the electrostatic field due to two equal circular coaxial conducting discs, where the distance separating the discs is d times their radius. It can be verified that

$$||K||_\infty = \underset{-1\leqslant x\leqslant 1}{\text{maximum}} \int_{-1}^{1} \left| \frac{d}{\pi[d^2 + (x-y)^2]} \right| dy = \frac{2}{\pi} \tan^{-1}\left(\frac{1}{d}\right) < 1,$$

and so the error bounds (4.5) can be applied to $||\epsilon_n^*||_\infty = ||f-f_n^*||_\infty$. For example, when $n = 6$ we have

$$0.70 \times 10^{-5} \leqslant \underset{-1\leqslant x\leqslant 1}{\text{maximum}} |f(x) - f_6^*(x)| \leqslant 0.41 \times 10^{-4} \,, \quad \text{for } d = \tfrac{1}{2}$$

and $$0.16 \times 10^{-6} \leqslant \underset{-1\leqslant x\leqslant 1}{\text{maximum}} |f(x) - f_6^*(x)| \leqslant 0.48 \times 10^{-6} \,, \quad \text{for } d = 1.$$

Example 3

Solve the Fredholm integral equation of the first kind

$$\int_0^1 e^{xy} f(y)\, dy = \begin{cases} g_3(x) \\ g_6(x) \\ g_{12}(x) \end{cases}, \quad 0 \leq x \leq 1, \tag{5.4}$$

where $g_d(x)$ represents $g(x) = \frac{e^{x+1} - 1}{x+1}$ to d significant figures.

Using the approximating function

$$f_n(x) = \sum_{j=1}^{n} \alpha_j x^{j-1}, \quad \text{for } n = 2,3,\ldots,6,$$

and discretizing [0,1] with m = 41 points defined by $x_i = 0.025(i-1)$ for $i = 1,2,\ldots,41$, the following results are obtained:

	3 sig. figs.	6 sig. figs.	12 sig. figs.
n	$\|\|\delta_n^*\|\|_\infty$	$\|\|\delta_n^*\|\|_\infty$	$\|\|\delta_n^*\|\|_\infty$
2	0.48×10^{-2}	0.38×10^{-3}	0.38×10^{-3}
3	0.47×10^{-2}	0.49×10^{-5}	0.67×10^{-6}
4	0.46×10^{-2}	0.47×10^{-5}	0.60×10^{-9}
5	0.46×10^{-2}	0.47×10^{-5}	0.46×10^{-11}
6	0.45×10^{-2}	0.47×10^{-5}	0.45×10^{-11}

Integral equations of the first kind are often ill-conditioned, and so when g(x) is known to only modest accuracy large errors can result from using any numerical method of solution. Equation (5.4) with $g(x) = \frac{e^{x+1} - 1}{x+1}$ has the exact solution $f(x) = e^x$, and our primary purpose in solving this problem with g(x) represented to different accuracies is to compare $||\delta_n^*||_\infty$ with $||\epsilon_n^*||_\infty = ||f - f_n^*||_\infty$.

Roughly speaking, although the sequence of values $||\delta_n^*||_\infty$ generated by increasing n converges to approximately the accuracy of g_d, the corresponding sequence $||\epsilon_n^*||_\infty$ decreases only until $||\delta_n^*||_\infty$ first reaches its limit. Thereafter, any further increase in n soon leads to large inaccuracies in the computed values of the parameters α_j^*, and consequently the sequence $||\epsilon_n^*||_\infty$ then diverges. For example, when n = 2 and n = 5 we obtain the following values:

		3 sig. figs.	6 sig. figs.	12 sig. figs.
n = 2	α_1^*	0.86022×10^0	0.85779×10^0	0.85780×10^0
	α_2^*	0.17155×10^1	0.17202×10^1	0.17202×10^1
n = 5	α_1^*	0.25639×10^3	0.30487×10^0	0.10001×10^1
	α_2^*	-0.48397×10^4	0.14166×10^2	0.99823×10^0
	α_3^*	-0.21022×10^5	-0.56650×10^2	0.51155×10^0
	α_4^*	-0.31878×10^5	0.86802×10^2	0.13818×10^0
	α_5^*	-0.15637×10^5	-0.42434×10^2	0.70217×10^{-1}

It is clear from these results that f_n^* can be a very poor approximation of f even when $||\epsilon_n^*||_\infty$ is small. Of course, we must be prepared for this when solving ill-conditioned problems. Equation (5.4) is used by Baker, Fox, Mayers, and Wright (1964) to illustrate a method for Fredholm equations of the first kind which obtains solutions as linear combinations of eigenfunctions of the kernel.

Example 4

Solve the Volterra integral equation of the second kind

$$f(x) - 17\int_0^x \frac{f(y)\,dy}{1+9y} = 1\ , \quad 0 \leqslant x \leqslant 1 \tag{5.5}$$

Using the approximating function

$$f_n(x) = \sum_{j=1}^{n} \alpha_j\, x^{j-1}\ , \text{ for } n = 3,4,\ldots,9,$$

and discretizing [0,1] with m = 51 points defined by $x_i = .02(i-1)$ for i = 1,2,...,51, the following results are obtained:

n	$\|\|\delta_n^*\|\|_\infty$	n	$\|\|\delta_n^*\|\|_\infty$
3	0.14×10^{0}	7	0.29×10^{-3}
4	0.18×10^{-1}	8	0.92×10^{-4}
5	0.37×10^{-2}	9	0.33×10^{-4}
6	0.96×10^{-3}		

Equation (5.5) is discussed by Watson (1973) who demonstrates that by minimizing an estimate for the error ϵ_n it is possible to obtain a more accurate approximate solution of the form f_n than the function f_n^* that we obtain by minimizing the residual δ_n. (However, considerable extra computation is involved in obtaining his estimate for ϵ_n prior to the minimization stage). The exact solution to (5.5) is $f(x) = \exp(\frac{17}{9} \log(1 + 9x))$, and so we can determine the error function $\epsilon_n^*(x) = f(x) - f_n^*(x)$ corresponding to each minimum residual function $\delta_n^*(x)$. For example, when n = 9 the error assumes the following values:

x	$\epsilon_9^*(x)$	x	$\epsilon_9^*(x)$
0.0	0.33×10^{-4}	0.6	-0.12×10^{-3}
0.1	0.22×10^{-4}	0.7	-0.99×10^{-4}
0.2	-0.50×10^{-4}	0.8	-0.16×10^{-3}
0.3	-0.30×10^{-4}	0.9	-0.22×10^{-3}
0.4	-0.26×10^{-4}	1.0	-0.26×10^{-3}
0.5	-0.95×10^{-4}		

As Watson (1973) observes, equation (5.5) gives errors that are predominantly one-signed when it is solved by residual minimizing techniques.

Since δ_n^* usually oscillates in sign at least n times, and in view of the relationship $\delta_n^* = L\,\epsilon_n^*$ from (4.2), in general the errors resulting from our method tend to be better distributed throughout the domain than might be suggested by this example.

6. Acknowledgements

It is a pleasure to acknowledge the assistance of Mr. C. Phillips in obtaining some of the numerical results of Section 5. The work of Chapters 5 and 8 was supported financially by the Science Research Council of Great Britain through SRC Grant No. B/SR/8872.

References

BAKER, C.T.H., FOX, L., MAYERS, D.F. and WRIGHT, K. (1964) Numerical solution of Fredholm integral equations of the first kind. Comput. J., vol.7, pp 141-148.

BARRODALE, I. and PHILLIPS, C. (1973) Solution of an overdetermined system of linear equations in the Chebyshev norm, Preprint, Dept. of Computational and Statistical Science, University of Liverpool.

BARRODALE, I. and YOUNG, A. (1970) Computational experience in solving linear operator equations using the Chebyshev norm, in Numerical Approximations to Functions and Data, Hayes, J.G. (ed.), The Athlone Press, London, pp 115-142.

LOVE, E.R. (1949) The electrostatic field of two equal circular coaxial conducting disks. Q.J. Mech. Appl. Math., vol.2, pp 428-451.

WATSON, G.A. (1973) On estimating best approximations of functions defined by integral equations. Comput. J. vol.16, pp 77-80.

CHAPTER 9 RAYLEIGH-RITZ-GALERKIN METHODS

L.M. Delves

University of Liverpool

1. Introduction

In Chapter 7, expansion methods for the solution of integral equations were discussed. An important class of such methods is that characterised by various titles: Rayleigh-Ritz, Galerkin, method of moments or variational; and in this chapter we give a brief discussion of this class*. §§2 and 3 define the methods for the linear eigenvalue problem and for linear inhomogeneous Fredholm equations respectively; their numerical behaviour is considered in §§4 and 5; while §6 gives a comparison with other expansion methods, and with quadrature methods. Finally, §7 considers the extension to non-linear problems.

Most of what we have to say is said most succinctly in terms of operators in an appropriate function space R, which we take to be a separable Hilbert space (see Chapter 4). The linear eigenvalue and inhomogeneous problems then take the form

$$\text{eigenvalue} \qquad Kf = \kappa f \tag{1.1}$$

$$\text{inhomogeneous} \qquad Lf = g \tag{1.2}$$

where K,L are linear operators in, and f,g elements of, R.

Viewed as Fredholm integral equations, K is the kernel operator; L=K (equation of the first kind) or L=I-K (equations of the second kind). We shall assume that (1.1) has at least one solution, and (1.2) only one solution, in R.

An expansion method introduces the <u>expansion set</u> $\{h_i\}$ which is assumed to be complete in R, and the <u>infinite</u> and <u>truncated</u> expansions

$$f = \sum_{i=1}^{\infty} b_i\, h_i \tag{1.3}$$

---

*The various names refer to different methods of deriving essentially equivalent approximation schemes. The domain of validity of the derivation varies with the name; the final scheme is often (especially in the Russian literature) referred to as "variational" irrespective of its derivation.

$$f_N = \sum_{i=1}^{N} a_i^{(N)} h_i \qquad (1.4)$$

and we start by considering the defining equations for the coefficients $a_i^{(N)}$.

2. The Eigenvalue Problem

This is the context in which Rayleigh-Ritz methods are best known, and perhaps the most used. We give the standard introduction here.

Suppose that K is Hermitian, and consider the functional

$$F[u] = (u,Ku)/(u,u) \quad (u,u) \neq 0 \qquad (2.1)$$

Let f satisfy (1.1) for some κ, and set $u = f + \epsilon$. Then clearly $F[f] = \kappa$, and

$$\begin{aligned} F[f+\epsilon] &= \left\{ (u,Kf) + (Kf,\epsilon) + (\epsilon,K\epsilon) \right\} /(u,u) \\ &= \left\{ \kappa(u,u-\epsilon) + \kappa(u-\epsilon,\epsilon) + (\epsilon,K\epsilon) \right\} /(u,u) \\ &= \kappa + (\epsilon,(K-\kappa)\epsilon)/(u,u). \end{aligned} \qquad (2.2)$$

From (2.2) several consequences follow:-

1) Since the second term (the <u>error term</u>) is quadratic in ϵ, $F[u]$ is stationary at the solution point f. This is made the basis of the numerical procedure below.

2) The eigenvalues of an Hermitian operator are real. If K is also bounded (as it is for an $\mathcal{L}^2$ Fredholm operator, for example) we may order the eigenvalues:-

$$||K|| \geq \kappa_1 \geq \kappa_2 \geq \ldots \geq \kappa_i \geq \kappa_{i+1} \ldots \geq -||K|| \qquad (2.3)$$

Equation (2.2) is valid for $\kappa_1, \kappa_2, \ldots$. If we set $\kappa = \kappa_1$, clearly $K - \kappa$ is negative definite, and hence

$$F[u] \leq \kappa_1 \quad \forall\, u \in R.$$

Similarly if the algebraically smallest eigenvalue is κ_{min} we find

$$F[u] \geq \kappa_{min} \quad \forall\, u \in R.$$

3) If we insert the form $u = f_N$ (see (1.4)) we obtain

$$F[f_N] = \mu^{(N)}(\underline{a}) = \underline{a}^+ \underline{K}\, a/\underline{a}^+ \underline{M}\, \underline{a} \qquad (2.4)$$

where $(\underline{K})_{ij} = (h_i, Kh_j)$ and $(\underline{M})_{ij} = (h_i, h_j)$ are Hermitian N×N matrices; M is clearly positive definite; and $(\underline{a})_i = a_i^N$. $F[f_n]$ is stationary with respect to the coefficients a_i if

$$\underline{K}\ \underline{a} - \mu^{(N)}(\underline{a})\ \underline{M}\ \underline{a} = 0 \tag{2.5}$$

where $\mu(\underline{a})$ is given by (2.4). Equation (2.5) is the defining equation for the Rayleigh-Ritz method. It is a standard N×N algebraic eigenvalue problem for $\mu^{(N)}$, and its eigenvalues $\mu_i^{(N)}$ indeed satisfy (2.4).

4) Since M is positive definite and symmetric, there exists a non-singular lower triangular matrix T such that

$$M = TT^T \tag{2.6}$$

Equation (2.5) can then be reformulated as

$$[K' - \mu^{(N)}\ I]\ \underline{a}' = 0 \tag{2.7}$$

$$K' = T^{-1}\ K\ T^{T-1} \quad : \quad \underline{a}' = T^T\ \underline{a}$$

The reduction (2.7) forms the basis of one standard numerical method for the solution of (2.4). It also leads to the following remark:

5) The N×N matrix $\underline{K}'$ is Hermitian; and since T, and hence T^{-1}, is triangular, the elements of K′ are independent of N: that is, $K'^{(N)}$ is the leading minor of $K'^{(M)}$ if M > N. Ordering the eigenvalues

$$\mu^{(N)} : \mu_1^{(N)} \geqslant \mu_2^{(N)} \geqslant \ldots \geqslant \mu_N^{(N)}$$

we therefore have the standard <u>separation theorem</u> (sometimes in this context called the <u>Hylleraas-Undheim</u> theorem):

$$\mu_1^{(N+1)} \geqslant \mu_1^{(N)} \geqslant \mu_2^{(N+1)} \geqslant \mu_2^{(N)} \geqslant \ldots \geqslant \mu_N^{(N+1)} \geqslant \mu_N^{(N)} \geqslant \mu_{N+1}^{(N+1)} \tag{2.8}$$

Thus, for fixed j, $\mu_j^{(N)}$ is an increasing function of N; and provided that

$$\mu_j^{(N)} \to \kappa_j \tag{2.9a}$$

we have the following extension of remark (2) above:-

$$\mu_j^{(N)} \leqslant \kappa_j \quad \forall_j,\ N \geqslant j. \tag{2.9b}$$

Conditions under which (2.9a) can be guaranteed are studied in, for example, Gould (1966), where it is shown that sufficient conditions are that K is compact, and $\left\{h_i\right\}$ complete, in R.

If we number the eigenvalues in increasing order we may similarly bound them from above (see Chapter 10, §3). This alternative bound is useful if K has negative eigenvalues; otherwise, we recall that for many Hermitian kernels, zero is an accumulation point of the spectrum, and in this case the lower bounds given here are more useful.

For a given integral equation there will usually be many possible choices of inner product; whether K is Hermitian depends on this choice. We take as an example the Fredholm kernel

$$(Kf)\ (x) = \int_a^b K(x,y)\ f(y)\ dy.$$

Then if K(x,y) is symmetric: $K(x,y) = K^*(y,x)$, (where * denotes complex conjugate) the kernel operator is Hermitian with the choice of inner product

$$(u,v) = \int_a^b u^*(x)\ v(x)\ dx.$$

It sometimes happens, however, that

$$K(x,y) = \overline{K}(x,y)\ w(y)$$

where $\overline{K}$ is symmetric. Such equations can always be symmetrised before solution, if $w(y) > 0$; alternatively, K may be made Hermitian by the choice of inner product

$$(u,v) = \int_a^b u^*(x)\ v(x)\ w(x)\ dx.$$

For a general non-Hermitian (but still linear) operator it is simplest to proceed as follows. We notice that (2.5) could have been derived by the method of moments (Galerkin method). In this method, we attempt to satisfy the approximate equations

$$K\ f_N \doteq \mu^{(N)}\ f_N \qquad (2.10a)$$

by taking their inner product with a set of functions h_i', i=1,2,...,N:-

$$(h_i',\ K\ f_N) = \mu^{(N)}\ (h_i',\ f_N). \qquad (2.10b)$$

Inserting the form (1.4) for f_N and identifying $h_i' \equiv h_i$ we regain (2.5). More generally, we notice

(1) We may choice the set $\{h_i'\}$ independently of $\{h_i\}$; we then have the unsymmetric method of moments or bi-variational method.

(2) The procedure does not require that K be Hermitian. The equation (2.5) remains invariant in form, with

$$(\underline{K})_{ij} = (h_i', Kh_j); \quad (\underline{M})_{ij} = (h_i', h_j) \tag{2.11}$$

and if $h_i' \equiv h_i$, the matrix M is still positive definite symmetric.

The same defining equations can also be reached by considering the stationary points of the modified functional

$$F[u^+,u] = (u^+Ku)/(u^+,u) \tag{2.12}$$

which is stationary with respect to variations in u^+ at the point u: $Ku = \kappa u$. The function u^+ represents the solution of the adjoint problem

$$K^+ u^+ = K u^+.$$

The convergence properties of these methods are considered in §§4 and 5 below.

3. Inhomogeneous Equations

The inhomogeneous equation (1.2) may be treated formally in a very similar manner. We do not assume that the operator ℓ is Hermitian. We proceed either directly by the method of moments, or by considering the functional $F_L(u^+,u)$:-

$$F_L(u^+,u) = (u^+, \ell u) - (g,u) - (u^+,g) \tag{3.1}$$

to produce the defining equations for the unsymmetric method of moments

$$(h_i', \ell x_N) = (h_i', g) \quad i=1,\dots,N. \tag{3.2}$$

Inserting the form (1.4) for f_N we find for the coefficients $a_i^{(N)}$:

$$L^{(N)} \underline{a}^{(N)} = \underline{g}^{(N)} \tag{3.3}$$

where $L^{(N)}$ is the leading N×N minor of the matrix L with

$$L_{ij} = (h_i', \ell h_j), \text{ and } g_i = (h_i', g).$$

As might be expected, the properties of (3.3) differ markedly depending on whether ℓ represents an integral operator of the second or first kind. If we set

$$\ell = I - \lambda K$$

and assume for simplicity that $\{h_i\} = \{h_i'\}$ is orthonormal, then

$$L_{ij} = \delta_{ij} - \lambda K_{ij}$$

Now $|K_{ii}| = |(h_i, K h_i)| \leq ||K||$; hence, at least if $|\lambda| \leq ||K||^{-1}$, we have $L_{ii} \neq 0$, and L positive definite. Of course, this condition is quite restrictive; but even if $|\lambda| > ||K||^{-1}$ the matrix L is usually well behaved and the solution of (3.3) straightforward numerically. None of the remarks is true if $\mathcal{L} = K$, giving an equation of the first kind; and indeed, we expect that the numerical behaviour of such equations will mirror their theoretical instability. We return to these points in the next two paragraphs.

4. Error Estimates

Error bounds and estimates for these procedures are available from a variety of sources. The first three can be considered classical. Let f_N denote any approximate solution of (1.2) and denote $f - f_N = \epsilon_N$.

1) Suppose that $\mathcal{L}$ is positive definite bounded below; that is, for some constant $\gamma > 0$

$$(u, \mathcal{L} u) \geq \gamma^2 (u,u) \quad \forall u \in R. \tag{4.1}$$

Then

$$||\epsilon||^2 = (\epsilon_N, \epsilon_N) \leq \frac{1}{\gamma^2} (\epsilon_N, \mathcal{L} \epsilon_N)$$

$$\leq \frac{1}{\gamma^2} ||\epsilon_N|| \, ||\mathcal{L} u - \mathcal{L} u_N||$$

i.e.

$$||\epsilon_N|| \leq \frac{1}{\gamma^2} || \mathcal{L} u_N - g|| \tag{4.2}$$

The L.H.S. of (4.1) can be evaluated a posteriori no matter how we arrive at the approximation u_N.

For an operator of the second kind: $\mathcal{L} = I - K$ we may take $\gamma^2 = 1 - ||K||$ provided that this is positive.

2) Suppose that in addition $\mathcal{L}$ is Hermitian. Then in the functional (2.12) we may set $u^+ = u$, and derive the identity

$$F[f_N] \equiv (f_N, \mathcal{L} f_N) - (f_N, g) - (g, f_N) = F[f_N] + (\epsilon_N, \mathcal{L} e_N) \tag{4.3}$$

The error term in (4.3) is positive, and hence the solution point f lies at a minimum of the functional. Suppose that for some δ

$$F[u] > \delta \quad \forall u \in R.$$

Then rewriting (4.3) we have

$$||\epsilon_N||^2_{\mathcal{L}} \equiv (\epsilon_N, \mathcal{L}\epsilon_N) = F[f_N] - F[f] \leq F[f_N] - \delta \qquad (4.4a)$$

We may then return to the "natural" norm by invoking (4.1) to obtain

$$||\epsilon_N||^2 \leq \frac{1}{\gamma^2} [F[x_N] - \delta] \qquad (4.4b)$$

3) Rather similar error bounds can be derived for the Hermitian eigenvalue problem in various ways. These bounds depend on producing a lower bound on the eigenvalue to adjoin to the upper bounds provided directly by (2.9b). We give one such bound (the Temple bound):- We introduce the operator $D=\mathcal{L}-\alpha I$. Then f satisfies the equation

$$Df = \frac{1}{\kappa-\alpha} D^2 f \qquad (4.5a)$$

for which the following functional G is the direct generalisation of (2.1):

$$G[u] = (u, Du) / (u D^2 u) \qquad (4.5b)$$

with the identity

$$G[f_N] = \frac{1}{\kappa-\alpha} + \frac{(\epsilon_N,(D-(\kappa-\alpha)^{-1}D^2)\epsilon_N)}{(f_N,D^2 f_N)} \quad . \qquad (4.5c)$$

Suppose that $\kappa_p > \alpha \geq \kappa_{p+1}$, and that we choose $\kappa=\kappa_p$ in (4.5c). Then the error term is positive, and we have

$$G[f_N] \geq \frac{1}{\kappa_p-\alpha}$$

whence we find

$$\kappa_p \leq \alpha + (G[f_N])^{-1} \quad \forall f_N: \; G[f_N] > 0. \qquad (4.6)$$

The bounds (4.6), (2.9b) allow us to bracket κ_p as closely as we please. Moreover, if we substitute the form (1.4) for f_N we can derive the defining equations for the coefficients $\underline{a}_i^{(N)}$ which maximise (4.5c):-

$$\begin{aligned} &\text{Sup } G[f_N] = \max \nu \; :- \\ &[D-\nu E] \, \underline{a} = 0 \\ &(D)_{ij} = (h_i,Dh_j); \; (E)_{ij} = (h_i,D^2h_j). \end{aligned} \qquad (4.7)$$

All of these bounds suffer from two disadvantages. First, they give no advance estimates of the likely error. Second, they contain no information on the dependence of the error on N - the convergence rate -

while in practice this is likely to be the crucial factor in determining the success or otherwise of the calculation. We sketch an alternative approach which concentrates on this aspect of the problem.

We note that for fixed N, the error in the approximate solution of either (1.1) or (1.2) via (2.5) or (2.15), depends only on the space spanned by the functions $\{h_i \dots h_N\}$ and not on the functions themselves; for the solution represents the stationary point of an appropriate functional within this space. Without essential loss of generality, we may therefore assume that the $\{h_i\}$ are orthonormal. Then the error norm may be written (see 1.3, 1.4):

$$(e_N, e_N) = \sum_{i=1}^{N} (b_i \dots a_i^{(N)})^2 + \sum_{i=N+1}^{\infty} b_i^2 = S_1(N) + S_2(N). \qquad (4.8)$$

In this identity the second term $S_2(N)$ depends only on the solution f, while the first depends also on the procedure for calculating the $a_i^{(N)}$. Both terms are positive; we may therefore <u>bound</u> the convergence rate by bounding $S_2(N)$; and this is a standard problem in approximation theory. The convergence rate of S_2 (the "Fourier" convergence rate) depends on the expansion set chosen, and on the analyticity properties of the solution f, and hence of the kernel K. Some comments on the choice of expansion set are made in Chapter 7. There are two possible types of set which might be considered. The first, of "finite element" type, uses piecewise polynomials: see Chapter 23 for a discussion of these in multi-dimensional calculations. The convergence rate of S_2 then depends primarily on the degree of continuity of the expansion set, and only weakly on the solution (not at all for sufficiently regular kernels). Alternatively we may use some set of orthogonal polynomials over the region involved; these may then be chosen so that the convergence is limited by the solution structure, and is very fast for smooth kernels. Estimates for various expansion sets are given in Mead and Delves (1973). Rapid convergence is particularly important in multi-dimensional problems.

If in addition $S_2(N)$ dominates $S_1(N)$, the estimates for S_2 become estimates of the convergence rate of $||e_N||^2$. In many situations this is the case; but a proof requires a detailed investigation of the numerical procedure used. Two approaches appear in the literature:-

(a) We first choose a norm in which it is simpler to discuss the error, leaving till later the question of converting if necessary to the natural norm. This "simpler" norm is then usually closely related to the norm $||e_N||_{\mathcal{L}}$ introduced in (4.4). From (4.3) we see that, if $\mathcal{L}$ is positive definite, the Rayleigh-Ritz procedure minimizes $||e_N||_{\mathcal{L}}$. Introducing a Hilbert space R with inner product $[x,y] = (x, \mathcal{L}\, y)$, it follows that the procedure produces a best approximation in this space and the problem reduces identically to one in approximation theory. Having solved this problem, the step back to $||e_N||$, the "natural" norm in R, may be made using (4.1) if applicable.

(b) The arguments of (a) return to the underlying Euclidean space in which the elements of R are functions, for the estimates of the convergence of some Fourier series. We may alternatively work wholly within the space R and seek estimates in terms of the structure of the equations to be solved. With an orthogonal expansion set $\{h_i\}$ in R, the matrix $\underline{K}$ of a Fredholm operator K has the following attributes

(i) the diagonal elements K_{ii} are decreasing in i;

(ii) for fixed i, K_{ij} is decreasing in j; for fixed j, K_{ij} decreases in i.

These attributes imply that for an equation of the second kind, the diagonal elements of the matrix L=I-K are in some sense dominant. By making appropriate assumptions on the form of L, it proves possible to characterise the inverse matrices $(L^{(N)})^{-1}$ in sufficient detail to bound the convergence rates of the coefficients b_i, and of $b_i - a_i^{(N)}$; and hence of $||e_N||$ and of $||e_N||_{\mathcal{L}}$. See Delves and Mead (1971); Freeman, Delves and Reid (1973); Delves and Freeman (1973) for details of the methods used.

5. Numerical Performance

Under this heading we consider briefly the various aspects of these methods which are important numerically.

(a) Implementation

The major effort in any implementation is expended in producing the matrix $\underline{K}$. For the integral equation

$$f(\underline{x}) = y(\underline{x}) + \int_{D(\underline{x})} K(\underline{x},\underline{y})\, f(\underline{y})\, d\underline{y}$$

the inner product (.) will be defined as an integral over some domain D: $D(\underline{x}) \in D$ for all $\underline{x}$ of interest:

$$K_{ij} = \int_D h_i(\underline{x}) \int_{D(\underline{x})} K(\underline{x},\underline{y})\ h_j\ (\underline{y})\ d\underline{y}\ d\underline{x} \tag{5.1}$$

For example, for a one-dimensional Volterra equation we have $D(x) = [a,x]$; $D = [a,b]$ and $a \leqslant x \leqslant b$. We see that to evaluate K_{ij} we have a double integral to perform. For a general kernel $K(x,y)$ this will typically need to be carried out numerically, although where the form permits it, analytic integration will usually be faster. Where numerical quadrature is used, the rule chosen should of course be tailored to the region $D \otimes D(s)$, and to any singular behaviour in the integrands $h_i(s)\ K(s,t)\ h_j(t)$. In "nice" circumstances: a Fredholm equation with high continuity kernel over a region $[a,b]$, for example, the domain is smooth (a square in this case) and a product Gauss rule is perhaps the best available choice. Whatever rule is chosen, however, economy suggests that it should be the same for each pair of suffices (i,j), since this minimises the number of evaluations of the functions involved. Moreover, a stability analysis (Delves (1973), p 93) suggests that the quadrature errors are then favourably correlated. Thus the use of, say, an adaptive integration procedure for individual elements is not recommended.

(b) <u>Rate of Convergence</u>

We have considered this formally in the previous section, and indicated that extremely rapid convergence rates can be achieved in practice. We illustrate this with an eigenvalue example relevant to Chapter 25. We consider the homogeneous equation

$$f(x) = \lambda \int_0^\infty K(x,y)\ f(y)\ dy \tag{5.2}$$

$$K(x,y) = k_0\ y^2(y^2+k^2)^{-1}\ v(x,y)$$
$$v(x,y) = (xy)^{-1}\ Q_0((x^2+y^2+\mu^2)/2xy) \tag{5.3}$$

where Q_0 is a Legendre function of the second kind, and k_0, k, μ are constants.

Table I shows the eigenvalue λ computed in three different ways. The first column gives the results using a quadrature method with a

Gauss-Rational rule (see Stroud and Secrest (1966) p 91), with N points. The second and third columns use a Rayleigh-Ritz procedure with N terms of the expansion sets:

$$A : h_i(x) = (x^2 + k^2)^{-i} \qquad , \; i = 1,2,\ldots$$

$$B : h_1(x) = v(x,k)$$

$$h_2(x) = (x^2 + k^2)^{-1}$$

$$h_i(x) = (x^2 + (i-2)\,\mu^2/2)^{-1} \quad i = 3,4,\ldots \tag{5.4}$$

In this table the constant k_0 has been chosen so that the exact eigenvalue is $\lambda = 1.0$.

TABLE I Numerically computed eigenvalues for the equation (5.2)

N	Quadrature	Set A	Set B
2		1.282	1.051
4		1.046	1.011
6	0.987	1.010	1.000
8	0.995	1.002	
10	0.998	1.000	
16	1.000		

At least in this example, the Rayleigh-Ritz calculations converge faster than the quadrature method; and set B converges extra-ordinarily fast. However, viewed in isolation the quadrature method also converges fast, and we comment further on the comparison later.

(c) Stability

The numerical errors in a variational calculation come from two sources:- setting up, and then solving, the defining equations. Of these sources, the former is usually dominant, especially if numerical quadrature is used to evaluate the matrix $\underline{K}$. Thus, we can usually assume that the computed equations are solved exactly. Even so, the total error from this source can grow without bounds as N increases, and the stability of the calculation is an important concept.

In the presence of errors in the computed matrices, equations (2.5) and (3.3) are replaced respectively by

$$[(K + k) - (\mu + \delta\mu)(M + m)][\underline{a} + \delta\underline{a}] = 0 \qquad (5.5)$$

$$(L + l)(\underline{a} + \delta\underline{a}) = (\underline{g} + \delta\underline{g}). \qquad (5.6)$$

For finite matrices, a standard treatment of (5.6) leads to the bound

$$||\delta\underline{a}|| \leqslant ||L^{-1}|| \left\{||\delta\underline{g}|| + ||l||\ ||\underline{a}||\right\} / \left\{1 - ||L^{-1}||\ ||l||\right\} \qquad (5.7)$$

provided the denominator is positive. A similar treatment can be given for the eigenvalue problem (Delves (1968); Delves (1973), p 85 et sequ) and we quote these results in two parts. First, if the expansion set is orthonormal we may set M=I, m=0 and a standard comparison theorem then yields

$$|\delta\mu| \leqslant ||k||. \qquad (5.8)$$

If the $\left\{h_i\right\}$ are not orthonormal we introduce the triangular decomposition (2.6) of M and obtain

$$|\delta\mu| \leqslant \frac{||T^{-1}||\ ||T^{-1T}||\left\{||k|| + |\mu|\ ||m||\right\}}{1 - ||T^{-1}||\ ||T^{-1T}||\ ||m||}. \qquad (5.9)$$

The bound (5.8) differs from (5.7) and (5.9) in two respects. First, it is cheap to compute in practice; the others involve the inversion of an N×N matrix. Delves (1968) gives an alternative approach to (5.9) which provides cheap estimates of $|\delta\mu|$ in the case M≠I. Second, and more important, we can read off from (5.8) the behaviour of the round-off errors as $N \to \infty$. For example, let us suppose that each element of K is computed to the same <u>relative accuracy</u> ϵ. Then for totally positive k

$$|k_{ij}| \leqslant \epsilon\ \underline{K}_{ij}$$

$$||k|| \leqslant \epsilon\ ||\underline{K}|| \leqslant \epsilon\ ||K||$$

so that the round-off errors are <u>bounded</u> in N. Such a calculation is <u>unconditionally stable</u>. To obtain a similar result from (5.7)or (5.9) requires that the norms $||L^{-1}||$, $||M^{-1}||$ be bounded as $N \to \infty$; and this is a very strong restriction on the expansion sets. The detailed analysis of (Mikhlin (1970)) takes this very strict definition of stability, and spells out these restrictions and their consequences in detail, chiefly in the context of differential equations. As one consequence of his definition, a given set $\{h_i\}$ may lead to a stable calculation, while the renormalised set $\{\alpha_i h_i\}$ does not. In most circumstances, such a re-scaling would make no difference numerically if, as is appropriate, the linear equations package used carried out an initial scaling of the equations before solution. A weaker definition of stability has been used by Freeman and Delves (1973) to give a similar analysis of the growth of round-off errors.

As mentioned above, all these bounds are pessimistic if a consistent numerical quadrature scheme has been used. See Delves (1973, p 93).

6. Extension to Nonlinear Equations

So far we have considered only linear equations; we sketch here the simplest extension of the Rayleigh-Ritz method to nonlinear equations.

Consider an inhomogeneous equation of the form

$$\mathcal{L}(f) = g \tag{6.1}$$

where $\mathcal{L}$ is now a nonlinear operator. We assume (temporarily) that there exists a decomposition of the form

$$\mathcal{L}(\phi) = \mathcal{L}^{(0)}\phi + \mathcal{L}^{(1)}(\phi) \quad \forall\ \phi \in R$$

where $\mathcal{L}^{(0)}$ is a linear operator. Then (6.1) may be written

$$\mathcal{L}^{(0)}f = g - \mathcal{L}^{(1)}(f)$$

which suggests the iterative scheme

$$\mathcal{L}^{(0)}f_0^{(n+1)} = g - \mathcal{L}^{(1)}\,(f^{(n)}). \tag{6.2}$$

We attempt a solution of (6.2) by the method of moments, replacing $f^{(n)}$ by $f_N^{(n)}$:-

$$f_N^{(n)} = \sum_{i=1}^{N} a_{i(n)}^{N}\, h_i \tag{6.3}$$

where the coefficients $a_{i(n)}$ depend on the iteration number. The Ritz-Galerkin or moment equations are:

$$(h_i, \mathcal{L}^{(0)}\, f_N^{(n+1)}) = (h_i, g) - (h_i, \mathcal{L}^{(1)}(f_N^{(n)})) \tag{6.4}$$

or in an obvious notation:

$$L^{(0)}\, \underline{a}_{(n+1)} = \underline{g} - \underline{L}^{(1)}\,(f_N^{(n)}) \tag{6.4a}$$

Thus each iteration in n involves the solution of a set of linear equations. Depending on the choice of $L^{(0)}$, these iterations may or may not converge. If they do, they clearly converge to a solution of the nonlinear equations

$$(h_i, \mathcal{L}(f_N)) = (h_i, g) \quad i = 1,2,\ldots,N. \tag{6.4b}$$

We could have started with these equations without introducing the decomposition $\mathcal{L} = \mathcal{L}^{(0)} + \mathcal{L}^{(1)}$; and numerically we may if we wish solve (6.4b) directly. However, in many problems a suitable $\mathcal{L}^{(0)}$ is readily available; and the form (6.4a) is very convenient numerically, since the matrix $L^{(0)}$ need be constructed once only. Further, the solution of the linear equations may be achieved very efficiently, the triangular reduction of $L^{(0)}$ being carried out once only.

Both of these remarks assume that $\mathcal{L}^{(0)}$ does not depend on the iteration n. If $\mathcal{L}$ has a Fréchet derivative, one possible form of $\mathcal{L}^{(0)}$ comes from writing

$$g = \mathcal{L}(f_N + \epsilon_N) = \mathcal{L}(f_N) + \mathcal{L}'(f_N)\,\epsilon_N + \mathcal{O}(||\epsilon_N||^2)$$

whence to first order we find the <u>correction equations</u>

$$\mathcal{L}'(f_N)\,\epsilon_N = g - \mathcal{L}(f_N) \tag{6.5}$$

This has the same structure as (6.4a), except that $\mathcal{L}'(f_N)$ <u>does</u> depend upon f_N and hence upon the iteration number n.

7. Comparison with other Methods

Finally, we attempt to compare variational with other available methods: alternative expansion methods, and quadrature methods. The conclusions reached in such a comparison depend partly on the use to which the solution will be put. For example, if the solution is to be repeatedly evaluated at a number of points in the domain, we shall favour a <u>short</u> expansion even if it is expensive to produce. The comparisons below exclude such conditions, and refer only to the cost of <u>producing</u> a solution of given accuracy.

(a) Comparison with quadrature methods

Methods based on collocation with a numerical quadrature formula form the natural counterpart to finite difference methods for differential equations. In that field, the relative advantages of expansion methods with a global (rather than finite element) type of expansion set are well known:- the rapid convergence rates attainable in favourable circumstances make possible the solution of high-dimensional problems not accessible in other ways. This comparison depends at least partly on the relatively slow ($\mathcal{O}(h^2)$ or $\mathcal{O}(h^4)$) convergence typical of a finite difference method. Quadrature methods for integral equations, on the other hand, can converge extremely fast (see the example in §5 above) if a high-order Gauss rule is used. This suggests:

Conclusion 1. Expansion methods show up relatively less well in the field of integral equations than in differential equations.

This conclusion is reinforced by the observation that an expansion method in an M dimensional problem requires an M dimensional quadrature for a differential equation, and a 2M dimensional quadrature for an integral equation:

Conclusion 2. Problems amenable to analytic evaluation of the appropriate integrals are better suited to expansion methods than those requiring numerical quadrature to produce $\underline{K}$.

Conclusion 3. If an expansion method is to be used, a differential equation formulation of a given problem may have advantages over the integral equation problem.

In favour of expansion methods, we add that analytic singularities in the kernel can often be better or more simply treated by the choice of expansion set, or of inner product, within an expansion method, than by a quadrature method.

Conclusion 4. The more difficult the problem the more worthwhile it is to look at expansion methods.

(b) Comparison with other expansion methods

There are many methods based on an expansion of either the kernel, or the solution, in some complete set of functions. We treat here only those which seem the most important, or commonly used

(i) Least squares methods

These introduce an expansion of the form (1.3), (1.4) for the solution, but compute the coefficients $a_i^{(N)}$ by minimising the norm of the residual, that is

$$\min_{a_i} ||Lx - g|| \quad \text{(inhomogeneous equation)} \qquad (7.1a)$$

or

$$\min_{a_i,\kappa} ||Kx-\lambda x|| \quad \text{(eigenvalue equation)} \qquad (7.1b)$$

although we group the methods together as "least squares" methods, the L_1 or L_∞ norm may be used; very often, over a discrete point set rather than the original space (see Chapters 5 and 8). For all such methods, (4.8) describes the error in the natural norm. If $S_2(N)$ dominates, as is likely for any reasonable method, the convergence rate attained will be identical with that of the Rayleigh-Ritz method. The amplitude of the error will not, however; least squares (L_2) methods for differential equations suffer from their introduction of the square of an unbounded operator. This is not the case for integral equations; here, we might expect comparable error amplitudes. For inhomogeneous equations, the comparison between methods reduces to deciding which will require least time to set up, and to solve, sets of equations of similar size. If a discrete point set is used for the residual minimisation, it is likely (to judge from the published examples) to be rather larger than would be needed for a numerical quadrature with the method of moments; but a high-order quadrature rule could equally well be used to generate the L_2 norms in (7.1).

Conclusion 5. The method of moments and least squares (L_2) methods are comparable for inhomogeneous integral equations. So may be L_1 and L_∞ methods.

For the eigenvalue problems, the above comparisons are probably still valid; least squares methods are also intrinsically more stable against integration errors (see, e.g. Delves 1973a). However, they have the disadvantage that the eigenvalue κ does not appear linearly, and this in practice seems overriding.

Conclusion 6. The method of moments is preferable for eigenvalue problems.

(ii) Methods based on a separable expansion of the kernel

These methods reduce the solution of the integral equation to that of a set of algebraic equations. The coefficients of these equations are defined as a single integral, compared with the double integral for a variational method; they are therefore relatively cheap to produce, provided that the coefficients of the separable expansion are viewed as free. If these must be produced separately, the cost is comparable to that of a complete Rayleigh-Ritz calculation, and the comparison changes abruptly. We also note that a calculation of this kind is formally identical with a Rayleigh-Ritz calculation with a particular choice of expansion set; and not necessarily a very appropriate one.

References

DELVES, L.M. (1968) Round-off Errors in Variational Calculations. Jnl. Comput. Phys. vol. 3, pp 17-28.

DELVES, L.M. (1973) Variational techniques in the nuclear three-body problem : a review. Adv. Nucl. Phys. vol. 5, pp 1-226.

DELVES, L.M. and FREEMAN, T.L. (1973) On round-off errors and stability in variational methods. To be published.

DELVES, L.M. and MEAD, K.O. (1971) On the convergence rates of variational methods I : Asymptotically diagonal systems. Maths. Comput. vol. 25, pp 699-716.

FREEMAN, T.L. DELVES, L.M. and REID, J.K. (1973) On the convergence rates of variational methods II : Systems of type B,C. Submitted to J. Inst. Math. & Its Appl.

FREEMAN, L. and DELVES, L.M. (1973) On the convergence rates of variational methods III : Eigenvalue problems. To be published.

GOULD, S.H. (1966) Variational methods for eigenvalue problems. Ontario: University of Toronto Press.

MEAD, K.O. and DELVES, L.M. (1973) The convergence rates of generalised fourier expansions. J. Inst. Math. & its Appl. To appear.

MIKHLIN, S.G. (1970) The numerical performance of variational calculations. Woltes-Noordhoff.

STROUD, A.H. and SECREST, O. (1966) Gaussian Quadrature Formulae. Prentice-Hall.

CHAPTER 10 NUMERICAL SOLUTION OF THE EIGENVALUE PROBLEM

C.T.H. Baker

University of Manchester

In this chapter we consider numerical methods for solving the equation

$$\int_a^b K(x,y)\, f(y)\, dy = \kappa\, f(x)$$

(for an eigenvalue κ and a corresponding eigenfunction $f(x)$) where a,b are finite constants and $K(x,y)$ is given. For the classical $\mathcal{L}^2$-theory, an eigenfunction is required to be square-integrable and non-null. There may be no such solutions, but, when there are, it is easy to establish that mild conditions on $K(x,y)$ guarantee the continuity of any eigenfunction which corresponds to a non-zero eigenvalue. An extension of such arguments can be used to determine the degree of differentiability of such an eigenfunction, and this information can be used to help motivate the selection of a numerical method. For a discussion, see Baker (to appear). This reference also gives further details of many of the topics covered in this chapter.

The available numerical methods are based on quadrature rules, product integration, collocation and Galerkin-type constructions, and the Rayleigh-Ritz method. With the exception of the quadrature method, the methods we describe for continuous kernels can be applied without modification in the weakly singular case where

$$K(x,y) = H(x,y)/|x-y|^{\alpha}$$

($0 < \alpha < 1$, $H(x,y)$ continuous).

If κ is an eigenvalue associated with $K(x,y)$ then there is at least one non-null function $\phi(x)$ such that

$$\int_a^b K(y,x)\, \phi(y) dy = \kappa\, \phi(x)$$

We call $\phi(x)$ a left-eigenfunction corresponding to κ. If κ is a simple eigenvalue, the accuracy obtainable in approximating κ by a numerical method is governed in part by the condition number

$$\mu(\kappa) = \frac{\left| \int_a^b f(x)\, \phi(x) dx \right|}{\|f(x)\|_2 \, \|\phi(x)\|_2}$$

which is invariant under scaling of $f(x)$ and $\phi(x)$. If $\mu(\kappa)$ is very small then κ is badly conditioned; for a simple eigenvalue of a Hermitian kernel $\mu(\kappa) = 1$. This situation is similar to that encountered in the matrix eigenvalue problem. Multiple eigenvalues are usually badly conditioned, unless $K(x,y)$ is Hermitian.

1. Methods Based on Quadrature Rules

Using a rule

$$\int_a^b \phi(y) dy \simeq \sum_{j=0}^{n} w_j \, \phi(y_j)$$

to replace the integral in

$$\int_a^b K(x,y)\, f(y) dy = \kappa\, f(x) \quad (a \leqslant x \leqslant b) \tag{1.1}$$

we obtain a functional equation

$$\sum_{j=0}^{n} w_j \, K(x,y_j)\, \tilde{f}(y_j) = \tilde{\kappa}\, \tilde{f}(x) \quad (a \leqslant x \leqslant b) . \tag{1.2}$$

If $y_0, y_1, \ldots, y_n \in [a,b]$, we can find a solution of (1.2) if we determine $\tilde{f}(y_0)$, $\tilde{f}(y_1), \ldots,$ $\tilde{f}(y_n)$ and $\kappa \neq 0$ such that

$$\sum_{j=0}^{n} w_j \, K(y_i,y_j)\, \tilde{f}(y_j) = \tilde{\kappa}\, \tilde{f}(y_i) \quad (i=0,1,\ldots,n). \tag{1.3}$$

The system (1.3) represents an eigenvalue problem for the matrix KD where

$$K = [K(y_i,y_j)] \ (0 \leqslant i,j \leqslant n) \quad \text{and} \quad D = \text{diag}\,(w_0,w_1,\ldots,w_n).$$

We obtain a finite eigensystem

$$\left\{\tilde{\kappa}_r,\ \tilde{f}_r = [\tilde{f}_r(y_0),\ \tilde{f}_r(y_1),\ldots,\tilde{f}_r(y_n)]^T \mid r=0,1,\ldots,n\right\}$$

on solving (1.3). When $\tilde{\kappa}_r \neq 0$, substitution in (1.2) can be employed to extend $\tilde{f}_r$ to a function $\tilde{f}_r(x)$, but it may sometimes be more convenient to obtain an approximate eigenfunction of the form

$$\hat{f}_r(x) = \sum_{i=0}^{n} \tilde{f}_r(y_i)\ \phi_i(x) \tag{1.4}$$

where $\left\{\phi_i(x)\right\}$ are, say, certain piecewise constant functions or the cardinal functions of Lagrangean interpolation.

There are computational and theoretical advantages in supposing $y_i \in [a,b]$ and $w_i > 0$ $(i=0,1,\ldots,n)$ in the quadrature rule; we assume this. Many such rules are Riemann sums (Baker(1968)). The theoretical analysis of (1.3) can be pursued by considering the eigenvalue problem for a degenerate kernel determined by $K(x,y)$ and the quadrature rule (see Bückner, (1962)) and this analysis is simplified slightly if we restrict attention to rules which are Riemann sums.

If $K(x,y) = \overline{K(y,x)}$, the eigenvalue problem

$$KD\ \tilde{f}_r = \tilde{\kappa}_r = \tilde{f}_r$$

can be recast as

$$(D^{\frac{1}{2}}\ KD^{\frac{1}{2}})\ \psi_r = \tilde{\kappa}_r\ \psi_r,$$

where $\psi_r = D^{\frac{1}{2}}\ \tilde{f}_r$, and $D^{\frac{1}{2}}\ KD^{\frac{1}{2}}$ is Hermitian (symmetric if K is real) and an algorithm designed for Hermitian matrices can be used to compute $\left\{\tilde{\kappa}_r,\ \tilde{f}_r\right\}$.

A good choice of quadrature rule is one for which the local truncation errors

$$\tau_r(x) = \sum_{j=0}^{n} w_j\ K(x,y_j)\ f_r(y_j) - \kappa_r\ f_r(x)$$

(corresponding to true eigensolutions which we assume exist) are small. Generally we need only consider the values $\tau_r(y_i)$ $(i=0,1,\ldots,n)$ for specific values of r. The quality of the quadrature rule can also be judged in terms of the quantity

$$\delta = \sup_{a \leq x, y \leq b} \left| \int_a^b K(x,z)\, K(z,y) dz - \sum_{j=0}^{n} w_j\, K(x,y_j)\, K(y_j,y) \right| \quad (1.5)$$

which can be used to provide a bound for $||\tau_r(x)||_2$.

For the repeated trapezium rule with step h the local truncation errors $\tau_r(y_i)$ are $O(h^2)$ as $h \to 0$, under mild conditions on $K(x,y)$. It can then be shown that if $\kappa \neq 0$ is a simple eigenvalue, there is an approximate eigenvalue $\tilde{\kappa}$ such that $|\kappa - \tilde{\kappa}| = O(h^2)$, as $h \to 0$. (For the repeated Simpson's rule we similarly obtain $|\kappa - \tilde{\kappa}| = O(h^4)$ under suitable differentiability conditions.) The assumption that $\kappa \neq 0$ is not a multiple eigenvalue is an essential hypothesis. If $\kappa \neq 0$ has algebraic multiplicity $m > 1$ then there will ultimately be m eigenvalues $\tilde{\kappa}_i (i=1,2,\ldots,m)$ converging as $h \to 0$ to κ, and in general we can only state that $|\kappa - \tilde{\kappa}_i| = O(h^{2/m})$. However, if we set

$$\hat{\kappa} = \sum_{r=1}^{m} \tilde{\kappa}_r \,/\, m$$

then $|\kappa - \hat{\kappa}| = O(h^2)$. Suitable differentiability conditions on $K(x,y)$ are also essential for such results and insight into both aspects can be obtained by considering certain degenerate kernels. Thus if

$$K(x,y) = \sum_{i=1}^{m} X_i(x)\; Y_i(y)$$

the non-zero eigenvalues κ and $\tilde{\kappa}$ are eigenvalues of two matrices of order m which are directly comparable. By choosing $\left\{X_i(x)\right\}$ and $\left\{Y_i(x)\right\}$, matrices can be constructed which exhibit various types of asymptotic behaviour of $\tilde{\kappa}$ as $h \to 0$; thus the results from matrix theory (Wilkinson (1965), Chapter 2) go over to the case of integral equations.

Treatment of a degenerate kernel simplifies the problem because the number of non-zero eigenvalues is at most finite (m). For a more general kernel, if we use (say) the trapezium rule with a step h which is progressively reduced, not only do the values of $\tilde{\kappa}_r (r=0,1,\ldots,n)$ change, but more approximate values are produced. If we wish to estimate the true eigenvalues as limit points of approximate eigenvalues and in particular if we use the deferred approach to the limit, described later) we may like to consider that a particular approximate eigenvalue moves along a smooth path to its limit point as $h \to 0$. To identify which path an eigenvalue lies on we must adopt a suitable indexing of the eigenvalues. In the case $\overline{K(x,y)} = K(y,x)$ (or whenever the true and approximate eigenvalues are always real) we index positive and negative eigenvalues separately so that

$$\kappa_r^+ \geq \kappa_{r+1}^+ \geq 0, \quad \kappa_r^- \leq \kappa_{r+1}^- < 0, \quad \text{and} \quad \tilde{\kappa}_r^+ \geq \tilde{\kappa}_{r+1}^+ \geq 0, \quad \tilde{\kappa}_r^- \leq \tilde{\kappa}_{r+1}^- < 0.$$

We can then show that as $h \to 0$, $\tilde{\kappa}_r^{\pm} \to \kappa_r^{\pm}$ (if $\lim \tilde{\kappa}_R^{+} = 0$). A more general convergence result is given in the following theorem.

<u>Theorem 1</u>

Every non-zero eigenvalue κ of a continuous kernel $K(x,y)$ is a limit point of approximate eigenvalues $\tilde{\kappa}$, and every non-zero cluster point of approximate eigenvalues is an eigenvalue κ.

The corresponding result for the eigenfunctions is not as strong. If κ is a simple eigenvalue and $\tilde{f}(x)$ an approximate eigenfunction corresponding to $\tilde{\kappa}$ where $\tilde{\kappa} \to \kappa \neq 0$, and if $\tilde{f}(x)$ is scaled so that $(f,\tilde{f}) = 1$, then

$$\lim \| f(x) - \tilde{f}(x) \|_\infty = 0.$$

More generally, if V_K is the linear space of eigenfunctions corresponding to κ,

$$\lim \inf_{\phi(x) \in V_K} \| \tilde{f}(x) - \phi(x) \|_\infty = 0.$$

Unfortunately, if κ is a multiple eigenvalue some of its eigenfunctions may arise from generalized eigenvectors (often called principal vectors) of KD.

From Theorem 1 we deduce that there is <u>some</u> indexing of the values $\{\tilde{\kappa}\}$, $\{\kappa\}$ such that if $\kappa_r \neq 0$, $\tilde{\kappa}_r \to \kappa_r$ as $h \to 0$. Stronger results are available, for, if κ_r is a simple eigenvalue

$$|\kappa_r - \tilde{\kappa}_r| = (b-a)\, \delta \left\{ 1 + O(1) \right\} / |\kappa_r \mu(\kappa_r)|$$

where δ is defined in (1.5), and under suitable differentiability conditions

$$\tilde{\kappa}_r = \kappa_r + \eta_{r,1} h^2 + \eta_{r,2} h^4 + \ldots + \eta_{r,N} h^{2N} + O(h^{2N+1}), \qquad (1.6)$$

where $\eta_{r,1}, \eta_{r,2}, \ldots, \eta_{r,N}$ are independent of h. Precise conditions for (1.6) are stated in Baker (1971).

A convincing demonstration of (1.6) is provided by the case

$$a=0,\ b=1, K(x,y) = x(1-y)\ (0 \leqslant x \leqslant y \leqslant 1),\ K(x,y) = K(y,x)\ (0 \leqslant y \leqslant x \leqslant 1).$$

Using the repeated trapezium rule with $h = 1/n$, we can establish that

$$\kappa_r^+ = \lim_{h \to 0} \tilde{\kappa}_r^+ ,$$

where $$\tilde{\kappa}_r^+ = h^2 \Big/ \left\{ 4 \sin^2\left(\frac{r\pi h}{2}\right) \right\}, \quad \text{for } r = 1,2,\ldots, n-1.$$

Thus for $r \neq n, n+1$,

$$\tilde{\kappa}_r^+ = \kappa_r^+ \left\{ 1 + \frac{1}{3}\left(\frac{\pi r h}{2}\right)^2 + \frac{1}{15}\left(\frac{\pi r h}{2}\right)^4 + \ldots \right\}$$

and $$\tilde{\kappa}_n^+ = \tilde{\kappa}_{n+1}^+ = 0.$$

Note (i) that zero is not a true eigenvalue of (1.1) though it is a "cluster point" of the approximate eigenvalues and (ii) the differentiability conditions of $K(x,y)$ are such that the local trunction errors $\tau_r^+(x)$ can be expanded in the form

$$\sum_{j=1}^{N} \upsilon_{r,j}(x)\, h^{2j} + O(h^{2N+1})$$

only when x has the form $x = ih$ $(i=0,1,\ldots,n)$.

For the kernel $K(x,y) = (xy)^{1/4}$ on $[0,1]$, $\tilde{\kappa} = \kappa + \eta h^{3/2} + O(h^2)$ because the kernel is not sufficiently differentiable for the validity of (1.6).

The validity of (1.6) provides a theoretical basis for the deferred approach to the limit which can be applied once a suitable indexing of the approximate eigenvalues is established. For the example following (1.6) we obtain $O(h^4)$ accuracy if we compute $\left\{ 4\tilde{\kappa}_r^+ \left(\frac{h}{2}\right) - \tilde{\kappa}_r^+(h) \right\} / 3$ for $r=1,2,\ldots,n-1$, and this process can be extended to obtain arbitrary high rders of accuracy. On the other hand, results obtained in this example using Simpson's rule have $O(h^2)$ accuracy, and we do not improve the order of accuracy if we apply the deferred approach to the limit to results computed using this rule.

The value of $\eta_{r,1}$ in (1.6) can be obtained theoretically. Thus if $\phi_r(x)$ is the left-eigenfunction, and if $K(x,y)$ has a continuous first derivative in the x- and in the y-variable, then

$$h^2 \eta_{r,1} = -\frac{1}{12} h^2 \left\{ \frac{\int_a^b \phi_r(x)[K(x,y)\, f_r'(y) + K_y(x,y)\, f_r(y)]_{y=a}^{y=b} dx}{\int_a^b \phi_r(x)\, f_r(x) dx} \right\}$$

(Equations governing $\eta_{r,j}$, $j>1$, can also be obtained.) The numerator in the expression for $h^2 \eta_{r,1}$ involves the dominant correction term for the trapezium rule and the role of the condition number $\mu(\kappa_r)$ is apparent. Instead of eliminating $\eta_{r,1}$ (as in the deferred approach) we can attempt to estimate $\eta_{r,1} h^2$ and add it as a correction to $\tilde{\kappa}_r$. This provides motivation for a method of deferred correction proposed for symmetric kernels by Fox and Goodwin (1953), and which we now extend slightly.

Suppose that values $\tilde{\kappa}$, $\tilde{\underset{\sim}{f}}$ are computed using the trapezium rule, so that

$$h \sum_{j=0}^{n}{}'' K(a+ih,\ a+jh)\ \tilde{f}(a+jh) = \tilde{\kappa}\ \tilde{f}(a+ih)\ , \quad i=0,1,\ldots,n,\ h=(b-a)/n.$$

We also require the vector $\tilde{\underset{\sim}{\phi}}$ of values $\tilde{\phi}(a+ih)$, where

$$h \sum_{j=0}^{n}{}'' K(a+jh,\ a+ih)\ \tilde{\phi}(a+jh) = \tilde{\kappa}\ \tilde{\phi}(a+ih).$$

When $K(x,y) = \overline{K(y,x)}$, $\tilde{\underset{\sim}{\phi}}$ is set to $\overline{\tilde{\underset{\sim}{f}}}$.

Gregory correction terms to the trapezium rule can be used to compute corrections $c_i(\tilde{f})$ to the trapezium rule approximation, thus:*

---

*The situation is complicated when jump discontinuities in the derivatives of $K(x,y)$ are present; these should be taken into account in forming $c_i(\tilde{f})$, but since $\underset{\sim}{f}$ can be extended to a smooth function it appears that we can risk the assumption that $\tilde{f}(x)$ is well-behaved.

$$\tilde{c}_i(f) \simeq \int_a^b K(a+ih, y)\, \tilde{f}(y)dy - h \sum_{j=0}^{n}{}'' K(a+ih, a+jh)\, \tilde{f}(a+jh)$$

(where $c_i(\tilde{f})$ involves differences in the y-direction).

A correction to $\tilde{\kappa}$ is now provided by computing the new approximation

$$\tilde{\kappa} + \frac{h \sum_{i=0}^{n}{}'' c_i(\tilde{f})\, \tilde{\phi}(a+ih)}{h \sum_{i=0}^{n}{}'' \tilde{f}(a+ih)\, \tilde{\phi}(a+ih)} \tag{1.7}$$

(in which h cancels in the quotient).

If we seek an iterative process we require corrected values of $\tilde{\underset{\sim}{f}}$, and these are less easily obtained. To see what could be done it is sufficient to consider Newton's method applied to the eigenvalue problem (see Collatz (1966) p.273 and p.325).

In an alternative process, based on the generalized Rayleigh quotient (§3) we can replace $\tilde{\underset{\sim}{\phi}}$ by $\tilde{\tilde{\underset{\sim}{f}}}$ in the expression for the "corrected" value (1.7).

2. Treatment of Discontinuities Using Methods Based on Approximate Integration

We discussed the methods of (§1) by reference to simple rules such as the repeated trapezium rule. When K(x,y) is smooth, higher-order rules may be employed and the convergence results and error estimates can be extended.

If the kernel has discontinuous derivatives there is little or no advantage* in using high-order rules. If K(x,y) is weakly singular the methods of (§1) fail.

To deal with ill-behaviour of K(x,y) on the line x=y (as in the case $K(x,y) = H(x,y) / |x-y|^{\alpha}$) the quadrature method can be modified.

*On the other hand, there may be no disadvantages in using a high-order rule in the presence of discontinuous derivatives, except when the deferred approach is being used.

We write

$$A(x) = \int_a^b K(x,y)dy$$

and set

$$f(x)\left\{\kappa - A(x)\right\} = \int_a^b K(x,y)\left\{f(y) - f(x)\right\}dy,$$

then replace the right-hand term using a quadrature rule to obtain

$$\tilde{f}(y_i)\left\{\tilde{\kappa} - A(y_i)\right\} = \sum_{\substack{j\neq i\\ j=0}}^{n} w_j\, K(y_i,y_j)\left\{\tilde{f}(y_j) - \tilde{f}(y_i)\right\} \qquad (2.1)$$

(The exclusion of the term $j=i$ in the summation is certainly justified if $\lim_{y\to x}\left\{f(x)-f(y)\right\}K(x,y) = 0$.) We now have the eigenvalue problem

$$M\tilde{\underset{\sim}{f}} = \tilde{\kappa}\,\tilde{\underset{\sim}{f}}$$

where $M_{ij} = w_j\, K(y_i,y_j)$ if $i\neq j$, and

$$M_{ii} = A(y_i) - \sum_{j\neq i} w_j\, K(y_i,y_j).$$

If $K(x,y) = \overline{K(y,x)}$ we can recover symmetry by considering the eigenproblem for $D^{\frac{1}{2}}MD^{-\frac{1}{2}}$.

We know how to establish convergence results which act as a justification for this method in the weakly-singular case, using certain rules such as the repeated trapezium rule and Simpson's rule. For continuous kernels with discontinuous derivatives, convergence results similar to those in §1 are valid.

Thomas (1971) used the repeated trapezium rule to apply the preceding method to the kernel

$$K(x,y) = -\log y \quad (0 \leqslant x \leqslant y \leqslant 1)$$

with

$$K(y,x) = K(x,y)$$

and then applied a process of repeated h^2-extrapolation. Though we have no theoretical justification for this process the computed results were excellent.

The modification described above is somewhat primitive, and to obtain high-order accuracy with weakly singular kernels it seems better to apply the method of product integration developed by Atkinson (1966).

Suppose that $K(x,y)$ is weakly singular and we write

$$K(x,y) = \sum_{r=0}^{N} L_r(x,y)\, M_r(x,y) \tag{2.2}$$

where $M_r(x,y)$ is weakly singular and $L_r(x,y)$ is continuous. Approximate integration formulae of the form

$$\int_a^b M_r(x,y)\, \phi(y)dy \simeq \sum_{j=0}^{n} v_{r,j}\,(x)\, \phi(z_j) \quad (r=0,1,\ldots,N)$$

with $z_i \in [a,b]$ for $i=0,1,\ldots,n$, can be used to give the equations

$$\sum_{r=0}^{N} \sum_{j=0}^{n} v_{r,j}(z_i)\, L_r(z_i,z_j)\, \tilde{f}(z_j) = \tilde{\kappa}\, \tilde{f}(z_i) \quad (i=0,1,\ldots,n) \tag{2.4}$$

for an approximate eigensystem. Unfortunately symmetry in the case $K(x,y) = \overline{K(y,x)}$ is usually lost in (2.4).

Convergence results similar to those which hold for the method of §1 have been established by Atkinson (1967), and in the case of a simple (non-zero) eigenvalue κ the accuracy obtainable is of the order of the local truncation error. Generalizations of the repeated trapezium rule and the repeated Simpson's rule give in general $O(h^2)$ and $O(h^3)$ (not $O(h^4)$) accuracy respectively, when approximations to simple eigenvalues are sought. Sometimes one can establish stronger results, see for example de Hoog and Weiss (1973). The decomposition of the kernel $K(x,y)$ in (2.2) affects the accuracy obtainable in a rather imponderable fashion, and various such decompositions can be tried (see Atkinson (1966)). Thus for $K(x,y) = \log|x-y|$ we may set

$$N=0,\ M_0(x,y) = |x-y|^{-\alpha}$$

and

$$L_0(x,y) = |x-y|^{\alpha} \log|x-y|$$

for <u>any</u> $\alpha \in (0,1)$, and the choice of α determines the accuracy.

3. Expansion Methods for the Eigenproblem

If we choose functions $\phi_0(x),\phi_1(x),\ldots,\phi_n(x)$ we can obtain approximate eigenfunctions of the form

$$\tilde{f}(x) = \sum_{i=0}^{n} \tilde{a}_i \, \phi_i(x).$$

Given such an approximation and an approximate eigenvalue $\tilde{\kappa}$, we find

$$\int_a^b K(x,y)\, \tilde{f}(y)dy - \tilde{\kappa}\, \tilde{f}(x) = \eta(x)$$

where, in general, $\eta(x)$ does not vanish. In general it is not possible to choose $\tilde{a}_0,\tilde{a}_1,\ldots,\tilde{a}_n$ and $\tilde{\kappa}$ to make $\eta(x)$ vanish, but various methods arise if we endeavour to choose these values to constrain $\eta(x)$ to be, in some sense, small. (In particular, if $\tilde{a}_0,\tilde{a}_1,\ldots,\tilde{a}_n$ have been chosen in any way, then assigning $\tilde{\kappa}$ the value of the generalized Rayleigh quotient $(K\tilde{f},\tilde{f})/(\tilde{f},\tilde{f})$ minimizes $||\eta(x)||_2$, when $||\tilde{f}(x)||_2 \neq 0$.)

In general our methods permit us to construct both $\tilde{f}(x)$ and $\tilde{\kappa}$ by imposing restrictions on $\eta(x)$. A disadvantage with these methods is the necessity to compute the functions

$$\psi_r(x) = \int_a^b K(x,y)\, \phi_r(y)dy$$

$(r=0,1,\ldots,n)$ either for specific values of x or for any $x\in[a,b]$. (It is often necessary to employ automatic numerical integration techniques to achieve this.) Thus choosing $z_0,z_1,\ldots,z_n\in[a,b]$ and requiring $\eta(z_i) = 0$ $(i=0,1,\ldots,n)$ we obtain the "collocation" equations

$$\sum_{j=0}^{n} \tilde{a}_j\, \psi_j(z_i) = \tilde{\kappa} \sum_{j=0}^{n} \tilde{a}_j\, \phi_j(z_i)$$

which can be written as a generalized eigenproblem

$$B\, \tilde{\underset{\sim}{a}} = \tilde{\kappa}\, A\, \tilde{\underset{\sim}{a}} \tag{3.1}$$

with $B_{ij} = \psi_j(z_i)$, $A_{ij} = \phi_j(z_i)$ and $\tilde{\underset{\sim}{a}} = [\tilde{a}_0,\tilde{a}_1,\ldots,\tilde{a}_n]^T$.

In the method of moments we choose linearly independent functions $\chi_0(x), \chi_1(x),\ldots,\chi_n(x)$ and require

$$\int_a^b \eta(x)\,\overline{\chi_i(x)}\,dx = 0 \quad \text{for} \quad i=0,1,\ldots,n.$$

We again obtain a generalized eigenproblem of the form (3.1) but now

$$B_{ij} = \int_a^b \psi_j(x)\,\overline{\chi_i(x)}\,dx$$

and

$$A_{ij} = \int_a^b \phi_j(x)\,\overline{\chi_i(x)}\,dx.$$

Thus

$$B_{ij} = (K\phi_j, \chi_i) = \int_a^b \int_a^b K(x,y)\,\phi_j(y)\,\overline{\chi_i(x)}\,dy\,dx\ ,$$

and the computation of this double integral will often cause practical difficulties. (We can, however, "justify" the use of a fixed quadrature rule with positive weights to evaluate each element A_{ij}, B_{ij}.)

The general form of the method of moments provides more variety than is usually required. The most commonly applied method is the classical Galerkin method in which we set $\chi_i(x) = \phi_i(x)$ $(i=0,1,\ldots,n)$. (A useful generalization of this classical method results if we choose $w(x) > 0$ for $x\in[a,b]$ and set $\chi_i(x) = w(x)\,\phi_i(x)$.) Moreover a simplification occurs if

$$\int_a^b \phi_j(x)\,\overline{\chi_i(x)}\,dx = \delta_{ij}$$

since A is then the identity matrix.

In the case $K(x,y) = \overline{K(y,x)}$ the classical Galerkin method reduces to the Rayleigh-Ritz method, preserving symmetry in A and B and giving one-sided bounds for the positive and for the negative eigenvalues of $K(x,y)$ respectively. See Schlessinger (1957) and Chapter 8.

In the collocation method we must choose $\phi_i(x)$ and z_i $(i=0,1,\ldots,n)$. It seems to be important to make a choice which ensures that A is well-conditioned with respect to inversion. Thus if A is orthogonal (or if A is, in particular, the identity) there appear to be computational advantages.

A popular choice is to take (with $[a,b] = [-1,1]$) $\phi_0(x) = \frac{1}{2}$,

$\phi_r(x) = T_r(x)$ $(r=1,2,\ldots)$ and for the set $\{z_i\}$ either the points of extrema of $T_n(x)$ or the zeros of $T_{n+1}(x)$. There also appears to be theoretical justification for taking the points z_i to be the zeros of the Legendre polynomial $P_{n+1}(x)$, and taking $\{\phi_r(x)\}$ to be the corresponding cardinal functions of polynomial interpolation. Low order spline functions with knots at, say, equally-spaced points z_i are also sometimes attractive.

As in the case of equations of the second kind (see Chapter 7), Chebyshev and Legendre polynomials and spline functions can also be used in the method of moments.

For many of the commonly used functions of approximation theory the theoretical convergence of the collocation and Galerkin methods can be established, and the order of accuracy depends on the smoothness of the eigenfunctions. It is sometimes possible to establish mean-square convergence for the eigenfunctions (though then it may be 'relatively uniform' - see Cochran (1972) p 137); however, the mean-square convergence is strengthened to uniform convergence if, given

$$\tilde{f}(x) = \sum_{i=0}^{n} \tilde{a}_i \, \phi_i(x)$$

we compute a new approximation to the eigenfunction in the form

$$K \tilde{f}(x) = \sum_{i=0}^{n} \tilde{a}_i \, \psi_i(x)$$

(at little extra cost in general). With these two functions, the generalized Rayleigh quotient $(K \tilde{f},\tilde{f})/(\tilde{f},\tilde{f})$ can be computed as a check on $\tilde{\kappa}$.

References

ANSELONE, P.M. and RALL, L.B. (1968) The solution of characteristic value-vector problems by Newton's method. Num. Math. vol. 11 pp 38-45.

ANSELONE, P.M. (1971) Collectively compact operator approximation theory and applications to integral equations. Prentice-Hall, Englewood Cliffs N.J.

ATKINSON, K.E. (1966) Extensions of the Nyström method for the numerical solution of linear integral equations of the second kind. MRC Tech. Report 686, Madison, Wisconsin.

ATKINSON, K.E. (1967) The numerical solution of the eigenvalue problem for compact integral operators. Trans. Am. Math. Soc. vol. 129, pp 458-465.

BAKER, C.T.H. (1968) On the nature of certain quadrature formulas and their errors. SIAM J. Numer. Anal. vol. 5, pp 783-804.

BAKER, C.T.H. (1971) The deferred approach to the limit for eigenvalues of integral equations. SIAM J. Numer. Anal. vol. 8, pp 1-10.

BAKER, C.T.H. (to appear) The numerical solution of integral equations. (Clarendon Press).

BRAKHAGE, H. (1961) Zur Fehlerabschatzung für die numerische Eigenwertbestimmung bei Integralgleichungen. Num. Math. vol. 3, pp 174-179.

BÜCKNER, H. (1952) Die praktische Behandlung von Integralgleichungen Springer-Verlag, Berlin.

BÜCKNER, H.F. (1962) Numerical methods for integral equations in Survey of Numerical Analysis (J. Todd, editor) McGraw-Hill,

COCHRAN, J.A. (1972) Analysis of linear integral equations. McGraw-Hill Book Co., New York.

COLLATZ, L.(1966) Functional analysis and numerical analysis. (transl. H. Oser) Academic Press, New York.

CRYER, C.W. (1967) On the calculation of the largest eigenvalue of an integral equation. Num. Math. vol. 10, pp 165-176.

DE HOOG, F. and WEISS, R. (1973) Asymptotic expansions for product integration. Math. Comput. vol. 27, pp 295-306.

FOX, L. and GOODWIN, E.T. (1953) The numerical solution of non-singular linear integral equations. Phil. Trans. R. Soc. (Lond.) vol. A 245, pp 501-534.

KELLER, H.B. (1965) On the accuracy of finite difference approximation to the eigenvalues of differential and integral operators. Num. Math. vol. 7, pp 412-419.

LINZ, P. (1970) On the numerical computation of eigenvalues and eigenvectors of symmetric integral equations. Math. Comput. vol. 24, pp 905-909.

MYSOVSKIH, I.P. (1959) On error bounds for approximate methods of estimation of eigenvalues of Hermitian kernels (Transl:) AMS Translations (Ser. 2) 1964, 35, pp 237-250.

MYSOVSKIH, I.P. (1959) On error bounds for eigenvalues calculated by replacing the kernel by an approximating kernel (Transl:) AMS Translations (Ser. 2) 1964, 35, pp 251-262.

SCHLESINGER, S. (1957) Approximating eigenvalues and eigenfunctions of symmetric kernels SIAM J. Numer. Anal. vol. 5 pp 1-14.

THOMAS, J. ALDYTH (1971) The numerical solution of certain eigenvalue problems in integral equations and ordinary differential equations. M.Sc. Thesis, University of Manchester.

VAINIKKO, G.M. (1964) Asymptotic evaluations of the error of projection methods for the eigenvalue problem (Transl:) USSR Comput. Math. and Math. Phys. vol. 4, pp 9-36.

VAINIKKO, G.M. (1965) Evaluation of the error of the Bubnov-Galerkin method in an eigenvalue problem. (Transl:) USSR Comput. Math. and Math. Phys. vol. 5, pp 1-31.

WIELANDT, H. (1956) Error bounds for eigenvalues of symmetric integral equations. AMS Proc. Symp. Appl. Math. vol. 6 pp 261-282.

WILKINSON, J.H. (1965) The algebraic eigenvalue problem. Clarendon Press, Oxford.

CHAPTER 11 VOLTERRA EQUATIONS OF THE SECOND KIND

D. Kershaw

University of Lancaster

1. Introduction

A convenient classification of Volterra integral equations is into those of the first kind typified by

$$\int_0^x K[x,y,f(y)]dy=g(x), \quad x>0, \tag{1.1}$$

and those of the second kind,

$$f(x)=g(x)+\int_0^x K[x,y,f(y)]dy, \quad x>0. \tag{1.2}$$

In each of these the functions g and K (the kernel) are given. Numerical methods for the solution of equations of the first kind will be dealt with in chapter 12; in the present chapter we shall be concerned solely with those of the second kind which we shall call 'Volterra equations' for brevity.

A general theoretical treatment of each kind of equation will be found in the recent monograph by Miller (1971). The book by Davis (1962) contains a brief introduction to the general non linear equation and for the theory of the linear equation (when $K[x,y,f]=K[x,y]f$) a convenient reference is Mikhlin (1957). A proof of the existence of a solution of (1.2) will be found in chapter 15. The theoretical aspects will not concern us; however we shall assume that g is continuous on some interval $[0,A]$ say, and K is continuous in $0\le y\le x\le A$ and that it satisfies a uniform Lipschitz condition in f, that is for all finite f_1, f_2

$$|K[x,y,f_1]-K[x,y,f_2]|\le L|f_1-f_2| \qquad \text{where L is a constant.}$$

There is clearly a strong similarity between Volterra equations and initial value problems for ordinary differential equations. Indeed any such problem can be reformulated as a Volterra equation.

For example

$$\text{"}f''(x)=K[x,f(x)],x>0,\ f(0),f'(0)\text{given"}$$

can be written as

$$f(x)=f(0)+xf'(0)+\int_0^x (x-y)K[y,f(y)]dy,\ x>0.$$

Consequently the methods for solving initial value problems are of considerable help in suggesting methods for dealing with Volterra equations.

The main references, apart from research papers, on the numerical solution of Volterra equations are the chapters of Mayers (1962) and Noble's (1963) comprehensive essay.

The methods for the solution of (1.2) fall into two classes, multistep and single step, and we shall give examples of each of these.

We shall adopt the following notation:

$f(rh)$ will indicate the value of the solution of (1.2) at the point rh,

f_r will denote the approximate solution at the same point.

2. Multistep methods

Let

$$\int_0^{rh} \varphi(y)dy=h\sum_{j=0}^{r} w_{rj}\varphi(jh)+E_r(\varphi) \tag{2.1}$$

represent an equal interval quadrature formula with remainder E_r, the weights $\{w_{rj}\}$ being supposed given or chosen. Then with the aid of (2.1) we can rewrite (1.2) as

$$f(rh)=g_r+h\sum_{j=0}^{r} w_{rj}K[rh,jh,f(jh)]+E_r(K). \tag{2.2}$$

Hence when $E_r(K)$ is negligible or ignored we obtain the following set of equations which one hopes determines a good approximation to f at the quadrature points rh, $r=1,2,\ldots,$:

$$f_r=g_r+h\sum_{j=0}^{r} w_{rj}K[rh,jh,f_j],\quad r=1,2,\ldots \tag{2.3}$$

If the quadrature formula is closed ($w_{rr}\neq 0$) then, assuming $f_0,f_1,\ldots,f_{n-1}$ are known, we have a non-linear equation to solve for f_n. This can be solved iteratively by a straightforward substitution process when $h|w_{nn}K_f[nh,nh,f_n]|<1$. Clearly for sufficiently small h this will be satisfied.

In the simplest case when (2.1) is the trapezoidal rule with remainder the equations of (2.3) have the simple form

$$f_r=g_r+h\sum_{j=0}^{r}{}'' K[rh,jh,f_j],\ r=1,2,\ldots. \tag{2.4}$$

Since $f_0=g_0$ we have an equation for f_1, which when solved will lead to an equation for f_2 and so on. However for more accurate computation one would naturally wish to use a higher order quadrature formula. In this case $f_1,f_2,\ldots,f_{n-1}$ will be required. The same situation arises in the numerical solution of ordinary differential equations and a special starting procedure is required. If the kernel is sufficiently regular it might be possible to find a power series expansion for f in the neighbourhood of the origin from which the necessary starting values can be found. An alternative is to use a Runge-Kutta type of single-step method- see §3. In a series of papers Day developes a number of starting procedures which are useful in a variety of circumstances. His approach is similar to that used by Rosser (1967) and Noble (1963). We give one of Day's methods here (Day(1968)); see also Noble (1963,p.252). The particular one we describe gives f_1,f_2,f_3 each with a truncation error of $O(h^4)$.

Day's starting procedure

Define

$$f_{11}=g_1+hK[h,0,g_0],$$

$$f_{12}=g_1+\tfrac{1}{2}h\left\{K[h,0,g_0]+K[h,h,f_{11}]\right\}$$

$$f_{13}=g_{\frac{1}{2}}+\tfrac{1}{4}h\left\{K[\tfrac{h}{2},0,g_0]+K[\tfrac{h}{2},\tfrac{h}{2},\tfrac{1}{2}g_0+\tfrac{1}{2}f_{12}]\right\}$$

Then

$$f_1=g_1+\tfrac{1}{6}h\left\{K[h,0,g_0]+4K[h,\tfrac{h}{2},f_{13}]+K[h,h,f_{12}]\right\}.$$

Next let

$$f_{21}=g_2+2hK[2h,h,f_1]$$

then

$$f_2=g_2+\frac{1}{3}h\left\{K[2h,0,g_o]+4K[2h,h,f_1]+K[2h,2h,f_{21}]\right\}.$$

Finally with

$$f_{31}=g_3+\frac{3}{2}h\left\{K[3h,h,f_1]+K[3h,2h,f_2]\right\}$$

we obtain

$$f_3=g_3+\frac{3}{8}h\left\{K[3h,0,g_o]+3K[3h,h,f_1]+3K[3h,2h,f_2]+K[3h,3h,f_3]\right\}.$$

Repeated Simpson's rule

A convenient and simple continuation of Day's starting procedure can be based on Simpson's rule in the following manner. (For this only f_o and f_1 are required.)

When r is even we can use repeated Simpson's rule immediately to give

$$f_r=g_r+\frac{h}{3}\sum_{j=o}^{r} w_{rj}K[rh,jh,f_j],\ r=2,3,\ldots. \tag{2.4a}$$

However when r is odd a different strategy is required and to maintain the local truncation error of $O(h^5)$ the 3/8ths rule is used at the upper end to give

$$f_r=g_r+\frac{h}{3}\sum_{j=o}^{r-3} w_{r-3j}K[rh,jh,f_j]+\frac{3}{8}h\left\{K[rh,\overline{r-3}h,f_{r-3}]\right.$$

$$\left.+3K[rh,\overline{r-2}h,f_{r-2}]+3K[rh,\overline{r-1}h,f_{r-1}]+K[rh,rh,f_r]\right\},r=3,5,\ldots$$

(2.4b)

The weights are given by $w_{po}=w_{pp}=1$, $w_{pj}=3-(-1)^j, 1\leq j\leq p-1$. It is tempting to believe that it is not important at which end the 3/8ths rule is used. Nevertheless we shall see later that it is of crucial importance to use it at the upper end.

Clearly other repeated rules can be used in place of Simpson's and equally clearly the more accurate the rule the more starting values it will require.

Another type of quadrature formula, Newton-Gregory, could also be used. In this as in the repeated rule a decision has to be taken as to the order of rule which is to be used. In the next section we consider the use of deferred correction in which the order of the quadrature formula is chosen appropriate to the equation and the point under consideration.

Newton-Gregory and deferred correction

We write the Newton-Gregory formula as

$$\int_0^{rh} \varphi(y)\,dy = h \sum_{j=0}^{r}{}'' \varphi_j + \frac{1}{12}[\Delta_o - \nabla_r]\varphi - \frac{1}{24}[\Delta_o^2 + \nabla_r^2]\varphi + \frac{19}{720}[\Delta_o^3 - \nabla_r^3]\varphi + \dots \quad (2.5)$$

(the coefficients can be found in Interpolation and Allied Tables, (1956) or can be computed from $-\int^1 \binom{t}{p+1} dt$, which is the multiplier of the p-th differences.) Mayers (1962) discusses the use of deferred correction for linear equations, and a more detailed study of this with examples will be found in Fox and Goodwin (1953).

With the aid of (2.5) we obtain the approximate equation

$$f_r = g_r + h \sum_{j=0}^{r}{}'' K[rh, jh, f_j] + \Delta_c \quad (2.5a)$$

where

$$\Delta_c = h \sum_{p=1}^{s} A_p [\Delta^p K[rh, x_o, f_o] + (-1)^p \nabla^p K[rh, x_r, f_r]] \quad (2.6)$$

is the correction. We assume that $f_o, f_1, \dots, f_{r-1}$ are known.

Initially we set $\Delta_c = 0$ and solve (2.5a) for $f_r^{(0)}$. This will allow us to set up a table of differences for K from which $\Delta_c^{(0)}$ can be calculated.

If $\Delta_c^{(0)}$ is negligible then $f_r^{(0)}$ is taken as the required value; on the other hand if it is not negligible then set $f_r^{(1)} = f_r^{(0)} + \Delta_c^{(0)}$.

With this new value we can compute the right hand side of (2.5a) to obtain $f_r^{(2)}$. This process is continued until convergence is reached.

After a few steps it will be apparent what order of difference can be neglected and the method could revert to a fixed quadrature formula with checks to make sure that the order is sufficient.

This method has the drawback that it is not easily programmed since it requires a decision to be made on the negligibility or otherwise of certain differences. It may be worthwhile to examine all the differences of a certain order before deciding to ignore them.

3. Runge-Kutta methods

As we have noted, a shortcoming of a multistep method is the need for starting values; another is the lack of flexibility for a change of step length. As with differential equations, a remedy for each of these is a single step Runge-Kutta method. Such a method has been developed by Pouzet (1960); see also Beltjukov (1966) and Loudet and Oules (1960).

We present a description of Pouzet's derivation in the case when the equation is linear.

We first recall that a Runge-Kutta method for the initial value problem

$$z'(x)=w[x,z(x)],z(0)=0$$

can be written as the sequence of operations defined by

$$\tilde{z}(\varphi_p h)=h\sum_{q=0}^{p-1} A_{pq}w[\varphi_q h,\tilde{z}(\varphi_q h)],p=1,2,\ldots,m,z(0)=0 \quad (3.1)$$

where the parameters $\{\varphi_q\}$ satisfy $0=\varphi_0\leq\varphi_1\leq\ldots\leq\varphi_m=1$ and the weights $\{A_{pq}\}$ satisfy

$$\sum_{q=0}^{p-1} A_{pq}=\varphi_p,\ p=1,2,\ldots,m.$$

The number $\tilde{z}(h)$ is the required $O(h^{m+1})$ approximation to $z(h)$ for $m\leq4$.

(For example the common Runge-Kutta formula with m=2 is given by

$\varphi_1=\varphi_2=\frac{1}{2}$, $\varphi_3=\varphi_4=1$, $A_{10}=A_{21}=\frac{1}{2}$, $A_{20}=A_{30}=A_{31}=0$, $A_{32}=1$,

$A_{40}=A_{43}=\frac{1}{6}$, $A_{41}=A_{42}=\frac{1}{3}$.)

Suppose that

$$K[x,y]=\sum_s u_s(x)v_s(y), \text{ then since}$$

$$f(x)=g(x)+\int_o^x K[x,y]f(y)dy, \quad x>0$$

we have

$$f(x)=g(x)+\sum_s u_s(x)z_s(x), \quad x>0$$

where

$$z_s'(x)=v_s(x)f(x), \quad z_s(0)=0.$$

We now apply the Runge-Kutta formula to this system of equations, to give

$$\tilde{z}_s(\varphi_p h)=h\sum_{q=o}^{p-1} A_{pq}v_s(\varphi_q h)f(\varphi_q h) \quad , \quad p=1,2,\ldots,m$$

which leads to

$$\tilde{f}(\phi_p h)=g(\varphi_p h)+\sum_s \tilde{z}(\phi_p h)u_s(\phi_p h)$$

$$=g(\varphi_p h)+h\sum_s u_s(\varphi_p h)\sum_{q=o}^{p-1} A_{pq}v_s(\varphi_q h)\tilde{f}(\varphi_q h)$$

$$=g(\varphi_p h)+h\sum_{q=o}^{p-1} A_{pq}K[\varphi_p h,\varphi_q h]\tilde{f}(\varphi_q h), \quad p=1,2,\ldots,m. \tag{3.2}$$

In this way we obtain $\tilde{f}(h)$ as the approximation to f_1.

We now state in full Pouzet's version in the case m=4 for the general non-linear equation.

Define

$$f_n^{(0)}=f_{n-1}^{(4)}, \quad f_o^{(0)}=g_o,$$

$$f_n^{(1)}=F_n(x_n+\tfrac{1}{2})+\tfrac{1}{2}hK[x_{n+\frac{1}{2}},x_n,f_n^{(0)}]$$

$$f_n^{(2)}=F_n(x_{n+\frac{1}{2}})+\tfrac{1}{2}hK[x_{n+\frac{1}{2}},x_{n+\frac{1}{2}},f_n^{(1)}]$$

$$f_n^{(3)}=F_n(x_{n+1})+hK[x_{n+1},x_{n+\frac{1}{2}},f_n^{(2)}]$$

$$f_n^{(4)}=F_n(x_{n+1})+\frac{h}{6}\Big\{K[x_{n+1},x_n,f_n^{(0)}]+2K[x_{n+1},x_{n+\frac{1}{2}},x_{n+\frac{1}{2}},f_n^{(1)}]+$$

$$+2K[x_{n+1},x_{n+\frac{1}{2}},f_n^{(2)}]+K[x_{n+1},x_{n+1},f_n^{(3)}]\Big\}$$

where

$$F_n(x)=g(x)+\frac{h}{6}\sum_{j=o}^{n-1}\Big\{K[x,x_j,f_j^{(0)}]+2K[x,x_{j+\frac{1}{2}},f_j^{(1)}]+$$

$$+2K[x,x_{j+\frac{1}{2}},f_j^{(2)}]+K[x,x_{j+1},f_j^{(3)}]\Big\},F_o(x)=g(x).$$

Pouzet states the result that $f(x_{n+1})=f_n^{(4)}+0(h^4)$.

The most obvious disadvantage of such a method is the number of function evaluations per step, of the order of $4n^2$, in contrast to the number of evaluations for repeated Simpson's rule, of the order of $\frac{1}{2}n^2$. Clearly the number of evaluations of K could be greatly reduced by the use of a multistep method such as given by Newton-Gregory for the calculation of $g_{n+1}+\int_o^{x_n}K[x_{n+1},y,f(y)]dy$ and indeed this is suggested by Pouzet.

Before leaving Runge-Kutta mention should be made of the approach by Beltjukov who derives his formulae from first principles. A number of these are given in his paper (Beltjukov (1966)) and we reproduce one such here. This is a third order method (none given by Beltjukov exceed this).

Define

$$k_o = hK[x_{\frac{1}{2}}, x_o, g_o]$$

$$k_1 = hK[x_1, x_1, g_1 + k_o]$$

$$k_2 = hK[x_1, x_{\frac{1}{3}}, g_{\frac{1}{3}} + \frac{1}{9}(k_o + 2k_1)],$$

then with $f_1 = g_1 + \frac{1}{4}[k_1 + 3k_2]$ we have $f(h) = f_1 + O(h^4)$.

This could be used to provide a start for repeated Simpson's rule.

We now turn to a description of a class of methods for which no starting values are required.

4. Block methods

The concept of a block method for integral equations seems to have been described first by Young (1954); a similar technique for use with differential equations was given by Milne (1953), see also Rosser (1967).

A block method is essentially an extrapolation procedure and has the advantage of being selfstarting. As we shall see, it produces a block of values at a time. One drawback of the method however is that it requires the kernel to be evaluated at points for which it may not be defined.

We shall follow Linz's (1967) description of the method and present his modification. For simplicity we confine ourselves to the simplest nontrivial case in which a block of two values is produced at each stage. The generalization will be obvious.

The method depends on the use of two three-point quadrature formulae. The first is Simpson's rule, and the second is given by

$$\int_o^h \varphi(y)dy = \frac{h}{12}[5\varphi_o + 8\varphi_1 - \varphi_2] + \frac{h^4}{24}\varphi^{(3)}(\xi). \qquad (4.1)$$

Suppose now that $f_o, f_1, \ldots, f_{2n}$ have been found (n may be zero), then since

$$f(\overline{2n+1}h)=g_{2n+1}+\int_{0}^{2nh} K[\overline{2n+1}h,y,f(y)]dy+\int_{2nh}^{\overline{2n+1}h} K[\overline{2n+1}h,y,f(y)]dy,$$

and

$$f(\overline{2n+2}h)=g_{2n+2}+\int_{0}^{2nh} K[\overline{2n+2}h,y,f(y)]dy+\int_{2nh}^{\overline{2n+2}h} K[\overline{2n+2}h,y,f(y)]dy,$$

we can use Simpson's rule and (4.1) to obtain the approximations

$$f_{2n+1}=g_{2n+1}+\frac{h}{3}\sum_{j=0}^{2n} w_j K[\overline{2n+1}h,jh,f_j]+\frac{h}{12}\Big\{5K[\overline{2n+1}h,2nh,f_{2n}]+$$

$$+8K[\overline{2n+1}h,\overline{2n+1}h,f_{2n+1}]-K[\overline{2n+1}h,\overline{2n+2}h,f_{2n+2}]\Big\} \qquad (4.2a)$$

$$f_{2n+2}=g_{2n+2}+\frac{h}{3}\sum_{j=0}^{2n+2} \overline{w}_j K[\overline{2n+2}h,jh,f_j], \qquad (4.2b)$$

where $\qquad w_0=w_{2n}=1,\ w_j=3-(-1)^j,\ 1\leq j\leq 2n-1$

and

$$\overline{w}_0=\overline{w}_{2n+2}=1,\ \overline{w}_j=3-(-1)^j,\ 1\leq j\leq 2n+1.$$

Thus we have a pair of nonlinear equations to solve for f_{2n+1} and f_{2n+2} and for sufficiently small h a simple iteration will give the solutions quickly.

It will have been noticed that in (4.2a) the kernel has to be evaluated at the point $[\overline{2n+1}h,\overline{2n+2}h,f_{2n+2}]$. This may not be feasible and Linz suggests the following alternative.

Replace

$$\int_{2nh}^{\overline{2n+1}h} K[\overline{2n+1}h,y,f(y)]dy$$

by

$$\frac{h}{6}\left\{K[2n\overline{+}1h,2nh,y_{2n}]+4K[2n\overline{+}1h,2n\overline{+}\tfrac{1}{2}h,\hat{y}_{2n+\frac{1}{2}}]+K[2n\overline{+}1h,2n\overline{+}1h,y_{2n+1}]\right\} \tag{4.3}$$

where

$\hat{y}_{2n+\frac{1}{2}}$ is an $O(h^3)$ approximation to $y((2n+\frac{1}{2})h)$ given by

$$\hat{y}_{2n+\frac{1}{2}} = \frac{1}{8}[3y_{2n}+6y_{2n+1}-y_{2n+2}].$$

If this is done then the kernel is evaluated at points within its domain.

5. Spline Approximations

There has recently been some work on the application of polynomial splines to the solution of Volterra equations. The results are somewhat disappointing although not surprising; see Hung (1970), El Tom (1971).

Let $x_o,x_1,\ldots,x_N$ be a set of points, called <u>knots</u>, which satisfy $0=x_o<x_1<\ldots<x_N$, then a spline of degree p with these knots is in $C^{p-1}[x_o,x_N]$ and its restriction to $[x_i,x_{i+1}]$ is a polynomial of degree at most p.

The most general form for such a spline is given by

$$\pi_p(x)+\sum_{j=o}^{N-1} c_jK_p^+(x-x_j)$$

where π_p is a polynomial of degree p and $K_p^+(z)=\frac{1}{p!}\left(\frac{z+|z|}{2}\right)^p$.

The splines which turn out to be useful for Volterra equations are quadratic splines.

We shall denote a quadratic spline with the knots $x_o,x_1,\ldots,$ by φ. Then in $[x_i,x_{i+1}]$ we have

$$\varphi(x)=A_i(x)\varphi_i+B_i(x)\varphi_{i+1}+C_i(x)\varphi_i' \tag{5.1}$$

where

$$A_i(x)=1-\left(\frac{x-x_i}{h}\right)^2,\quad B_i(x)=1-A_i(x),\; C_i(x)=\frac{1}{h}(x-x_i)(x-x_{i+1}).$$

$(h=x_{i+1}-x_i)$

For the continuity of φ' we require that

$$\varphi_i'+\varphi_{i-1}' = \frac{2}{h}(\varphi_i-\varphi_{i-1}),\quad i=1,2,\ldots \tag{5.2}$$

Suppose now that $\varphi_o,\varphi_1,\ldots,\varphi_{r-1},\varphi_o',\varphi_1',\ldots,\varphi_{r-1}'$ are known, then

$$\varphi_r=g_r+\sum_{j=o}^{r-2}\int_{x_j}^{x_{j+1}} K[x_r,y,A_j(y)\varphi_j+B_j(y)\varphi_{j+1}+C_j(y)\varphi_j']dy + \tag{5.3}$$

$$+\int_{x_{r-1}}^{x_r} K[x_r,y,A_{r-1}(y)\varphi_{r-1}+B_{r-1}(y)\varphi_r+C_r(y)\varphi_{r-1}']dy.$$

This is an equation for φ_r which is solved by iteration. From φ_r and (5.2) we can calculate φ_r' and so proceed to the calculation of φ_{r+1}. The integrals in (5.3) are replaced by a quadrature formula.

It will be clear that this approach can be used to generate a spline solution of any required degree; however as we shall see in the next section it happens that for cubic splines the method is unstable. (Hung (1970) shows that the quadratic spline method is weakly stable.)

More recent work suggests that if the continuity conditions are relaxed for a higher order spline then the method may be useful for these.

6. Convergence and Stability

We now consider the twin problems of the convergence and the stability of multistep methods for the solution of the Volterra equation

$$f(x)=g(x)+\int_o^x K[x,y,f(y)]dy,\quad x>0. \tag{6.1}$$

A detailed examination of each problem will be found in Kobayasi (1966) and more conveniently in Noble (1969). Mayers (1962) gives some attention to the problem of stability for the linear equation, and this reference provides a good introduction. Linz (1968) also considers these problems, and we shall outline Linz's proof of convergence.

A. Convergence

The numerical method is <u>convergent</u> if the solution of the approximating set of equations converges to the solution of the exact problem as the interval length h decreases to zero; that is if

$$\lim_{h\to o} |f(rh)-f_r|=0 \quad \text{with} \quad rh \text{ fixed.}$$

To examine the conditions under which convergence can be guaranteed, we make the necessary and obvious assumption that the quadrature formula converges in the sense that if

$$\int_o^{rh} \varphi(y)dy=h\sum_{j=o}^{r} w_{rj}\varphi(jh)+E_r(\varphi) \qquad (6.2)$$

then, with rh fixed, $E_r(\varphi)\to 0$ as $h\to 0$. When this is so the quadrature formula is said to be <u>consistent</u>.
It follows from (6.1) and (6.2) that

$$f(rh)=g_r+h\sum_{j=o}^{r} w_{rj}K[rh,jh,f(jh)]+E_r(K); \qquad (6.3)$$

the equations of the multistep method are

$$f_r=g_r+h\sum_{j=o}^{r} w_{rj}K[rh,jh,f_j], \quad r=n,n+1,.... \qquad (6.4)$$

Hence we obtain

$$f(rh)-f_r=h\sum_{j=o}^{r} w_{rj}\Big\{K[rh,jh,f(jh)]-K[rh,jh,f_j]\Big\}+E_r(K),\ r=n,n+1,... \qquad (6.5)$$

We now use the assumption that K satisfies a uniform Lipschitz condition in f to give the result that

$$|e_r| \leq hL \sum_{j=o}^{r} |w_{rj}||e_j|+|E_r(K)|, \quad r=n,n+1,\ldots$$

where $e_j=f(jh)-f_j$ and L is the Lipschitz constant.
Thus for sufficiently small h we have

$$|e_r| \leq \frac{hLw}{1-hLw} \sum_{j=o}^{r-1} |e_j| + \frac{1}{1-hLw} |E_r(K)| \qquad (6.6)$$

where $w=\max_{r,j} |w_{rj}|$. This type of inequality is familiar from the theory of multistep methods for ordinary differential equations, see Henrici (1962), and it is not difficult to deduce that

$$|e_r| \leq \frac{[|E_r(K)|+hLw(n-1)e]}{1-hLw} \exp\left\{\frac{wL}{1-hLw} rh\right\} \qquad (6.7)$$

where $e=\max_{o\leq j\leq n-1} |e_j|$.

Consequently $|e_r|\to 0$ as $h\to 0$ with rh fixed.

A more illuminating form of this result is

$$|e_r|=0(h^p)+0(h^q) \qquad (6.8)$$

where the error in the quadrature formula is $0(h^p)$ and the error in the starting values is $0(h^{q-1})$.
When Newton-Gregory is used as the quadrature formula the following result is true,

$$|e_r|\leq e \exp\left\{2rhL\right\}+h^{k+2} \frac{M_{k+2}}{L} |A_{k+1}| \left\{\exp(2rhL)-1\right\} \qquad (6.9)$$

where k is the order of difference used, and M_{k+2} is a constant.
Details of this will be found in Phillips (1971).

B. Stability

The solution of the differential equation $f'(x)=K[x,f(x)]$ with a given initial condition by a multistep method can be described in the form of the recurrence

$$\sum_{j=0}^{k} \alpha_j f_{n+j} = h \sum_{j=0}^{k} \beta_j K[x_j, f_{n+j}] \qquad (6.10)$$

where $\alpha_0, \alpha_1, \ldots, \alpha_k$, $\beta_0, \beta_1, \ldots, \beta_k$ are constants with $\alpha_k=1$. It is well known (see Henrici (1962) or Lambert (1973)) that the stability or otherwise of such a method depends on the location and quality of the zeros of the polynomial

$$\sum_{j=0}^{k} \alpha_j t^j - h \sum_{j=0}^{k} \beta_j t^j.$$

When this polynomial has a zero outside the unit disc then there will be a component in the solution of the difference equation which diverges. Consequently any errors which are made in the starting values for the method, including rounding errors, will be amplified as the integration proceeds.

This may not be important if the required solution increases more rapidly than the error does, however if the solution decreases then this component will dominate it.

With regard to a multistep method for a Volterra equation the situation is more complicated since in place of (6.10) we have the difference equations

$$f_r = g_r + h \sum_{j=0}^{r} w_{rj} K[rh, jh, f_j], \quad r=n, n+1, \ldots \qquad (6.11)$$

in which the weights will usually vary from step to step. The use of a repeated rule in (6.11) will give some connection between weights in successive formulae and it is this situation which is most easily dealt with. We define <u>stability of the multistep method</u> of (6.11) to mean that if the required solution of the difference equations tends to ero then the effect of errors in the starting values also tend to ero. Clearly we have here excluded the possibility of relative stability in which the solution increases at a faster rate than does the propogation of errors.

Mayers (1962) examined the stability of repeated Simpson's rule with trapezoidal rule at one or other of the ends when the number of intervals was odd. He concluded that for the integral equation $f(x)=1-a\int_0^x f(y)dy$, $a>0$ the method is stable when the correction was applied at the upper end (for small enough h) and unstable otherwise.

Later Kobayasi (1966) considered the general nonlinear equation; however his results seem intractable as far as deciding on the merits of a given quadrature formula. Linz (1968) gave a simplified version of Kobayasi's results and Noble (1969) produced a rigorous and intuitive approach to Kobayasi's analysis.

In order to motivate Noble's result we present Mayer's analysis for the equation

$$f(x)=1-a\int_0^x f(y)dy \qquad \text{where } a>0. \tag{6.12}$$

(a) Trapezoidal rule and repeated Simpson's rule

For an even number of intervals we have

$$f_{2r}=1-\frac{ah}{3}\sum_{j=o}^{2r} w_{2rj}f_j \tag{6.13}$$

where

$$w_{2ro}=w_{2r2r}=1,\ w_{2rj}=3-(-1)^j,\ 1\leq j\leq 2r-1.$$

When the number of intervals is odd we use trapezoidal rule for the first interval followed by repeated Simpson's rule,

$$f_{2r+1}=1-\frac{ah}{2}[f_o+f_1]-\frac{ah}{3}\sum_{j=1}^{2r+1} w_{2r+1j}f_j \tag{6.14}$$

It is easily shown that

$$(1+\frac{ah}{3})f_{r+1}+\frac{4}{3}ahf_r+(1-\frac{ah}{3})f_{r-1}=0 \qquad (6.15)$$

which is Milne-Simpson and known to be weakly stable (that is, stable in the sense given above only at h=0). The zeros of the associated polynomial are approximately e^{-ah}, $e^{ah/3}$, consequently for nonzero h there will be an error component which increases.

(b) Repeated Simpson's rule + trapezoidal rule

When there is an even number of intervals we obtain again (6.13), but when there is an odd number the following equation replaces (6.14),

$$f_{2r+1}=1-\frac{ah}{3}\sum_{j=o}^{2r} w_{2rj}f_j-\frac{ah}{2}[f_{2r}+f_{2r+1}] \qquad (6.16)$$

from which it can be shown that in general

$$f_r=\left(\frac{6-7ah+3a^2h^2}{6+5ah+a^2h^2}\right)f_{r-1}. \qquad (6.17)$$

It is easily shown that for $0<ah<6$ the multiplier of f_{r-1} is always less than unity in modulus. Consequently the method is stable for sufficiently small h.

We now examine in more detail these two methods. For the first we see that f_{2j} is multiplied by 2 and f_{2j+1} by 4 in the even interval case but that the reverse holds in the odd interval case. Noble suggested that the even and odd indexed values should be regarded as deriving from different functions. Let $\{p_j\}$ and $\{q_j\}$ be such that

$$p_{2j}=f_{2j} \text{ and } q_{2j+1}=f_{2j+1} \qquad (6.18)$$

Then from (6.13) we have

$$p_{2r}=1-2\frac{ah}{3}\sum_{j=o}^{r}{}'' p_{2j}-4\frac{ah}{3}\sum_{j=1}^{r} q_{2j-1}, \qquad (6.19)$$

and from (6.14)

$$q_{2r+1}=1-\frac{ah}{2}[p_o+q_1]-2\frac{ah}{3}\sum_{j=o}^{r}{}'' q_{2j+1}-4\frac{ah}{3}\sum_{j=1}^{r} p_{2j} \qquad (6.20)$$

Associate with $p_o, p_2, \ldots, p_{2r}$ the function p and with $q_1, q_3, \ldots, q_{2r+1}$ the function q; then the sums in (6.19) can be regarded as repeated trapezoid rule applied to the integral $\int_o^{rh} p(y)dy$ and repeated midpoint rule applied to $\int_h^{2\bar{r}-1h} q(y)dy$, each with interval length 2h.

Consequently (6.19) would be the result of discretizing an equation of the form

$$p(x)=1+\epsilon_1-\frac{a}{3}\int_o^x p(y)dy-\frac{2}{3}a\int_o^x q(y)dy, \quad x>0 \qquad (6.21)$$

by these methods.

Similarly (6.20) would be the result of discretizing

$$q(x)=1+\epsilon_2-\frac{a}{3}\int_o^x q(y)dy-\frac{2}{3}a\int_o^x p(y)dy, \quad x>0 \qquad (6.22)$$

where ϵ_1 and ϵ_2 are errors introduced by the starting values. These equations can be uncoupled to give

$$p(x)+q(x)=2+\epsilon_1+\epsilon_2-a\int_o^x [p(y)+q(y)]dy$$

and

$$p(x)-q(x)=\epsilon_1-\epsilon_2+\frac{a}{3}\int_o^x [p(y)-q(y)]dy. \qquad (6.23)$$

The solutions of these equations are given by

$$(p+q)(x)=(2+\epsilon_1+\epsilon_2)e^{-ax} \text{ and } (p-q)(x)=(\epsilon_1-\epsilon_2)e^{\frac{ax}{3}}.$$

Thus we see that there is an unstable component in the solution. The situation for the second method is quite different for in this case we have

$$(p+2q)(x)=2+\epsilon_1+2\epsilon_2-a\int_0^x (p+2q)(y)dy$$

$$(p-q)(x)=\epsilon_1-\epsilon_2, \tag{6.24}$$

and we conclude that the method is stable.

The arguments which have been used to reach these conclusions are obviously heuristic, however the results agree with those arrived at by considerations of the difference equations.

Noble uses the same device to analyse the use of repeated Simpson's rule together with the 3/8ths rule on the equation

$$f(x)=g(x)+\int_0^x k(x,y)f(y)dy$$

and shows that the same conclusion holds, namely that the 3/8ths rule applied at the upper end leads to a stable method.

Noble's general conclusion is as follows. Let us define the repetition factor of a repeated rule (this term was introduced by Linz (1968)).

The repetition factor of the quadrature rules defined by the array $\left\{w_{ro}, w_{r1}, \ldots, w_{rr}\right\}$, $r=1,2,3,\ldots$ is the smallest integer K such that $w_{r+kj}=w_{rj}$ for $j=\alpha,\alpha+1,\ldots,r-\beta$ where α and β are integers independent of r.

Thus for Simpson + trapezoidal rule the repetition factor is 1 and for trapezoidal rule + Simpson the repetition factor is 2.

Then (Noble (1969)) rules with a repetition factor of 1 are stable, however those for which it is greater than unity tend to be unstable. For more details of these results Noble's paper should be consulted.

7. A-stability and the cubic spline method

In the field of ordinary differential equations the concept of A-stability is important. Hung (1970) and Weiss (1972) have suggested the following definition of A-stability for Volterra equations:

A numerical method is A-stable if all solutions tend to zero as $j\to\infty$ with h fixed when the method is applied to the integral equation

$$f(x)=1-\lambda\int_0^x f(y)dy,\ Re(\lambda)>0. \qquad (7.1)$$

This definition is an equivalent statement of Dahlquists' definition of A-stability for methods for ordinary differential equations. It is obviously convenient, however its true relevance to the integral equation situation does not appear to have been thoroughly examined.

We have seen that repeated Simpson rule + trapezoidal rule is A-stable in this sense and we shall now investigate the A-stability of the cubic spline method.

Let φ be a cubic spline with the knots 0, h, 2h,... In the interval $[jh,\overline{j+1}h]$ we have the representation from Hermite's two point formula

$$\varphi(x)=A_j(x)\varphi_j+B_j(x)\varphi_{j+1}+C_j(x)\varphi'_j+D_j(x)\varphi'_{j+1},\ j=0,1,\ldots$$

The imposition of the conditions of the continuity of ϕ'' at the knots gives the relations

$$\varphi'_{j-1}+4\varphi'_j+\varphi'_{j+1}=\frac{3}{h}[\varphi_{j+1}-\varphi_{j-1}]. \qquad (7.2)$$

We substitute φ for f in the integral equation with x=rh to give

$$\varphi_r=1-a\int_0^{rh}\varphi(y)dy=1-ah\sum_{j=o}^{r}{}''\varphi_j+\frac{ah^2}{12}[\varphi'_r-\varphi'_o]$$

by the Euler-Maclaurin sum formula.
Hence

$$\varphi_{r+1}-\varphi_r=-\frac{ah}{2}[\varphi_{r+1}+\varphi_r]-\frac{ah^2}{12}[\varphi'_{r+1}-\varphi'_r]. \qquad (7.3)$$

The derivatives can be eliminated between (7.2) and (7.3) to give

$$(1+\frac{ah}{4})\varphi_{r+1}+(3+\frac{11}{4}ah)\varphi_r+(-3+\frac{11}{4}ah)\varphi_{r-1}+(-1+\frac{ah}{4})\varphi_{r-2}=0.$$

The associated polynomial is

$$(1+\frac{ah}{4})t^3+(3+\frac{11}{4}ah)t^2-(3-\frac{11}{4}ah)t-(1-\frac{ah}{4})$$

which for small h has a zero in the vicinity of $-2-\sqrt{3}$. Consequently the recurrence is unstable and so the cubic spline method is not A-stable.

This result was proved by Hung (1970).

References

BELTJUKOV, B.A. (1966) An analogue of the Runge-Kutta method for the solution of nonlinear integral equations of Volterra type. Differential equations. Vol. 1 pp.417-26.

DAVIS, H.T. (1962) Introduction to nonlinear differential and integral equations. Dover.

DAY, J.T. (1968) On the numerical solution of Volterra integral equations. BIT Vol. 8 pp.134-7.

EL TOM,(1971) An application of spline functions to Volterra integral equations. J. Inst. Math. and it's Appl. Vol. 8 pp.354-57

FOX, L., and GOODWIN, E.T. (1953) The numerical solution of non-singular linear integral equations. Phil. Trans. R. Soc. Vol. 254 pp.501-34.

HENRICI, P. (1962) Discrete variable methods in ordinary differential equations. Wiley.

HUNG, H.S. (1970) The Numerical Solution of Differential and Integral Equations by Spline functions. M.R.C. Tech. Rpt. 1053.

KOBAYASI, M. (1966) On numerical solution of the Volterra integral equations of the second kind by linear multistep methods. Rep. Stat. Appl. Res. JUSE Vol. 13 pp.1-21.

LAMBERT, J.D. (1973) Computational methods in ordinary differential equations. Wiley.

LINZ, P. (1968) The numerical solution of Volterra integral equations by finite difference methods. M.R.C. Tech. Rept. 825.

LOUDET, M., and OULES, H. (1960) Sur l'integration numerique des equations integrales du type de Volterra. In Symposium on the Numerical treatment of Ordinary differential equations, integral and integro-differential equations. Birkauser Verlag. Basel. pp.117-21.

MAYERS, D.F. (1962) In Numerical solution of ordinary and partial differential equations. Ed. L. Fox. Pergamon.

MIKHLIN, S.G. (1957) Integral Equations. Pergamon.

MILLER, R.K. (1971) Nonlinear Volterra Integral Equations. Benjamin.

MILNE, W.E. (1953) Numerical solution of differential equations. Wiley.

NOBLE, B. (1969) In Conference on the numerical solution of Differential equations Springer Lecture Notes 109.

PHILLIPS, G.M. (1971) An error estimate for Volterra integral equations. B.I.T. Vol. 11 pp.181-186.

POUZET, P. (1963) Method d'integration numerique des equations integrales et alegro differentielles du type de Volterra de seconde espece. Formules de Runge-Kutta. in LOUDET and OULES (1960) pp.362-368

ROSSER, J.B. (1967) A Runge-Kutta for all seasons. SIAM Review Vol. 9 pp.417-52.

WEISS, R. (1972) Numerical Procedures for Volterra Integral Equations Ph.D. thesis Canberra.

YOUNG, A. (1954) The application of product integration to the numerical solution of integral equations. Proc. R. Soc. A. 224 pp.561-73.

CHAPTER 12 METHODS FOR VOLTERRA EQUATIONS OF FIRST KIND

C.T.H. Baker

University of Manchester

1. Introduction

In this chapter we consider Volterra equations of the first kind, that is, equations of the form

$$\int_0^x K(x, y)\, f(y)dy = g(x) \quad (X \geqslant x \geqslant 0) \tag{1.1}$$

in which the kernel $K(x, y)$ and the term $g(x)$ are given. In §§3-6 we discuss numerical methods for the solution of (1.1). We begin by discussing certain difficulties associated with the theory and solution of equation (1.1); these difficulties sometimes lead us to consider the reformulation of the problem as indicated in §2. A number of applications of (1.1) are mentioned by authors whom we refer to later.

The treatment of the equation will depend on the smoothness of $K(x, y)$ and $g(x)$; as an example in which $K(x, y)$ is unbounded at $x = y$ consider the equation of Abel type:

$$\int_0^x \frac{f(y)}{(x^p - y^p)^\alpha}\, dy = g(x) \quad (0 < \alpha < 1,\ p = 1, 2). \tag{1.2}$$

Methods which are suitable for such equations are mentioned in §6, and Edels, Hearne and Young (1962), Linz (1967b) and Minerbo and Levy (1969) have considered Abel's equation in detail. Unless otherwise stated we shall suppose that $K(x, y)$ and $g(x)$ have continuous derivatives of the order required.

Integral equations of the first kind (see also Ch.13) are generally suspected of being ill-conditioned or ill-posed. In such circumstances small changes in $g(x)$ or $K(x, y)$ may have a large effect in the solution $f(x)$ of the problem; if the original problem had a solution the perturbed equation may have no solution, and vice-versa. Whereas it is recognized that difficulties associated with ill-posedness occur in Fredholm equations of the first kind, the view sometimes expressed is that Volterra equations of the first kind are less difficult to deal with than their counterparts of Fredholm type. We believe that there is an element of truth in this.

To illustrate that Volterra equations of the first kind can be ill-posed, however, consider the simple equation

$$\int_0^x f(y)dy = g(x). \tag{1.3}$$

We note immediately that there is no solution unless $g(0) = 0$ and $g'(x)$ exists. The solution is then $f(x) = g'(x)$; the solution is integrable, continuous or differentiable depending on the corresponding properties of $g'(x)$. If we change $g(x)$ to $g(x) + \varepsilon(x)$, there is a solution only if $\varepsilon(0) = 0$ and $\varepsilon'(x)$ exists. Moreover the continuity properties of any solution of the perturbed equation depend upon the properties of $\varepsilon'(x)$. Further, observe that the approximate solution for this problem can be obtained directly by approximating $g'(x)$ numerically, and this process is known to be sensitive to cancellation errors when tabular values of $g(x)$ are differenced. Such difficulties appear in other numerical methods for solving this integral equation.

2. Conversion to Equations of the Second Kind.

The technique employed to solve (1.3) can frequently be employed to produce a more tractable equation from the equation of the first kind. Assuming that the differentiations can be performed, we find, on differentiating (1.1), that

$$K(x, x)\ f(x) + \int_0^x K_x(x, y)\ f(y)dy = g'(x) \quad (X \geq x \geq 0). \tag{2.1}$$

If $K(x, x) \neq 0$ for $x \geq 0$ we have

$$f(x) + \int_0^x K_1(x, y)\ f(y)dy = g'(x), \tag{2.2}$$

where $K_1(x, y) = K_x(x, y)/K(x, x)$, and equation (2.2) is a Volterra equation of the second kind for $f(x)$. This equation can be solved numerically by the techniques described in Chapter 11, provided the derivatives involved can be obtained fairly accurately. If $K(x, x) \equiv 0$ then (2.1) assumes the form of (1.1) and the process can be repeated. The process terminates if $[(\partial/\partial x)^r\ K(x, y)]_{y=x} \neq 0$ for some r, but the idea does not seem to be of practical relevance unless r is small.

It is of interest to note that on integrating by parts we find

$$\int_0^x K(x, y)\ f(y)dy = K(x, x)\ F(x) - \int_0^x K_y(x, y)\ F(y)dy, \tag{2.3}$$

where

$$F(x) = \int_0^x f(y)dy, \tag{2.4}$$

and if $K(x, x) \neq 0$ we obtain from (1.1) an equation of the second kind for $F(x)$ of the form

$$F(x) - \int_0^x K_2(x, y)\, F(y)dy = g(x), \qquad (2.5)$$

where $K_2(x, y) = K_y(x, y)/K(x, x)$. The function $F(x)$ may be differentiated to provide the solution $f(x)$ for (1.1), when it exists. (In practice $F'(x)$ could be approximated numerically.)

3. Numerical Methods of Solution

For certain equations of the form (1.1) we would recommend conversion to an equation of the second kind and the application of numerical methods to the resulting equation. A more direct treatment is frequently simpler computationally, and methods which have been suggested for Volterra equations of the first kind are generally analogues of methods for equations of the second kind. To summarize the possibilities briefly, these are methods based on quadrature formulae, methods using product integration, collocation, and Galerkin-type methods. Not all such methods are satisfactory. Regularization methods similar to those proposed for Fredholm equations of the first kind have also been proposed for Volterra equations. We do not discuss these but refer to Chapter 13 and Schmaedeke (1968).

4. Use of Quadrature Rules in the Equation of the First Kind.

Equation (1.1) with $x = rh$, $h > 0$, $r = 0, 1, \dots,$ assumes the form

$$\int_0^{rh} K(rh, y)\, f(y)dy = g(rh).$$

A quadrature rule of the form

$$\int_0^{rh} \phi(y)dy \simeq \sum_{j=0}^{r} w_{r,j}\, \phi(jh)$$

can be employed to discretize the equation; we obtain the equations $g(0) = 0$ and

$$\sum_{j=0}^{r} w_{r,j}\, K(rh, jh)\, \tilde{f}(jh) = g(rh),\ r = 1, 2, \dots, \qquad (4.1)$$

for the approximate values $\tilde{f}(jh) \simeq f(jh)$, $j = 0, 1, 2, \dots$. Collatz (1960) suggests the use of the r-times repeated trapezium rule using function values at step h, so that equation (4.1) becomes

$$h \sum_{j=0}^{r}{}'' K(rh, jh)\, \tilde{f}(jh) = g(rh) \qquad (4.2)$$

for $r = 1, 2, \dots$.

At first sight (4.2) appears to provide a system of equations for $\tilde{f}(0), \tilde{f}(h), \dots, \tilde{f}(rh)$ which can be solved by forward substitution. On closer examination we observe that corresponding to $r = 0$ there is no information about $\tilde{f}(0)$ and a value for $\tilde{f}(0)$ must be computed by other means. We also see that the r-th equation provides no information about $\tilde{f}(rh)$ if $K(rh, rh) = 0$. It is therefore important to assume that $K(x, x) \neq 0$ for $0 \leq x \leq X$, if we require approximate values of $f(x)$ with $0 \leq x \leq X$, and with this assumption we obtain $\tilde{f}(0) \simeq f(0) = g'(0)/K(0, 0)$ from equation (2.1). If $g'(0)$ is not known we must approximate this value; a suitable accuracy is obtained by setting $g'(0) \simeq (g(h) - g(-h))/(2h)$. (We shall suppose either that the correct value of $f(0)$ is used or that this approximation is used and $g''(x)$ is continuous in $[-h, h]$.) When $\tilde{f}(0)$ is obtained we can employ (4.2) to obtain

$$\tilde{f}(rh) = \left\{g(rh) - h \sum_{j=0}^{r-1}{}' K(rh,jh)\ \tilde{f}(jh)\right\} / (h\ K(rh,rh))$$

for $r = 1, 2, \ldots, n$.

Jones (1961) studied the behaviour of the values $\tilde{f}(rh)$ as approximations to $f(rh)$ in the case where $K(x, y)$ is a difference kernel, of the form $K(x, y) = k(x - y)$. Linz (1967a) generalized the theory of Jones and established the following result.

Theorem 1. If $g(0) = 0$, $K(x, x) \neq 0$ for $0 \leq x \leq X$, and $K(x, y)$, $g(x)$ and the true solution $f(x)$ of (1.1) are "sufficiently smooth" then $f(rh) - \tilde{f}(rh) = O(h^2)$ as $h \to 0$ with $rh \leq X$ fixed. (Precise smoothness conditions can be determined from Linz (1967a) and Kobayasi (1967).)

Linz (1967a) also established a similar result when the values of $\tilde{f}(x)$ are computed using the mid-point rule: thus

$$h \sum_{j=0}^{r-1} K(rh,(j + \tfrac{1}{2})h)\ \tilde{f}((j + \tfrac{1}{2})h) = g(rh) \qquad (4.3)$$

for $r = 1, 2, \ldots$. This scheme provides approximate values $\tilde{f}(\tfrac{1}{2}h)$, $\tilde{f}(3h/2)$, $\tilde{f}(5h/2)$ ..., with accuracy $O(h^2)$. The scheme can be incorporated in (4.1) on replacing $\tfrac{1}{2}h$ by h.

The above convergence results are the sort to be hoped for, but the fears of pessimists are justified by a negative result, also given by Linz (1967a).

Theorem 2. Suppose in (4.1) that the Gregory rule is employed with corrections including second differences for $r \geq 2$. Then, in general, if $h \to 0$ and rh is fixed, $\tilde{f}(rh) \not\to f(rh)$.

To reinforce this theorem with a numerical example, Linz considers the equation $\int_0^x \cos(x - y)\, f(y)dy = \sin(x)$ (with solution $f(x) = 1$) and employs a fourth-order Gregory method. With $h = 0.1$ he computes $\tilde{f}(2) = 8.4$ and with $h = 0.05$, $\tilde{f}(2) = 1.4 \times 10^7$.

The disturbing result of Theorem 2 does not appear to be confined to the Gregory rule. From computational and theoretical work, Linz concluded that some other direct high-order methods (including block-by-block methods analogous to those for equations of the second kind) can display "non-convergence". In §5, however, we give more encouraging results.

The mid-point rule provides a better tool than the trapezoidal rule in some respects, in particular since the former has better stability properties and the latter produces "oscillations" in the values $\tilde{f}(rh)$ (and also requires a starting value). The oscillations are noted by Jones and by Linz, and it has been suggested that one should smooth the results by forming, in place of the values $\tilde{f}(h)$, $\tilde{f}(2h)$, ..., the approximations

$$\tilde{\phi}(rh) = \frac{1}{4}\{\tilde{f}(r-1)h) + 2\tilde{f}(rh) + \tilde{f}((r+1)h)\} \qquad (4.4)$$

for $r = 1, 2, \ldots$. Such observations are compatible with the work of Kobayasi (1967), who gives a result which shows that if

$$\int_0^x K(x, y)\, e(y)dy = -\frac{1}{12}\Big[(\partial/\partial y)K(x, y)\, f(y)\Big]_0^x \quad (x \geqslant 0)$$

and

$$K(x, x)\, d'(x) + K_y(x, x)\, d(x) = 0 \qquad (x \geqslant 0)$$

(where $d(0)$ is appropriately defined) then

$$\tilde{f}(rh) - f(rh) = h^2 e(rh) + (-1)^r h^2 d(rh) + O(h^3). \qquad (4.5)$$

(Kobayasi's method of smoothing the oscillations in $f(x)$ is similar in form to that of (4.4).) For the mid-point rule, Linz (who also considers the result for the trapezium rule) obtains a similar result, of the form

$$\tilde{f}((r+\tfrac{1}{2})h) - f((r+\tfrac{1}{2})h) = h^2\, \varepsilon((r+\tfrac{1}{2})h) + O(h^3) \qquad (4.6)$$

where $\varepsilon(x)$ satisfies a Volterra equation of the second kind.

Equation (4.6) provides a theoretical basis for an application of the deferred approach to the limit, to obtain approximations with accuracy $O(h^3)$. If we compute $\tilde{f}_0(h_0/2)$, $\tilde{f}_0(3h_0/2)$, $\tilde{f}_0(5h_0/2)$, ... using the mid-point method with $h = h_0$ and repeat with $h = h_0/3$ to obtain $\tilde{f}_1(sh_0/6)$ for $s = 1, 3, 5, 7, 9, 11, 13, \ldots$ then

$\tilde{f}_0(rh) - \{(9\,\tilde{f}_1\,(rh_0) - f_0(rh_0))/8\} = O(h_0^3)$.

For an application of the deferred approach using the trapezium rule it is necessary to compute, say, $\tilde{\phi}(rh)$ $(r = 1, 2, \ldots)$ defined by (4.4), because of the term $(-1)^r$ in (4.5). Thus, from (4.5),

$$\tilde{\phi}(rh) - \tfrac{1}{4}\{f((r-1)h) + 2f(rh) + f((r+1)h)\}$$

$$= \tfrac{1}{4}h^2\{e((r-1)h) + 2\,e(rh) + e((r+1)h)\}$$

$$+ \tfrac{1}{4}h^2\{2d(rh) - [d((r-1)h) + d((r+1)h)]\} + O(h^3).$$

Under suitable differentiability conditions we see (on employing Taylor series) that

$$\tilde{\phi}(rh) - f(rh) = \tfrac{1}{4}h^2\eta(rh) + O(h^3),$$

where $\eta(x) = f''(x) + e(x)$. If we calculate $\tilde{\phi}_0(h_0)$, $\tilde{\phi}_0(2h_0)$, $\tilde{\phi}_0(3h_0)$, ... using a step $h = h_0$, and then halve h to compute $\tilde{\phi}_1(h_0/2)$, $\tilde{\phi}_1(h_0)$, $\tilde{\phi}_1(3h_0/2)$, $\tilde{\phi}_1(2h_0)$, ..., we have

$$\{4\,\tilde{\phi}_1(rh_0) - \tilde{\phi}_0(rh_0)\}/3 - f(rh_0) = O(h_0^3).$$

(There seems to be less wastage of function evaluations if we use $\tilde{\phi}(x)$ instead of values of $\tilde{f}(x)$ produced with the mid-point rule using steps h_0 and $h_0/3$, but stability must also be taken into account.)

5. Higher Order Accuracy.

The use of the trapezium rule or the mid-point rule produces $O(h^2)$ accuracy in the approximations $\tilde{f}(rh) \simeq f(rh)$, assuming sufficient differentiability, etc. A single h^2-extrapolation (or alternatively a deferred correction, see Linz, 1967a) can be used to 'improve' the accuracy, but, in view of the negative result embodied in Theorem 2, a careful approach is required for high-order accuracy.

Alternative methods have been devized by Linz (1967a) and by de Hoog and Weiss (1971). The methods of de Hoog and Weiss involve a direct treatment of (1.1); the method of Linz is closely related to the results in §2 and will be discussed later.

For equation (1.1), de Hoog and Weiss set $x_{ij} = (i+u_j)h$ where $0 \leq u_0 < u_1 < \ldots < u_n = 1$ and employ certain quadrature rules

$$\int_{ih}^{(i+1)h} \phi(y)dy \simeq h \sum_{k=0}^{n} w_k\,\phi(ih + u_k h) \tag{5.1}$$

$$\int_{ih}^{(i+u_j)h} \phi(y)dy \simeq h \sum_{k=0}^{n} \omega_{jk}\,\phi(ih + u_k h). \tag{5.2}$$

In our exposition we suppose that $u_0 = 0$, although a choice $u_0 > 0$ seems to be better in practice.
The quadrature rules (5.1) and (5.2) are constructed to be interpolatory, in the sense that the degree of precision of (5.1) and (5.2) is n and the 'weights' w_k and ω_{jk} can be constructed by integrating the Lagrangian form of the polynomial of degree n having values $\phi(ih + u_k h)$ at $x = ih + u_k h$, $k = 0, 1, \ldots, n$. Constraints on a suitable choice of $u_1, u_2, \ldots, u_{n-1}$ are mentioned later.

The approximations (5.1) and (5.2) are used to produce equations approximating

$$\int_0^{x_{rs}} K(x_{rs}, y)\, f(y)dy = g(x_{rs}),$$

with $r = 0, 1, 2, \ldots,$ $s = 0, 1, \ldots, n$. Two such schemes are produced, the first giving

$$\sum_{i=0}^{r-1} \sum_{j=0}^{n} hw_j\, K(x_{rs}, x_{ij})\, \tilde{f}(x_{ij}) + \sum_{j=0}^{n} h\, \omega_{sj}\, K(x_{rs}, x_{rj})\, \tilde{f}(x_{rj}) = g(x_{rs}). \tag{5.3}$$

This scheme is produced by writing

$$\sum_{i=0}^{r-1} \int_{ih}^{(i+1)h} K(x_{rs}, y)\, f(y)dy + \int_0^{x_{rs}} K(x_{rs}, y)\, f(y)dy = g(x_{rs}) \tag{5.4}$$

and using (5.1) to approximate the first terms and (5.2) to approximate the last term in the left-hand side. We require $\tilde{f}(0)$ to be obtained independently, as before. In (5.3) we require values of $K(x, y)$ for $y > x$, which may not be available. Accordingly de Hoog and Weiss proposed a scheme based on (5.4) as before, but approximating the integral on $[rh, x_{rs}]$ in a different way. They write

$$\int_{rh}^{x_{rs}} K(x_{rs}, y)\, f(y)dy \simeq \int_{rh}^{x_{rs}} K(x_{rs}, y)\, p_r(y)dy,$$

where $p_r(x)$ is the polynomial of degree n with values $\tilde{f}(x_{rs})$ when $x = x_{rs}$, for $s = 0, 1, \ldots, n$, and they employ the quadrature rule (5.1) to approximate the latter integral. The numerical scheme thus produced is

$$\sum_{i=0}^{r-1} \sum_{j=0}^{n} h\, w_j\, K(x_{rs}, x_{ij})\, \tilde{f}(x_{ij}) + \sum_{j=0}^{n} \sum_{k=0}^{n} h\, w_k\, u_s\, K(x_{rs}, (i+u_s u_k)h) \times \ell_j(u_s u_k)\, \tilde{f}(x_{rj}) = g(x_{rs}), \tag{5.5}$$

where the functions $\ell_j(x)$ are the coefficients in the Lagrangian form $p_r(x) = \sum_{j=0}^{n} \ell_j(x) \tilde{f}(x_{rj})$. The schemes (5.3) and (5.5) can be solved, if $u_0 = 0$, for $\tilde{f}(u_0h)$, $\tilde{f}(u_1h)$, ... $\tilde{f}(h)$; $\tilde{f}((1 + u_1)h)$, $\tilde{f}((1 + u_2)h)$, ... in a block-by-block fashion, and de Hoog and Weiss (1971) give some numerical examples of their application. For (1.3) the corresponding scheme with $u_0 > 0$ is equivalent to a method of approximate differentiation of $g(x)$.

de Hoog and Weiss (1971) have presented some interesting results concerning the convergence and stability of the computed approximations. They assume that $f(x)$ is $(n + 4)$ times continuously differentiable on $[0, X]$ and that $K(x, y)$ is $(n + 4)$ times continuously differentiable for $0 \leq y \leq x + \delta$, $0 \leq x \leq X$, where for (5.5) $\delta = 0$ and for (5.3) $\delta > 0$ is fixed. They also suppose $K(x, x) \neq 0$ for $0 \leq x \leq X$. Among other results, de Hoog and Weiss (1971) give the following theorem.

Theorem 3. Let $\eta = \prod_{k=0}^{n-1} (1 - u_k) / \prod_{k=1}^{n} (-u_k)$. If $\eta \in [-1, 1]$ the schemes (5.3), (5.5) are convergent, the errors being $O(h^{n+1})$ if $\eta \in [-1, 1)$, and $O(h^n)$ if $\eta = 1$, on $[0, X]$. (More powerful results are given for the case $u_0 > 0$ by de Hoog and Weiss, 1973.)

Such convergence results do not guarantee the success of the methods with arbitrary step-size h. Linz (1967a) considered an application of (5.3) to a simple equation in which the method behaved disastrously with $h = 0.3$ and with $h = 0.15$, and it is interesting to see the explanation of de Hoog and Weiss (1971) that this behaviour is due to too large a choice of h. The errors they obtain with smaller h are fairly satisfactory, and much improved when they employ (5.5).

Linz (1967a) constructs high-order approximations to the solution of (1.1) using a finite difference analogue of (2.2). His technique can be extended to situations in which $K(x,x) = 0$ for $0 \leq x \leq X$ but, say, $[K_x(x,y)]_{y=x}$ is non-vanishing. Linz employs a p-th order method for solving the integral equation (2.2), of the form

$$f(rh) - \sum_{j=0}^{n} w_{rj} K_1(rh,jh) f(jh) = g'(rh),$$

but he replaces $g'(x)$ and $K_x(x, y)$ (in the definition of $K_1(x, y)$) by finite difference approximations with accuracy $O(h^p)$. It is then not difficult to show that $\tilde{f}(rh) - f(rh) = O(h^p)$ as $h \to 0$ with rh

fixed. de Hoog and Weiss (1973) have some comments on the relative merits of this approach and their own schemes.

6. Product Integration.

The methods of §5 depend on general quadrature formulae, but it is also possible (and sometimes necessary) to use product-integration approximations, of the form

$$\int_{\alpha}^{\beta} K(x, y)\ \phi(y)dy \simeq \sum_{j=0}^{n} \nu_j^{\alpha,\beta}(x)\ \phi(z_j^{\alpha,\beta})$$

say, which are tailor-made to the kernel $K(x, y)$. Thus with $n = 0$ and the approximation

$$\int_{\alpha}^{\beta} K(x, y)\ \phi(y)dy \simeq \phi\left(\frac{\alpha + \beta}{2}\right) \int_{\alpha}^{\beta} K(x, y)dy$$

(which we call the modified mid-point rule) we can produce a composite or repeated version to give

$$\int_{0}^{rh} K(x, y)\ \phi(y)dy \simeq \sum_{j=0}^{r-1} \phi((j+\tfrac{1}{2})h) \int_{jh}^{(j+1)} K(x, y)dy,$$

which leads to the scheme

$$\sum_{j=0}^{r-1} \tilde{f}((j+\tfrac{1}{2})h) \int_{jh}^{(j+1)h} K(rh, y)dy = g(rh),\ r = 0, 1, \ldots$$

The modified trapezium rule is based on an approximation

$$\int_{\alpha}^{\beta} K(x, y)\ \phi(y)dy \simeq \nu_0^{\alpha,\beta}(x)\ \phi(\alpha) + \nu_1^{\alpha,\beta}(x)\ \phi(\beta),$$

which is exact if $\phi(x)$ is linear in $[\alpha, \beta]$. For such rules applied to Volterra equations of the first kind see Anderssen and White (1971), Weiss and Anderssen (1971), Atkinson (1971), Linz (1967a).

The use of product integration is particularly appropriate when $K(x, y)$ is badly behaved but the solution $f(x)$ is not; in particular for Abel's equation and equations with weakly-singular kernels, e.g. $\int_0^x \frac{H(x, y)}{(x - y)^{\mu}} f(y)dy = g(x)$ where $0 \leqslant \mu < 1$ and $H(x, y)$ is smooth. Weiss and Anderssen (1971) analyse the convergence of product integration techniques for such an equation, whilst Atkinson (1971) (with some involved analysis) obtains convergence results for Abel's equation (1.2) with $\alpha = \frac{1}{2}$ and $p = 2$.

Equations with weakly singular kernels occur quite frequently in practice and the techniques of this section are then particularly useful, although their theoretical justification is not yet complete.

References

ANDERSSEN, A.S. and WHITE, E.T. (1971) Improved numerical methods for Volterra integral equations of the first kind. Comput. J. Vol. 14, pp.442-443.

ATKINSON, K.E. (1971) The numerical solution of an Abel integral equation by a product trapezoidal method. Computer Centre, Australian National University, Canberra.

CAMPBELL,G.M. (1970) On the numerical solution of Volterra integral equations. Ph.D. Thesis, Pennsylvania State University, U.S.A.

CAMPBELL, G.M. and DAY, J.T. (1970) An empirical method for the numerical solution of Volterra integral equations of the first kind. Pennsylvania State University, U.S.A.

COLLATZ, L. (1960) The Numerical Treatment of Differential Equations. Springer-Verlag, Berlin (3rd edition, 1966).

DE HOOG, F. and WEISS, R. (1971) High order methods for first kind Volterra integral equations. Computer Centre, The Australian National University, Canberra.

DE HOOG, F. and WEISS, R. (1973) On the solution of Volterra integral equations of the first kind. Num. Math. Vol. 21, pp.22-32.

EDELS, H., HEARNE, K. and YOUNG, A. (1962) Numerical solution of the Abel integral equation. J. Math. and Phys. Vol. 41, pp.62-75.

FETTIS, H.G. (1964) On the numerical solution of equations of the Abel type. Maths. Comput. Vol. 18, pp.491-496.

JONES, J.G. (1961) On the numerical solution of convolution integral equations and systems of such equations. Maths. Comput. Vol. 18, pp.491-496.

KOBAYASI, M. (1967) On numerical solution of the Volterra integral equations of the first kind by trapezoidal rule. Rep. Stat. Appl. Res. JUSE Vol. 14, pp.65-78.

LINZ, P. (1967a) The numerical solution of Volterra integral equations by finite difference methods. MRC Tech. Report 825,Madison, Wisconsin.

LINZ, P. (1967b) Applications of Abel transforms to the numerical solution of problems in electrostatics and elasticity. MRC Tech. Report 826,Madison, Wisconsin.

LINZ, P. (1971) Product integration methods for Volterra integral equations of the first kind. Nord. Tidskr. Inf. Behandl. (BIT)

Vol. 11, pp.413-421.

MINERBO, G.N. and LEVY, M.E. (1969) Inversion of Abel's integral equation by means of orthogonal polynomials. SIAM J. numer. Anal. Vol. 6, pp.598-616.

NOBLE, B. (1971) A bibliography on "Methods for solving integral equations" - subject listing. MRC Tech. Report 1177 Madison, Wisconsin, U.S.A.

SCHMAEDEKE, W.W. (1968) Approximate solutions of Volterra integral equations of the first kind. J. math. Anal. and Appl. Vol. 23, pp.604-613.

SQUIRE, W. (1969) Numerical solution of linear Volterra equations of the first kind. Aerospace Eng. TR-15, West Virginia University.

WEISS, R. and ANDERSSEN, R.S. (1971) The convergence of a product integration method for a class of singular first kind Volterra equations. Computer Centre, Australian National University, Canberra.

Note The reports originating from the Australian National University carry no date. In the majority of instances these reports have been submitted for publication in the journals.

Appendix

In this appendix we report on some numerical experiments performed using a method of de Hoog and Weiss (1971). For a simple test equation of the form

$$\int_0^x K(x, y)\ f(y)dy = g(x) \tag{A1}$$

we formulated and solved the equations (5.5).

These equations were produced using choices of $u_0 = 0, u_1, \ldots, u_n = 1$ which lead, respectively, to various Newton-Cotes rules and a Lobatto rule in (5.1). For such rules, with $u_0, u_1, \ldots, u_n$ symmetrically placed about the mid-point of the interval $[0, 1]$, it follows immediately that $|\eta| = 1$ and (according to Theorem 3) convergence should be obtained as $h \to 0$ for suitable equations.

Problem 1

In equation (A1) we set $K(x, y) = 0.5 + x^4 - y^4$, $g(x) = 0.5x + 0.8x^5$, and the solution is $f(x) = 1$. Linz (1967) and de Hoog and

Weiss consider this example. The exact starting value $f(0) = 1$ is supplied. Blocks of values of the approximate solution are determined by solving systems of linear equations, and we also give the maximum condition number (in the uniform norm) of the matrices encountered in the numerical solution.

(i) Rule: 3/8-ths rule: $u_0 = 0$, $u_1 = 1/3$, $u_2 = 2/3$, $u_3 = 1$, with $\eta = -1$. The solution is computed on $[0, 1.5]$.

ERRORS

x	Step h = 0.15		Step h = 0.0375
x = 0	0	B	0
0.05	-1.39×10^{-7}	L	2.28×10^{-8}
0.10	-3.89×10^{-6}	O	-1.52×10^{-8}
0.15	-1.79×10^{-5}	C	8.57×10^{-11}
		K	
0.20	5.87×10^{-6}		2.30×10^{-8}
0.25	-9.77×10^{-6}		-1.47×10^{-8}
0.30	3.69×10^{-7}		1.30×10^{-9}
....	....		....
1.25	-2.17×10^{-5}		1.60×10^{-7}
1.30	1.39×10^{-5}		1.44×10^{-7}
1.35	-1.46×10^{-4}		9.22×10^{-7}
1.40	8.05×10^{-5}		5.31×10^{-7}
1.45	-4.75×10^{-5}		5.98×10^{-7}
1.50	4.89×10^{-4}		4.70×10^{-6}
	Maximum condition number 105		Maximum condition number 49

Here it should be noted that for each x, $K(x, y)f(y)$ is a quartic in y, and the 3/8 rule is exact for cubics.

(ii) If we use the 5-point Lobatto rule we expect the correct answer for this problem. Since, however, the weights and abscissae of this rule are irrational numbers, rounding errors are introduced; we found that these errors propagated fairly slowly over the range of the solution, the maximum condition number with $h = 0.15$ being 166.

(iii) If we use the 9-point Newton-Cotes rule (which has negative weights) any errors are again due to rounding. The maximum condition number recorded was 5800 and rounding errors were significantly worse than in (ii).

Problem 2

The preceding methods were also applied to the equation

$$\int_0^u k(v + 3 - u)\,\phi(v)dv = \gamma(u) \qquad (0 \leqslant u \leqslant 6)$$

where $k(z) = 1 + \cos(\pi z/3)$ if $|z| \leqslant 3$ and $\gamma(u)$ is given. This equation was obtained by reformulating a (now classic) Fredholm equation of the first kind studied by Phillips (1962) (see the references for Chapter 13). The methods produced disastrous results for this equation, which can be explained by the vanishing of $k(z)$ when $z = 3$. Repeated differentiation of the Volterra equation of the first kind does however lead to an equation of the second kind which can be solved numerically.

The calculations reported here were performed by D.K. Sayers, using a CDC7600.

CHAPTER 13 FREDHOLM EQUATIONS OF THE FIRST KIND

G.F. Miller

National Physical Laboratory

. Introduction

The linear Fredholm equation of the first kind is defined by

$$\int_a^b K(x, y)\, f(y)\, dy = g(x) \quad (c \leqslant x \leqslant d) \tag{1.1}$$

or, briefly, in operator form,

$$Kf = g. \tag{1.2}$$

For simplicity we consider only linear equations and suppose that all the functions concerned are real. Notice that the x-range (c, d) does not necessarily coincide with the range of integration (a, b). An equation of first kind is distinguished by the fact that the unknown function f(y) appears only under the sign of integration. The consequences of this are profound; the equation has properties which are quite different from an equation of second kind and poses special difficulties.

First we propose to give a broad general view of the problem and its difficulties; then we will analyse it in a little more detail in terms of singular functions (eigenfunctions in symmetric cases), and finally consider some methods of numerical solution.

2. Nature of the Problem

The problem has much in common with that of solving a system of algebraic equations $A\underline{x} = \underline{b}$, in which the matrix A is severely ill-conditioned (or even singular) owing to the near linear dependence of its adjacent rows and columns. Furthermore, since A corresponds to a completely continuous integral operator it may be supposed to possess a cluster of small eigenvalues; consequently any attempt to solve the system directly is likely to produce a wildly oscillating solution.

In the case of an integral equation of the second kind, on the other hand, the corresponding algebraic system will be of the form $(I - \lambda A)\underline{x} = \underline{b}$, and unless λ is very large, the near-singularity of A is not a source of difficulty.

However, there are features related to the infinite-dimensionality of the integral operator which are not fully reflected in the matrix operator. First let us note that by no means every function $g(x)$ can be represented in the form prescribed by (1.1); we can express this by saying that g must lie in the range of the operator K.

Suppose, for example, that

$$K(x, y) = \cos x \cos y.$$

Then if we are given any integrable $f(x)$ whatsoever and we evaluate the integral, the resulting $g(x)$ must clearly be a multiple of $\cos x$. It follows that if we are given a $g(x)$ which is not of this form we have no hope of finding a solution $f(y)$ of (1.1).

This example may appear extreme but it does serve to illustrate the way in which $g(x)$ may be restricted by the character of $K(x, y)$ as a function of x; with an equation of second kind degeneracy of the kernel has no comparable effect. In general $g(x)$ must not contain components which are not matched by a corresponding component in $K(x, y)$.

Closely bound up with this consideration of compatibility (especially in symmetric cases) is the question of uniqueness. In the above example there is clearly an infinity of functions $\phi(y)$ such that $K\,\phi = 0$ and hence the solution, even if it exists, cannot be unique.

There is another compatibility requirement on $g(x)$. It is important to note that K has the character of a smoothing operator, its effect being to smooth out any discontinuity or roughness in the $f(y)$ on which it operates; the degree of smoothing depends a good deal on the type of kernel. In order that $g(x)$ may lie in the range of K it must possess a sufficient degree of "smoothness" relative to K. (This question is associated with the convergence condition in §3.)

From a computational standpoint, however, perhaps the most troublesome feature is that the problem is ill posed (or incorrectly or improperly posed), by which we mean that the solution f does not depend continuously on the data function g; in other words a very small perturbation on g can give rise to an arbitrarily large perturbation in f. To show this let us observe that, according to the celebrated Riemann-Lebesgue theorem, if $K(x, y)$ is absolutely integrable (as a function of y for each x) then

$$h_n(x) \equiv \int_a^b K(x, y) \cos ny \, dy \to 0 \quad (c \leqslant x \leqslant d)$$

as $n \to \infty$. Therefore by choosing n large enough we can ensure that, given ε however small,

$$\max_{c \leqslant x \leqslant d} |h_n(x)| \leqslant \varepsilon.$$

Now suppose that the equation $Kf = g$ has a unique solution f, and let g be given a small perturbation $\delta g = C\, h_n(x)$. The corresponding change in f is $\delta f = C \cos nx$. Clearly by choosing a sufficiently large n we can make the ratio $||\delta f|| / ||\delta g||$ as large as we like. (For a general account of the theory of ill-posed problems and some applications, see Lavrentiev, 1967.)

3. Singular Function Analysis

Let the kernel be given in the form of an expansion in terms of the singular functions (s.f.'s) $u_i(x)$, $v_i(y)$, and singular values κ_i:

$$K(x, y) = \sum_i \kappa_i\, u_i(x)\, v_i(y). \tag{3.1}$$

In the case of symmetry this reduces to an eigenfunction expansion - but this case appears not to be particularly common or significant in relation to equations of first kind. The s.f. expansion provides the natural extension in the general case. It is not necessary that the expansion should converge pointwise; all we require, for an L_2 kernel, is that

$$\lim_{n\to\infty} \iint \{K(x, y) - \sum_{i=1}^{n}\}^2\, dx\, dy = 0,$$

for which it is necessary and sufficient that the series $\sum_i \kappa_i^2$ converge. (For a comprehensive theoretical treatment see Smithies, 1958, Chapter 8).

The sequences $\{u_i(x)\}$ and $\{v_i(y)\}$ constitute orthonormal systems with the properties

$$K\, v_i = \kappa_i\, u_i,\; K^T\, u_i = \kappa_i\, v_i. \tag{3.2}$$

Hence

$$K\, K^T\, u_i = \kappa_i^2\, u_i,\quad K^T K\, v_i = \kappa_i^2\, v_i, \tag{3.3}$$

so that u_i and v_i are eigenfunctions for the symmetric (self-adjoint) operators $K\,K^T$ and $K^T K$ respectively, associated with eigenvalue κ_i^2. The κ_i^2 are necessarily real and positive and we can arrange that

$\kappa_i > 0$ also.

The effect of operating with K on an arbitrary (L_2-integrable) function f is readily seen to be

$$Kf = \sum_i \kappa_i f_i u_i(x), \tag{3.4}$$

where

$$f_i = (f, v_i) = \int_a^b f(y) v_i(y) dy. \tag{3.5}$$

Now let us suppose that the function $g(x)$ in (1.1) can be expanded in the form

$$g(x) = \sum_i g_i u_i(x), \tag{3.6}$$

with

$$g_i = (g, u_i) = \int_c^d g(x) u_i(x) dx. \tag{3.7}$$

(We can always form the expansion but there is in general no guarantee of equivalence.) If the equation $Kf = g$ is to be satisfied, the two expansions (3.4) and (3.6) must agree; hence, in virtue of the orthogonality of the $u_i(x)$, we require

$$\kappa_i f_i = g_i. \tag{3.8}$$

From this simple relation we can derive most of the properties which interest us.

4. Fundamental Theorems

The above analysis suggests as a tentative solution of (1.1)

$$f(y) = \sum_i \frac{g_i}{\kappa_i} v_i(y). \tag{4.1}$$

We now enquire: under what conditions is this solution valid and when is it unique? The answers to these questions are briefly as follows:

A. Compatibility. For any functions u such that $K^T u = 0$ we must have $(g, u) = 0$. (This may be thought of as the case of a zero singular value; thus $\kappa_i = 0$ implies $g_i = 0$.)

B. Convergence. For the existence of an L_2 solution we clearly require $\sum_i (g_i/\kappa_i)^2 < \infty$. (This may be regarded as a "smoothness" condition; obviously the more rapidly the κ_i decrease the more severely is g restricted.)

A and B together represent necessary and sufficient conditions for existence.

C. Uniqueness. Suppose there is a function v such that $Kv = 0$. Then if f is a solution, so is $f + Cv$. Moreover, any two solutions differ by a function v with this property. Hence uniqueness occurs if and only if the equation $Kv = 0$ has no nontrivial solution.

The ill-posed character of the solution is also apparent from the singular function analysis; we have

D. Instability. Suppose that a solution exists and that it is unique (if it is not unique the problem is certainly ill posed). By the Hilbert-Schmidt theory the sequence of singular values $\{\kappa_i\}$ has zero as its only possible limit point; the assumed uniqueness implies that the sequence is infinite, and so zero is indeed a limit point. Thus by choosing i sufficiently large we can find a κ_i as small as we like. Perturb $g(x)$ by changing the coefficient g_i by an amount δg_i. This causes a perturbation $\delta f = (\delta g_i/\kappa_i)\, v_i(y)$ in $f(y)$. Hence the ratio $\|\delta f\|/\|\delta g\|$ can be made arbitrarily large.

5. Applications; Noise; Need for Filtering

The fact that a problem is "ill posed" does not necessarily mean that it has been badly or wrongly formulated. A great many physical problems (e.g. in spectroscopy, X-ray scattering, chemical analysis, queueing theory, optimal control and in the calculation of atmospheric temperature profiles) do give rise to equations of first kind and it is essential to investigate methods for dealing with them. Reviews of applications, with extensive bibliographies, are to be found in Tihonov (1967), Turchin, Kozlov & Malkevich (1971), and Nedelkov (1972).

A commonly occurring situation is that in which the solution function f corresponds to some actual physical phenomenon that one wishes to observe or measure, the operator K corresponds to the action of the measuring instrument, while g represents the set of observations or data on which the estimation of f is to be based. Limitations in the resolving power of the instrument will lead to loss of some of the fine detail in f; high-frequency components become attenuated, as is indicated by the relation (3.2). Some such loss is inevitable. What we must try to avoid, however, is to treat small oscillatory components of $g(x)$ which are really in the nature of noise as if they were due to large oscillations in $f(y)$. The situation is further complicated by the fact that additional errors resembling noise are liable to be introduced in the course of computation.

The difficulty is not easily resolved, and in fact the remedy can only lie in a revision of the problem itself. Because of its instability (property D of §4) the fact that we find a function f^* which makes $||Kf^* - g||$ small by no means implies that f^* is close to the "true" f. In order to overcome this defect it is necessary to restrict the field of admissible functions f^* by invoking some additional property of the solution (such as smoothness) which is not contained in the original equation (1.1). Usually one seeks a "best" solution within a certain class and according to some specific criterion. If the latter is suitably chosen the effect will be that of a filtering process which discriminates against excessive contributions from the higher singular functions, and so prevents the appearance of wild oscillations in the solution.

6. Method of Expansion in Singular Functions

One method of filtering which immediately suggests itself is to calculate successive terms in the s.f. expansion and to terminate at a suitable point. Such a method was proposed by Baker, Fox, Mayers and Wright (1964) for the symmetric case, while extensions to the general case have been considered by Hanson (1971, 1972), Vainshtein (1972a, b) and Crone (1972).

By matrix methods we may calculate approximations to the $v_i(y)$ and coefficients g_i and hence form the sequence of partial s.f. expansions

$$f^{(m)}(y) = \sum_{i=1}^{m} \frac{g_i}{\kappa_i} v_i(y), \quad m = 1, 2, \ldots \tag{6.1}$$

At the same time we may, if we wish, form the corresponding residual sum of squares, given by

$$R^{(m)} \equiv ||Kf^{(m)} - g||^2 = ||g||^2 - \sum_{i=1}^{m} g_i^2. \tag{6.2}$$

Each time we increase m by one we satisfy one more of the equations

$_i f_i = g_i$, thereby adding a term to $f^{(m)}$ and eliminating one from $R^{(m)}$.

A kind of law of diminishing returns operates here; as m increases each reduction in $R^{(m)}$ requires a proportionately large change in $f^{(m)}$, while in practice the problem is accentuated by the difficulty of computing the higher singular functions sufficiently accurately. If g contains noise the terms must eventually begin to increase in size and it seems natural to stop in the region of the smallest term.

The method is theoretically attractive for the insight it provides into the problem, but has the practical disadvantage that it is rather laborious and expensive to apply.

7. Use of Other Expansions

A rather simpler procedure is to choose some other basic system of functions, $\{p_j(y)\}$ say, and to approximate f in the form

$$f^{(m)}(y) = \sum_{j=1}^{m} c_j^{(m)} p_j(y). \tag{7.1}$$

We then operate on it with K and seek to satisfy the equation

$$\sum_{j=1}^{m} c_j (Kp_j)(x) = g(x) \tag{7.2}$$

as closely as possible, e.g. in a least-squares sense, using an appropriate quadrature formula to approximate the quantities Kp_j.

This method is not to be despised. It can work out quite well, especially if we happen to know a set of functions $p_j(y)$ which is particularly well suited to the problem, i.e. such that $f(y)$ can be represented by means of only a few terms. Undoubtedly many ill-posed problems have been dealt with in this way with fair success, and in some cases without the solver ever being aware of the inherent difficulties! The instability is held at bay so long as m is kept small. As soon as we increase m it begins to appear in the form of ill-conditioning of the resulting matrix operator, so that there is a limit to the accuracy which can be obtained by increasing the order of the approximation. Another disadvantage is that the approximate solution may depend strongly on m and on the choice of the system $\{p_j(y)\}$. Some discussion is included in Turchin et al. (1971) (an admirable review paper covering many aspects of the problem).

8. Method of Regularization

Another method of attack which has received a good deal of attention in recent years is that of regularization. It was proposed independently by Phillips (1963) in the U.S.A. and Tihonov (1963, 1964) in the U.S.S.R., and has been subsequently developed and extended by Twomey (1964, 1965), Ribière (1967) and many others especially in the U.S.S.R. (Bakushinskii, Arsenin, Ivanov, Morozov).

The method essentially consists in replacing the ill-posed problem by a stable minimization problem involving a small positive parameter α: instead of attempting to solve the equation $Kf = g$ directly we seek to minimize the quadratic functional

$$||Kf - \tilde{g}||^2 + \alpha||L f||^2, \tag{8.1}$$

where L is some linear operator and $\tilde{g}$ denotes $g + \delta g$. If L is suitably chosen, the second term has a smoothing or stabilizing effect on the solution. We may, for example, take $Lf = f$, f' or f''; if the kth derivative is selected the process is termed kth order regularization. More generally the term $||Lf||^2$ can be replaced by a linear combination of the squared norms of several derivatives. Another choice, which is often appropriate if an initial estimate $\bar{f}$ of f is available, is $Lf = f - \bar{f}$.

It is of interest to note that there are two distinct ways of arriving at this formulation. Either we can say:

(i) It is impracticable (and unwise) to try to solve equation (1.1) exactly since the data is only approximate. Let us therefore replace it by $||Kf - \tilde{g}|| \lesssim \varepsilon$, where ε is chosen to correspond to the noise level in g, and let us minimize $||Lf||$ subject to this restriction;

or we can say:

(ii) We should like our solution to have the property $||Lf|| \lesssim B$. Let us restrict our search to this class and determine within it the function f which most closely satisfies the integral equation, in the sense that $||Kf - \tilde{g}||$ is minimized. (This f is termed a "quasi-solution" relative to L.)

In either case we are led, by a theorem in the calculus of variations, to the minimization of (8.1) for some particular value of α which must be determined to correspond with the assigned ε or B.

Subject to certain restrictions on L relative to K (essentially there must be no functions ϕ of unit norm such that $||K\phi||$ and $||L\phi||$ are small simultaneously), there exists a unique minimizing function for (8.1), f_α say. An increase in α produces an increase in $||Kf_\alpha - g||$ and a decrease in $||Lf_\alpha||$, and conversely. It is a question of balancing the magnitude of the residual against smoothness of the solution as measured by $||Lf_\alpha||$.

The regularized solution f_α has the desirable property that if $\delta g = 0$ and equation (1.1) possesses a unique solution f then $||f - f_\alpha|| \to 0$ as $\alpha \to 0$, subject to suitable conditions on L and provided that $||Lf|| < \infty$. Moreover, the same is true even with the use of an approximate $\tilde{g}$ provided that $||\delta g||^2 \to 0$ no less rapidly than α. (See e.g. Ribière, 1967, for a detailed account.)

Let us examine the method in slightly more detail in the simple case when $Lf = f$. The solution of the minimization problem (8.1) is then obtained as the solution of the linear equation

$$(K^T K + \alpha I) f_\alpha = K^T \tilde{g}. \qquad (8.2)$$

The operator acting on f_α is clearly positive-definite when $\alpha > 0$ and hence has a bounded inverse. Solving in terms of singular functions we obtain

$$f_\alpha(y) = \sum_i \frac{\kappa_i g_i}{\kappa_i^2+\alpha} v_i(y). \qquad (8.3)$$

Comparing this with the exact expansion (4.1) we see that the effect of regularization has been to insert a <u>filter factor</u> $\kappa_i^2/(\kappa_i^2 + \alpha)$. This is close to unity so long as κ_i is large compared with α but tends to zero as $\kappa_i \to 0$, the rate of transition depending on α. Different choices of L (as well as α) - and other methods generally - lead to different filter factors. Our particular choice $L = I$ has produced a rather mild "roll-off". This is in strong contrast to the method of s.f. expansion, in which the series is simply truncated at a selected point.

If we split $\tilde{g}$ into $g + \delta g$ the expression (8.3) becomes

$$f_\alpha = \sum_i \frac{\kappa_i g_i}{\kappa_i^2+\alpha} v_i(y) + \sum_i \frac{\kappa_i \delta g_i}{\kappa_i^2+\alpha} v_i(y). \qquad (8.4)$$

As regards the first term it is advantageous to make α small in order to reduce the error due to regularization; by contrast the second term, which consists wholly of error, is made small by taking α large. Thus there is a conflict, and we would like to achieve the best compromise. The question of the choice of an optimum α has been discussed by many authors (see, for example, Ribière, 1967) but there appears to be no simple rule. However, from the consideration that the maximum value of the function $\kappa/(\kappa^2 + \alpha)$ is $1/(2\alpha^{\frac{1}{2}})$ we may deduce that the maximum error in f due to an error δg in $\tilde{g}$ has a maximum norm of approximately $\|\delta g\| /(2\alpha^{\frac{1}{2}})$. Hence in order to keep the error from this source small α must certainly exceed $\frac{1}{4}\| \delta g \|^2$.

Regularization methods have several attractive features:

(i) They proceed by converting an ill-posed problem into a well-posed one.

(ii) They allow more flexibility than the expansion methods in that the quality criterion is independent of the mode of representation of the solution.

(iii) They are economical and comparatively easy to apply.

9. Iterative Methods

These provide yet another approach to the problem. The simplest method, variously ascribed to Landweber (1951) and Fridman (1956), has the form

$$f^{(n+1)} = f^{(n)} + \beta K^T (g - K f^{(n)}), \quad n = 0, 1, \ldots, \qquad (9.1)$$

starting with $f^{(0)} = 0$. It is not difficult to show that, in terms of singular values and functions,

$$f^{(n)} = \sum_i \{1 - (1 - \beta \kappa_i^2)^n\} \frac{g_i}{\kappa_i} v_i(y) \qquad (9.2)$$

and that this converges to the true solution (or, in case of non-uniqueness, to the solution of minimum norm) provided that $0 < \beta < 2/\kappa_1^{\ 2}$.

The method can be generalized in various ways, notably by inserting an additional operator to the left of K^T in (9.1), in order to obtain a filter function with improved characteristics; see Marchuk and Vasilev (1970) and Strand (1973). It was first shown by Bakushinskii (1967) that the iterative process can be regarded as a regularization procedure with $1/n$ playing the role of the parameter α.

Another possibility is to apply the regularization method iteratively as considered by Shaw (1972) and Strand (1973). If, for example, we take $L f = f - \bar{f}$, where $\bar{f}$ is some prior estimate of f, the linear equation determining f_α (cf. equation (8.2)) becomes

$$(K^T K + \alpha I) f_\alpha = K^T g + \alpha \bar{f}. \tag{9.3}$$

In the absence of an initial estimate we may start with $\bar{f} = 0$ and solve repeatedly, each time replacing $\bar{f}$ by the last computed f_α. Thus we have the iterative process

$$(K^T K + \alpha I) f^{(n)} = K^T g + \alpha f^{(n-1)}, \quad n = 1, 2, \ldots, \tag{9.4}$$

and we can show that

$$f^{(n)} = \sum_i \left\{1 - \left(\frac{\alpha}{\kappa_i^2 + \alpha}\right)^n\right\} \frac{g_i}{\kappa_i} v_i(y) . \tag{9.5}$$

The effect of increasing n is to admit more of the high-frequency components but to produce a steeper roll-off. In view of this it is sometimes advantageous to take a larger α than would be appropriate for a single application of the process, and to perform several iterations. However, if the process is carried too far the benefit of regularization is lost and instability sets in.

10. Probabilistic Methods

Finally we briefly mention another class of methods in which the problem is viewed from a probabilistic or statistical standpoint. An essential feature is that the functions f and g are now replaced by stationary random processes - in recognition of the fact that an actual experiment usually involves both sampling of f and random errors of observation in g. Given certain statistical information (mean and covariance matrix) concerning g at each point, or in each component, the object is to determine similar information about f. However, it turns out that in order to make any progress here we need also to have certain *a priori* information about f itself; this is tantamount to saying that the stochastic problem needs to be regularized just as much as does the deterministic one. The probabilistic methods may in fact be regarded as an extension of the regularization methods, and the actual process of computation is somewhat similar. For details see Turchin et al. (1971) (which in turn refers to important earlier work of Turchin), Strand and Westwater (1968) and Franklin (1970).

References

BAKER, C.T.H., FOX, L., MAYERS, D.F. and WRIGHT, K. (1964). Numerical solution of Fredholm integral equations of first kind. Comput.J. Vol. 7, pp.141-147.

BAKUSHINSKII, A.B. (1967). A general method of constructing regularizing algorithms for a linear incorrect equation in Hilbert space. USSR Comput. Math. and Math. Phys. Vol. 7, pp.279-287.

CRONE, L. (1972). The singular value decomposition of matrices and cheap numerical filtering of systems of linear equations. J. Franklin Inst. Vol. 294, pp.133-136.

FRANKLIN, J.N. (1970). Well-posed stochastic extensions of ill-posed linear problems. J. math. Anal. & Applic. Vol. 31, pp.682-716.

FRIDMAN, V.M. (1956). The method of successive approximations for Fredholm integral equations of the first kind. Usp. Mat. Nauk Vol. 11, pp.233-234.

HANSON, R. (1971). A numerical method for solving Fredholm integral equations of the first kind using singular values. SIAM J. numer. Anal. Vol. 8, pp.616-622.

HANSON, R. (1972). Integral equations of immunology. Communs. Ass. comp. Mach. Vol. 15, pp.883-890.

LANDWEBER, L. (1951). An iteration formula for Fredholm integral equations of the first kind. Am. J. Math. Vol. 73, pp.615-624.

LAVRENTIEV, M.M. (1967). Some Improperly Posed Problems of Mathematical Physics. Springer-Verlag.

MARCHUK, G.I. and VASILEV, V.G. (1970). On an approximate solution for operator equations of the first kind. Soviet Math. Doklady Vol. 11, pp.1562-1566.

NEDELKOV, I.P. (1972). Improper problems in computational physics. Comput. Phys. Commun. Vol. 4, pp.157-164.

PHILLIPS, D.L. (1962). A technique for the numerical solution of certain integral equations of the first kind. J. Ass. comput. Mach. Vol. 9, pp.84-96.

RIBIÈRE, G. (1967). Régularization d'opérateurs. Rev. Franç. Inf. Rech. Opér. Vol. 1, pp.57-79.

SHAW, C.B. Jr. (1972). Improvement of the resolution of an instrument by numerical solution of an integral equation. J. math. Anal. & Appl. Vol. 37, pp.83-112.

SMITHIES, F. (1958). Integral Equations. Cambridge Tract No. 49. Cambridge University Press.

STRAND, O.N. (1973). Theory and methods related to the singular function expansion and Landweber's iteration for solving integral equations of the first kind. SIAM J. numer. Anal. (to appear)

STRAND, O.N. and WESTWATER, E.R. (1968). Statistical estimation of the numerical solution of a Fredholm integral equation of the first kind. J. Ass. comput. Mach. Vol. 15, pp.100-114.

TIHONOV, A.N. (1963). On the solution of incorrectly posed problems and the method of regularization. Soviet Math. Vol. 4, pp.1035-1038.

TIHONOV, A.N. (1964). Solution of nonlinear integral equations of the first kind. Soviet Math. Vol. 5, pp.835-838.

TIHONOV, A.N. (1967). Methods for the solution of incorrect problems (Russian). Vych. Metody i Program. Vol. 8, pp.3-33.

TURCHIN, V.F., KOZLOV, V.P. and MALKEVICH, M.S. (1971). The use of mathematical-statistics methods in the solution of incorrectly posed problems. Soviet Phys. Usp. Vol. 13, pp.681-702.

TWOMEY, S. (1963). On the numerical solution of Fredholm integral equations of the first kind by the inversion of the linear system produced by quadrature. J. Ass. comput. Mach. Vol. 10, pp.97-101.

TWOMEY, S. (1965). The application of numerical filtering to the solution of integral equations encountered in indirect sensing measurements. J. Franklin. Inst. Vol. 279, pp.95-109.

VAINSTEIN, L.A. (1972a). Filtering of noise in a numerical solution of integral equations of the first kind. Soviet Phys. Dokl. Vol. 17, pp.519-521.

VAINSTEIN, L.A. (1972b). Numerical solution of integral equations of the first kind using a priori information on the function to be determined. Soviet Phys. Dokl. Vol. 17, pp.532-534.

CHAPTER 14 METHODS FOR INTEGRO-DIFFERENTIAL EQUATIONS

C.T.H. Baker
University of Manchester

. Introduction. Volterra Integro-differential Equations

The types of integro-differential equation which can arise are so aried that there is no satisfactory unifying theory. Our purpose here ill be to discuss numerical methods under the assumption that the quation considered has a (locally unique) solution. (Clearly ill-osed problems can arise, as in integral equations, but we shall not onsider such cases.) For all types of equations considered in this hapter, the numerical methods borrow from those developed for integral quations, but they may also be regarded as extensions of methods for he numerical solution of differential equations.

Probably the type of equation to be treated in the literature in ost generality is the integro-differential equation for a function of ne variable. Consider in particular the first-order equation

$$f'(x) = F(x,\ f(x),\ Vf(x)), \tag{1.1}$$

here

$$Vf(x) = \int_0^x K(x,\ t,\ f(t))dt, \tag{1.2}$$

ith an initial condition $f(0) = f_0$. Here V is a Volterra operator, cting on $f(x)$ in (1.1). If we write $Vf(x) = v(x)$ then (1.1) and 1.2) have the appearance of a *coupled* pair of differential-integral quations. Linz (1969) states Lipschitz conditions sufficient to nsure a solution, and quotes a well-known example of (1.1) of the form

$$f'(x) = \alpha f(x) - \beta\ \{f(x)\}^2 + f(x)\int_0^x K(x-y)\ f(y)dy.$$

simple non-linear recurrence relation for the approximate solution of quations (1.1), (1.2) is obtained as follows, where $\tilde{f}(rh)$ is the pproximate value of $f(rh)$,

$$\frac{1}{h}\{\tilde{f}((r+1)h) - \tilde{f}(rh)\} = F(rh,\ \tilde{f}(rh),\ h\ \sum_{j=0}^{r}{}'' K(rh,\ jh,\ \tilde{f}(jh)), \tag{1.3}$$

or $r = 0, 1, 2, \ldots$, with $\tilde{f}(0) = f_0$.

More generally, we can apply a numerical method for the step-by-step solution of the differential equation (1.1) if we approximate $v(x) = Vf(x)$ numerically. Thus Linz (1969) discusses an extension of multistep methods for first-order ordinary differential equations to deal with (1.1), (1.2). If we write $Vf(x) = v(x)$ in (1.1), a multistep method for (1.1) has the form

$$\sum_{r=0}^{k} \alpha_r \tilde{f}(x_{n+r}) = h \sum_{r=0}^{k} \beta_r F(x_{n+r}, \tilde{f}(x_{n+r}), \tilde{v}(x_{n+r})), \tag{1.4}$$

where $x = kh$ for some fixed $h > 0$ and where $\alpha_0, \alpha_1, \ldots, \alpha_k \neq 0, \beta_0, \beta_1, \ldots, \beta_k$ are parameters determining the method. We see that in (1.4) we require values of $\tilde{v}(x_{n+r})$ $(r = 0, 1, \ldots, k)$ and to obtain these we write

$$\tilde{v}(x_{n+r}) = \sum_{j=0}^{n+r} w_{n+r} K(x_{n+r}, x_j, \tilde{f}(x_j)), \; r = 0, 1, \ldots, k, \tag{1.5}$$

using a quadrature rule to replace (1.2) (as in the treatment of Volterra integral equations). If $K(x, t, f(t))$ is badly behaved we may require some modification of (1.5); an ordinary quadrature rule may not suffice.

For a first-order equation the initial condition prescribes only the value of $f(0)$, and (1.4) cannot be used if $k > 1$. Let us suppose for the present that $\tilde{f}(0) = f(0)$, and $\tilde{f}(h), \ldots, \tilde{f}((k-1)h)$ have been obtained by some other means. Then setting $n = 0$ in (1.4) and substituting from (1.5) we have an equation for $\tilde{f}(kh)$, thus

$$\sum_{r=0}^{k} \alpha_r \tilde{f}(rh) = h \sum_{r=0}^{k} \beta_r F(rh, \tilde{f}(rh), \sum_{j=0}^{r} w_{rj} K(rh, jh, \tilde{f}(jh))), \tag{1.6}$$

and the equation is explicit if β_k is zero. If the equation is non-linear and (1.6) is not explicit, it is necessary to solve (1.6) iteratively (by an extension, say, of the predictor-corrector method used for differential equations, see Lambert, 1973). Setting $n=1,2,\ldots$ in (1.4), (1.5) we obtain equations, comparable with (1.6), for $\tilde{f}((k+1)h), \tilde{f}((k+2)h), \ldots$.

The choices of $\alpha_0, \alpha_1, \ldots, \alpha_k, \beta_0, \beta_1, \ldots, \beta_k$ and of the weights w_{rj} determine the numerical properties of the algorithm indicated above. Linz (1969) analyses conditions on these parameters required to extend the theory of Dahlquist (1956) for stability (asymptotic stability as $h \to 0$), and for consistency and convergence. To the extent that the conditions involve $\{\alpha_r\}$, $\{\beta_r\}$ the conditions are the same as for the differential equation arising when $K(x, u, v) \equiv 0$, and sensible choices of these parameters may be obtained from Lambert (1973). For the weights w_{rj} we ask that $|w_{rj}| \leq W$ for all $r, j \leq r$ and that for any continuous function $\phi(x)$

$$\int_0^{rh} \phi(y)dy - \sum_{j=0}^{r} w_{rj}\ \phi(jh) = \theta(h),$$

where $\theta(h) \to 0$ as $h \to 0$ and $rh = x$ is fixed. This condition is satisfied if we construct the weights w_{rj} using a combination of, say, Simpson's rule and the trapezium rule:

	j=0	1	2	3
r=1	$\frac{1}{2}h$	$\frac{1}{2}h$		
r=2	$\frac{1}{3}h$	$\frac{4}{3}h$	$\frac{1}{3}h$	
r=3	$\frac{1}{3}h$	$\frac{4}{3}h$	$\frac{5}{6}h$	$\frac{1}{2}h$

In practice the propagation of errors with fixed $h > 0$ is determined in part by these parameters and there is ample scope for testing practical schemes. The local truncation errors are determined both by the coefficients α_r, β_r and the weights w_{rj}. In the choice of weights w_{rj} we would expect a similar theory to that which holds for Volterra equations of the second kind.

For the purposes of discussion we supposed that for $k > 1$ the values $\tilde{f}(h), \tilde{f}(2h), \ldots, \tilde{f}((k-1)h)$ had been obtained. In practice we require a "starting procedure" to obtain these values, and Linz (1969) suggests a scheme based on Simpson's rule and quadratic interpolation. Alternatively, explicit Runge-Kutta type formulae can be used to provide starting formulae or to achieve the entire numerical solution. In particular the ALGOL procedure of Pouzet (1970; p.201) for the treatment of the equation

$$f'(x) = F(x, f(x), \int_0^x k(x, t, f(t), f'(t))dt)$$

with $f(0) = f_0$ can be applied to the more special equations (1.1), (1.2).

The methods discussed by Feldstein and Sopka (1973) are 'one-step' methods in the special sense that $k = 1$ in the general framework above. As with ordinary differential equations, change of step-size h with $k = 1$ is simpler than with $k > 1$, although the quadrature rule may require modification. (The terminology 'k-step' is not particularly meaningful and does not appear to apply to the method of Mocarsky (1971) who treats the integrated form of (1.1) as an integral equation.) Feldstein and Sopka call their Algorithm 1 a 'p-th order perturbed Taylor algorithm' and a particular case is the Runge-Kutta formula. Their method of investigation appears promising; they are primarily concerned with obtaining classes of methods of a given local truncation error by approximating the terms of a local Taylor expansion for $f(x)$. The true derivatives of $f(x)$ are expressible in terms of integrals which are to be approximated by quadrature, and the order of approximation required decreases as the order of derivative increases.

Let us consider briefly integro-differential equations of higher order, say

$$f^{(p)}(x) = F(x, f(x), \ldots, f^{(p-1)}(x)) + \int_0^x R(x, t, f(t), \ldots, f^{(p)}(t))dt$$

with initial conditions defining $f(0), f'(0), \ldots, f^{(p-1)}(0)$. Pouzet (1960, p.365) presents a Runge-Kutta method for this equation (in which we note the integral term appears linearly). It would further appear that the methods analyzed by Linz (1969) can also be extended to higher order equations, by generalising the way in which a high-order differential equation can be reduced to a first-order system.

There has been little or no investigation in the published literature of the efficient *automatic* implementation of methods for Volterra integro-differential equations. The stability properties of various methods with practical choices of $h > 0$ and the estimation of a suitable step-size h merit further consideration. The issues to be investigated are similar to those for initial-value problems in ordinary differential equations, but in the implementation of a method there is the added complication that all 'past' values of $\tilde{f}(rh)$, $r = 0, 1, \ldots, k-1$ are used in the calculation of $\tilde{f}(kh)$, because they enter the approximation to the integral term.

2. Boundary Value Problems in Integro-differential Equations

In the present context, a boundary value problem is one in which here are two or more constraints on the values of the solution at the nd points a,b of an interval in which the solution $f(x)$ is sought. here is no clear distinction between boundary-value and initial-value roblems, and there is no reason to suppose that the integral in the ntegro-differential equation is over the interval $[a,b]$, since it may n fact reduce to a Volterra integral. (In this case a shooting method ased on the methods of §1 may be feasible.) In general, the numerical ethods for boundary-value problems are extensions of those applied to oundary-value problems in ordinary differential equations and to redholm equations of the second kind. Thus a plausible method is to eplace derivatives by differences and integrals by sums (using uadrature rules), as in Adachi (1956).

Consider for example the equation

$$f''(x) + \lambda\int_0^1 K(x, y)\, f(y)dy = g(x) \quad (0 \leqslant x \leqslant 1), \tag{2.1}$$

ith boundary conditions

$$f(0) = f(1) = 0 \tag{2.2}$$

here $K(x, y)$ and $g(x)$ are continuous $(0 \leqslant x, y \leqslant 1)$. This articular equation can be converted to a Fredholm integral equation of he second kind,

$$f(x) + \lambda\int_0^1 H(x, y)\, f(y)dy = \gamma(x), \tag{2.3}$$

here $H(x, y) = \int_0^1 G(x, y)\, K(z, y)dy$, $\gamma(x) = \int_0^1 G(x, z)\, g(z)dz$, with

$(x, y) = -x(1-y)$ $(0 \leqslant x \leqslant y \leqslant 1)$, $G(y, x) = G(x, y)$ (i.e. $G(x, y)$ is Green's function). The existence of a solution $f(x)$ is assured for ll but a countable set of values of λ, the characteristic values of (x, y). (In practice the existence of a Green's function may be known ut its form may be difficult to obtain, so that this approach provides heoretical insight only.)

A direct finite-difference approach to (2.1)-(2.2) using central ifferences and the trapezium rule, with a step $h = {}^1/n$, provides pproximations satisfying the equations

$$\left.\begin{aligned} &\tilde{f}(0) = 0, \\ &\frac{1}{h^2}\{\tilde{f}((i-1)h) - 2\tilde{f}(ih) + \tilde{f}((i+1)h)\} \\ &\qquad + \lambda h \sum_{j=0}^{n}{}'' K(ih, jh)\tilde{f}(jh) = g(ih) \\ &\qquad (i = 1, 2, \ldots, n-1), \\ &\tilde{f}(nh) = 0. \end{aligned}\right\} \qquad (2.4)$$

In vector-matrix notation we can write

$$(h^{-2}\, T + \lambda h\, K)\underline{\tilde{f}} = \underline{g} \qquad (2.5)$$

where (in the present case) $\underline{\tilde{f}} = [\tilde{f}(h), \tilde{f}(2h), \ldots, \tilde{f}(1-h)]^T$,
$\underline{g} = [g(h), g(2h), \ldots, g(1-h)]^T$,

$$K = \begin{bmatrix} K(h,h), & K(h,2h), \ldots, & K(h,1-h) \\ K(2h,h), & K(2h,2h), \ldots, & K(2h,1-h) \\ \cdot & \cdot & \cdot \\ \cdot & \cdot & \cdot \\ \cdot & \cdot & \cdot \\ \cdot & \cdot & \cdot \\ K(1-h,h), & K(1-h,2h), \ldots, & K(1-h,1-h) \end{bmatrix}, \quad T = \begin{bmatrix} -2 & 1 & & & \\ 1 & -2 & 1 & & \\ & 1 & -2 & 1 & \\ & \cdot & \cdot & \cdot & \\ & & \cdot & \cdot & \cdot \\ & & \cdot & \cdot & 1 \\ & & & 1 & -2 \end{bmatrix}$$

Now the inverse of $h^{-2}T$ is the matrix hG with entries $hG(ih, jh)$ $(1 \leq i, j \leq n-1)$ so that

$$(I + \lambda h^2 GK)\underline{\tilde{f}} = hG\underline{g}, \qquad (2.6)$$

and hence the method we propose is equivalent to using the trapezium rule in the quadrature method to solve (2.3), having first replaced $\int_0^1 G(x, z)\, K(z, y)dz$ by $h \sum_{j=0}^{n}{}'' G(x, jh)\, K(jh, y)$ and $y(x)$ by $h \sum_{j=0}^{n}{}'' G(x, jh)\, g(jh)$. The solvability of the system when λ is not a characteristic value of $H(x, y)$ is assured for h sufficiently small, and the convergence of the approximate solution as $h \to 0$ can also be established, using approximation theory for integral equations.

Let us consider the more general implications of the preceding example, without being too rigorous. (For rigour, see also Vainikko,

967.) Suppose that $\mathcal{D}\{\ \}$ is a *linear* differential operator and we seek the solution of

$$\mathcal{D}\{f(x)\} + \int_a^b F(x, y; f(y))dy = 0 \tag{2.7}$$

with linear boundary conditions

$$\mathcal{B}\{f(x)\} = 0. \tag{2.8}$$

The boundary conditions will thus have the form

$$(d_i f)(a) + (\delta_i f)(b) = 0 \quad (i = 1, 2, \ldots, k)$$

where d_i and δ_i are (differential) operators acting on $f(x)$.

If there is a Green's function $G(x, y)$ for the problem $\mathcal{D}\{\phi(x)\} = 0$, and $\mathcal{B}\{\phi(x)\} = 0$ and if the equation

$$f(x) + \int_a^b \int_a^b G(x, z)F(z, y; f(y))dzdy = 0$$

has a unique solution, then the boundary value problem for the integro-differential equation has a unique solution. (The Green's function is defined so that if $\mathcal{D}\{f(x)\} = \phi(x)$, $\mathcal{B}\{f(x)\} = 0$, then $f(x) = \int_a^b G(x, y)\phi(y)dy$.) Suppose for our numerical scheme that we approximate the values of $\mathcal{D}\{f(x)\}$ at $x = ih$, $i = 1, 2, \ldots, n-1$ $(h = 1/n)$, using a difference scheme which is strongly stable (Stetter, 1973) so that the inverse of the resulting matrix "approximates" the integral operator with the Green's function $G(x, y)$ as kernel. Suppose also that we approximate the integral by a convergent quadrature rule using values of the integrand at equal step h. Then the resulting scheme is convergent as $h \to 0$. Methods of deferred approach to the limit and of deferred correction can also be applied, given sufficient differentiability.

In our treatment of (2.1)-(2.2) we chose the trapezium rule to replace the integral, to match the $O(h^2)$ local truncation error of the difference approximation to $f''(x)$. Since the matrix $h^{-2}T + h\lambda K$ is in any case a full matrix, we can (with little additional computational expense) use as many of the values $\tilde{f}(0), \tilde{f}(h), \ldots, \tilde{f}(nh)$ as required to approximate the derivative term to higher order accuracy, and it is then reasonable to employ a higher-order quadrature rule.

In a practical computation along similar lines to those above, Robertson (1956) considers the numerical solution of the homogeneous equation

$$f''(x) + \left\{k^2 - \frac{p(p+1)}{x^2} + V(x)\right\} f(x) = \int_0^\infty K(x,y)f(y)dy \quad (0 < x < \infty), \qquad (2.9)$$

$$f(0) = 0.$$

In this equation the integral is over an infinite range and for computational purposes Robertson replaces it by an integral on $[0, X]$ (X "large"). The integral is then approximated by the Lagrangian form of a Gregory formula, and the central-difference approximation

$$f''(x) \simeq \{f(x-h) - 2f(x) + f(x+h)\}/h^2$$

is used to produce a computational scheme, with deferred correction also using central differences.

3. Expansion Methods for Integro-differential Equations

If we seek an approximation of the form

$\tilde{f}(x) = \sum_{r=0}^{n} \tilde{a}_r \phi_r(x)$ to our integro-differential equation (equation (2.7) say) the expansion methods of Chapter 7 may be readily adapted. As in the treatment of ordinary differential equations (Collatz, 1960, §4.1) there is now slightly more flexibility.

We obtain (i) an "interior method" if the functions $\phi_r(x)$ $(r = 0, 1, \ldots, n)$ are chosen so that $\tilde{f}(x)$ automatically satisfies the boundary conditions. In particular, when the boundary conditions are linear this is achieved if each of the functions $\phi_r(x)$ satisfies the boundary conditions. We obtain (ii) a "boundary method" in those (rare) circumstances where $\tilde{f}(x)$ automatically satisfies the integro-differential equation but not necessarily the boundary conditions. We have (iii) a "mixed method" if $\tilde{f}(x)$ is not constrained to satisfy either the boundary conditions or the equation.

If we treat (2.1)-(2.2) as an example, substitution of $\tilde{f}(x)$ in the equation and boundary conditions gives, for $0 \leq x \leq 1$,

$$\eta(x) = \tilde{f}''(x) + \lambda\int_0^1 K(x,y)\tilde{f}(y)dy - g(x), \qquad (3.1)$$

and the errors in the boundary conditions are

$$\varepsilon_0 = \tilde{f}(0) - 0, \quad \varepsilon_1 = \tilde{f}(1) - 0, \tag{3.2}$$

all as functions of $\tilde{a}_0, \tilde{a}_1, \ldots, \tilde{a}_n$. For convenience, we assume that $\phi_r''(x)$ exists for $r = 0, 1, \ldots, n$. If we write $K\phi_r(x) = \psi_r(x)$, then

$$\eta(x) = \sum_{r=0}^{n} \tilde{a}_r\{\phi_r''(x) + \lambda\psi_r(x)\} - g(x).$$

Referring to the classification (i)-(iii) given above, in the interior method $\phi_r(a) = \phi_r(b) = 0$ and we obtain approximations $\tilde{f}(x)$ satisfying the boundary conditions if we choose $\tilde{a}_0, \tilde{a}_1, \ldots, \tilde{a}_n$ to ensure that $\eta(x)$ in (3.1) is "small"; we obtain a collocation method if we require $\eta(z_i) = 0$ $(i = 0, 1, \ldots, n)$ where $a \leqslant z_0 < z_1 < \ldots < z_n \leqslant b$, and we obtain a classical Galerkin method if we require $\int_a^b \eta(x)\, \overline{\phi_r(x)}dx = 0$, $r = 0, 1, \ldots, n$. In the example of (2.1)-(2.2) it is readily seen that the Galerkin method is equivalent to a method of moments applied to (2.3) using functions

$\chi_r(x) = \int_0^1 G(x, y)\, \phi_r(y)dy$ (see Chapter 7).

In the method of collocation we seek $\tilde{f}(x)$ such that $\varepsilon_0 = \varepsilon_1 = 0$ and $\eta(\xi_i) = 0$ for $i = 0, 1, \ldots, n-2$, $a < \xi_0 < \xi_1 < \ldots < \xi_{n-2} < b$. For a mixed Galerkin method a number of generalizations of the interior Galerkin method could be suggested but we would not advocate them in the present state of our knowledge.

Rayleigh-Ritz methods can be developed for certain integro-differential equations. The construction of these methods proceeds differently from the "error distribution principles" used to obtain the collocation and Galerkin methods. To study an example we consider the equation

$$\gamma^2 f''(x) = \int_a^b (K^*K)(x, y)f(y)dy - \int_a^b K^*(x,y)g(y)dy, \tag{3.3}$$

with boundary conditions

$$f'(a) = f'(b) = 0 \tag{3.4}$$

and with $\gamma > 0$. The boundary-value problem (3.3)-(3.4) arises in a method for regularizing a Fredholm equation of the first kind, and we have written

$$(K^*K)(x, y) = \int_a^b K^*(x, z)K(z, y)dz, \qquad (3.5)$$

where $K^*(x, y) = \overline{K(y, x)}$. We suppose that $K(x, y)$ is continuous for $a \leqslant x, y \leqslant b$ and that there is no non-trivial continuous function $\psi(x)$ such that $\int_a^b K(x, y)\, \psi(y)dy = 0$.

If we consider the functional

$$J[\psi] = \gamma^2\int_a^b |\psi'(x)|^2dx + \int_a^b |K\psi(x) - g(x)|^2dx, \qquad (3.6)$$

where $K\psi(x) = \int_a^b K(x, y)\, \psi(y)dy$, we can show that $J[\psi]$ attains a stationary value (a minimum) as $\psi(x)$ varies in $C^2[a,b]$. For, consider $\psi(x)$, $\phi(x)$ in $C^2[a,b]$ and $\varepsilon = \zeta + i\eta$ where ζ, η are real. Then $J[\psi + \varepsilon\phi]$ has a stationary value at $\psi(x)$ if

$$(\partial/\partial\zeta)J[\psi + \varepsilon\phi]_{\varepsilon=0} = (\partial/\partial\eta)J[\psi + \varepsilon\phi]_{\varepsilon=0} = 0 \qquad (3.7)$$

for all $\phi(x) \in C^2[a,b]$. Now,

$$J[\psi + \varepsilon\phi] = J[\psi] + \varepsilon\gamma^2\int_a^b \overline{\psi'(x)}\phi'(x) + \bar{\varepsilon}\gamma^2 \int_a^b \psi'(x)\overline{\phi'(x)}dx$$

$$+ \varepsilon\int_a^b \overline{(K\psi(x)-g(x))}\, K\phi(x)dx$$

$$+ \bar{\varepsilon}\int_a^b (K\psi(x)-g(x))\, \overline{K\phi(x)}dx + O(|\varepsilon|^2).$$

In compact notation,

$$J[\psi + \varepsilon\phi] = J[\psi] + (\zeta + i\eta)\{\gamma^2(\phi', \psi') + (\phi, K^*K\psi - K^*g)\}$$

$$+ (\zeta - i\eta)\{\gamma^2(\psi', \phi') + (K^*K\psi - K^*g, \phi)\} + O(|\varepsilon|^2),$$

so that

$$(\partial/\partial\zeta)J[\psi + \varepsilon\phi]_{\varepsilon=0} = 2\mathrm{Rl}\{(\psi',\phi')\gamma^2 + (K^*K\psi - K^*g, \phi)\}$$

and

$$(\partial/\partial\eta)J\left[\psi + \varepsilon\phi\right]_{\varepsilon=0} = -2\mathrm{Im}\{(\psi',\phi')\gamma^2 + (K^*K\psi - K^*g, \phi)\}.$$

If both of these last terms vanish for all $\phi(x) \in C^2[a,b]$ then

$$\gamma^2(\psi', \phi') + (K^*K\psi - K^*g, \phi) = 0 \tag{3.8}$$

for all such $\phi(x)$. Now

$$(\psi', \phi') \equiv \int_a^b \psi'(x)\overline{\phi'(x)}dx = \left[\psi'(x)\overline{\phi(x)}\right]_{x=a}^{x=b} - \int_a^b \psi''(x)\overline{\phi(x)}dx$$

so that the condition (3.8) can be written

$$\gamma^2\left[\psi'(x)\overline{\phi(x)}\right]_{x=a}^{x=b} - \int_a^b \eta(x)\overline{\phi(x)}dx = 0 \quad (\forall\, \phi(x) \in C^2[a,b]), \tag{3.9}$$

where

$$\eta(x) = \gamma^2\psi''(x) - K^*K\psi(x) + K^*g(x). \tag{3.10}$$

If $\psi(x)$ satisfies the integro-differential equation, $\eta(x) = 0$, while if $\psi(x)$ satisfies the boundary conditions, $\psi'(a) = \psi'(b) = 0$, and (3.9) vanishes for all $\phi(x) \in C^2[a,b]$. Thus for $\psi(x) = f(x)$, $J[\psi]$ has a stationary value. (It is not difficult to show that there is only one stationary value which is necessarily a minimum.)

Observe that since $\phi'(a)$, $\phi'(b)$ are unconstrained, the function $\psi(x) + \varepsilon\phi(x)$ is not required to satisfy the boundary conditions (3.4) while the stationary value is being sought. Such boundary conditions are said to be suppressible or 'natural', whilst any boundary conditions which must be satisfied by the function in the variational functional are called essential.

For further example, in which some of the boundary conditions are essential, consider the equation (assumed real)

$$f^{iv}(x) - f''(x) - f(x) + \lambda\int_{-1}^{1} f(y)dy = 0$$

with the boundary conditions

$$f(1) = f(-1) = f''(1) = f''(-1) = 0.$$

A corresponding variational problem is to find a stationary value of

$J[\psi]$ subject to the conditions $\psi(1) = \psi(-1) = 0$, where

$$J[\psi] = \int_{-1}^{1} \{[\psi''(x)]^2 + [\psi'(x)]^2 - [\psi(x)]^2\}dx + \lambda\left\{\int_{-1}^{1} \psi(x)\,dx\right\}^2. \quad (3.11)$$

(Here the stationary value is again an extreme value.)

Collatz (1960, p.489) provides a fairly general approach to the task of relating an integro-differential equation to a variational problem. The integro-differential equations considered have, in general, derivatives of even order $2n$ and $2n$ boundary conditions.

When the integro-differential equation and its boundary conditions have been expressed as a solution of a stationary value problem for a functional $J[\psi]$ subject to certain essential boundary conditions, a variational method proceeds in the usual way. Thus we find a function $\tilde{f}(x) = \sum_{r=0}^{n} \tilde{a}_r \phi_r(x)$ which is required to lie in the class of functions in which $\psi(x)$ varies and to satisfy the essential boundary conditions of the variational problem. Among such functions $\tilde{f}(x)$, we choose one which gives $J[\tilde{f}]$ a stationary value. When the boundary conditions of the variational formulation are linear this is most easily achieved by choosing $\phi_r(x)$ $(r = 0, 1, \ldots, n)$ to satisfy these boundary conditions, and solving the equations $(\partial/\partial\tilde{a}_r)\, J[\tilde{f}] = 0$ $(r = 0, 1, \ldots, n)$. We then take the corresponding function $\tilde{f}(x)$ as an approximation to $f(x)$.

4. Partial Integro-differential Equations

There is a wide variety of types of partial integro-differential equations, most of which are virtually untreated in the literature. In this section, we discuss some particular examples, to indicate the types of method which have been used in practice.

(i) The first example is a 'simple' first order equation

$$\left.\begin{aligned} \mu\frac{\partial\psi(t,\mu)}{\partial t} &= \psi(t,\mu) - \tfrac{1}{2}\int_{-1}^{1} \psi(t,\nu)\,d\nu - e^t \\ &\qquad (0 \leqslant t \leqslant 1,\ |\mu| \leqslant 1), \end{aligned}\right\} \quad (4.1)$$

with boundary conditions $\psi(0, \mu) = 0$ $(\mu < 0)$, $\psi(1, \mu) = 0$ $(\mu > 0)$, treated by Anselone (1971) to exemplify the discrete-ordinates method, apparently due originally to Wick and Chandrasekhar (after whom it is sometimes named). References in the literature are also found under "Discrete P_{n-1}-method". This equation can be solved by finding the solution of a linear integral equation

$$f(x) - \tfrac{1}{2}\int_0^1 E_1(|x-y|)\, f(y)\,dy = e^{-x},$$

giving

$$\psi(t, \mu) = -\int_0^t \frac{e^{(t-\tau)/\mu}}{\mu} f(\tau)\,d\tau, \quad \mu < 0, \tag{4.2}$$

with a similar expression for $\mu > 0$ (Anselone, 1971). ($E_1(z)$ is the exponential integral function, with a logarithmic singularity at $z = 0$.) However, a direct numerical treatment is available if we use a quadrature rule

$$\int_{-1}^1 \phi(\nu)\,d\nu \simeq \sum_{j=0}^n w_j\, \phi(\nu_j) \quad (\nu_j \in [-1, 1]) \tag{4.3}$$

to replace the integral in (4.1). We obtain a system of ordinary differential equations for the functions $f_i(t) = \tilde{\psi}(t, \nu_i)$, where (for $|\mu| \leq 1$, $t \in [0, 1]$)

$$\mu \frac{d}{dt}\tilde{\psi}(t, \mu) = \tilde{\psi}(t, \mu) - \tfrac{1}{2}\sum_{j=0}^n w_j\, \tilde{\psi}(t, \nu_j) - e^{-t} \tag{4.4}$$

and $\tilde{\psi}(0, \mu) = 0$ if $\mu < 0$, $\tilde{\psi}(1, \mu) = 0$ if $\mu > 0$.

We set $\mu = \nu_i (i = 0, 1, \ldots, n)$ to obtain the system of $(n + 1)$ differential equations, which can be solved numerically for the values of $f_i(t) = \tilde{\psi}(t, \nu_i)$. To define the values $\tilde{\psi}(t, \nu)$ for all $|\nu| \leq 1$ we can perform conventional interpolation or use a discrete analogue of (4.2) (Anselone, 1971). The quadrature rule (4.3) used originally in the literature was the Gauss-Legendre rule, but repeated Gauss rules have been used with greater success for certain similar equations.

(ii) The equation (4.1) above arises from a relatively simple transport problem. Integro-differential equations arise fairly frequently in the theory of scattering, transfer and transport problems in physics. For example, a first-order equation which has received considerable attention is the neutron transport equation, which in cartesian co-ordinates has the form

$$\left.\begin{aligned} &(\partial/\partial t + v_x\, \partial/\partial x + v_y\, \partial/\partial y + v_z\, \partial/\partial z)\, \psi(\underset{\sim}{x}, \underset{\sim}{v}, t) \\ &\quad = -v\sigma(v)\, \psi(\underset{\sim}{x}, \underset{\sim}{v}, t) + \int_{-\infty}^{\infty}\int_{-\infty}^{\infty}\int_{-\infty}^{\infty} v'\sigma(v')\, K(\underset{\sim}{v}, \underset{\sim}{v}') \\ &\qquad \times \psi(\underset{\sim}{x}, \underset{\sim}{v}', t)\, dv'_x\, dv'_y\, dv'_z, \end{aligned}\right\} \tag{4.5}$$

with suitable boundary conditions (Richtmyer and Morton, 1967, p.219ff). Equation (4.5) reduces to simpler forms on making special assumptions (e.g. that of symmetry). For a discussion in the literature see, for example, Richtmyer and Morton (1967), Clark and Hansen (1964), Bennett (1964) and his references, Alder, Fernbach and Rotenberg (1963), Underhill and Russell (1962), the PICC proceedings of 1960, and, more recently, the SIAM-AMS proceedings on "Transport Theory" (1969), and Erdos, Haley, Marti, and Mennig (1970).

If we consider the transport equation (4.5) for a slab with symmetry and isotropic scattering we can obtain (Richtmyer and Morton, 1967, p.233) for $|z| \leqslant a,\ t \geqslant 0$ the equation

$$\left(\frac{1}{v}\frac{\partial}{\partial t} + \mu\frac{\partial}{\partial z} + \sigma\right)\Psi(z, \mu, t) = \frac{\sigma(1+g)}{2}\int_{-1}^{1}\Psi(z, \nu, t)\,d\nu, \qquad (4.6)$$

with boundary conditions $\Psi(z, \mu, t) = 0$ for $z = a,\ \mu < 0$ and for $z = -a,\ \mu > 0$. Compared with (4.1), (4.6) is complicated by the partial derivatives with respect to both z and t. In the discrete-ordinates method for (4.6) we now write

$$\left(\frac{1}{v}\frac{\partial}{\partial t} + \mu\frac{\partial}{\partial z} + \sigma\right)\tilde{\Psi}(z, \mu, t) = \frac{\sigma(1+g)}{2}\sum_{j=0}^{n} w_j\,\tilde{\Psi}(z, \nu_j, t),$$

and setting $\mu = \nu_i$ $(i = 0, 1, \ldots, n)$

$$\left(\frac{1}{v}\frac{\partial}{\partial t} + \nu_i\frac{\partial}{\partial z} + \sigma\right)f_i(z, t) = \frac{\sigma(1+g)}{2}\sum_{j=0}^{n} w_j\,f_j(z, t), \qquad (4.7)$$

where $f_i(z, t) = \tilde{\Psi}(z, \nu_i, t)$ and

$$f_i(z, t) = 0 \quad \text{if} \quad z = a,\ \nu_i < 0,$$

$$f_i(z, t) = 0 \quad \text{if} \quad z = -a,\ \nu_i > 0,$$

(If $\nu_i = 0$ the term $(\partial/\partial z)\,f_i(z, t)$ is missing in (4.7).)

The system (4.7) can be written in vector-matrix notation as

$$\left(\frac{1}{v}\frac{\partial}{\partial t} + M\frac{\partial}{\partial z} + \sigma\right)\underline{f} = \sigma(1+g)\ S\underline{f} \qquad (4.8)$$

where $\underline{f} = [f_0(x, t), f_1(x, t), \ldots, f_n(x, t)]^T$ and S, M are matrices, M being diagonal.

The system (4.7) is a coupled system of linear first-order partial differential equations with constant coefficients, and Richtmyer and

Morton (1967, p.237ff) discuss a variety of difference schemes for the approximate solution of this system; with the exception of a scheme which they attribute to Carlson, there are stability conditions to be imposed on the mesh-steps of the discretization. ("Carlson's scheme" is stable for any ratio of mesh-steps in t and z and though the scheme is implicit it appears to be easily solved.)

For the transport equation of a spherical system with isotropic scattering, the transport equation has a term $\frac{\partial}{\partial\mu}\psi(\;)$ which is to be discretized. The "S_n-method" of Carlson for treating this case is also discussed by Richtmyer and Morton (1967, p.244).

(iii) We next consider the related equation

$$\left(\frac{1}{v}\frac{\partial}{\partial t} + \mu\frac{\partial}{\partial z} + \sigma\right)\Psi(z, \mu, t) = \frac{\sigma(1+g)}{2}\int_{-1}^{1}\Psi(z, \nu, t)\,d\nu . \qquad (4.9)$$

In the "spherical harmonic method" (Richtmyer and Morton 1967, p.244) we seek an approximate solution of the form

$$\tilde{\Psi}(z, \mu, t) = \sum_{r=0}^{n}\sqrt{r+\tfrac{1}{2}}\,P_r(\mu)\,\phi_r(z, t). \qquad (4.10)$$

We proceed somewhat differently from Richtmyer and Morton to demonstrate a general principle. If we substitute $\tilde{\Psi}$ in the equation for Ψ we obtain a residual $\eta = \eta(z, \mu, t)$ and we can choose the functions $\phi_r(z, t)$ so that

$$\sqrt{s+\tfrac{1}{2}}\int_{-1}^{1}\eta(z, \mu, t)\,P_s(\mu)\,d\mu = 0$$

for $s = 0, 1, \ldots, n$. Thus, we apply a classical Galerkin method in the μ-direction. The equations derived in this way are given below; we obtain these results using the orthogonality properties of the functions $\sqrt{r+\frac{1}{2}}\,P_r(\mu)$ for $\mu \in [-1, 1]$, $r = 0, 1, \ldots$. For the equation (4.9) above we obtain

$$\frac{1}{v}\frac{\partial}{\partial t}\phi_s(z, t) + \alpha_s\frac{\partial}{\partial z}\phi_{s-1}(z, t) + \beta_s\frac{\partial}{\partial t}\phi_{s+1}(z, t) \qquad (4.11)$$

$$+ \sigma\phi_s(z, t) = \sigma(1 + g)\,\delta_{s0}\,\phi_0(z, t) \quad (s = 0, 1, \ldots, n)$$

where δ_{s0} is the Kronecker delta and α_s, β_s are simple rational functions of s (arising from the relations amongst the Legendre polynomials). This system reduces to the form (4.8). For a slightly different equation of a similar type see Richtmyer and Morton (p.234).

If we solve the coupled system of first order equations (4.11) numerically (Richtmyer and Morton p.228ff.) we may construct $\tilde{\Psi}(z, \mu, t)$ in (4.10). Stability of the numerical method is again an important factor.

The spherical harmonic method is equivalent for the equation above to the method of discrete ordinates, for which convergence proofs are available in the literature. A similar approach to that employed for the spherical harmonic method in the treatment of the transport equation has been employed (Buckingham and Burke, 1960) in the treatment of a second-order partial integro-differential equation. The equation is thus reduced to an approximating ordinary integro-differential equation of second-order, of the type considered by Robertson (1956).

(iv) To conclude this section we note that a second-order "parabolic" integro-differential equation

$$\frac{\partial^2\psi(x, t)}{\partial x^2} = F(x, t, \psi, \frac{\partial\psi}{\partial x}, \frac{\partial\psi}{\partial t}, v(x, t)) \tag{4.12}$$

with $v(x, t) = \int_0^t K(x, t, \tau, \psi(x, \tau), \frac{\partial}{\partial x}\psi(x, \tau))d\tau$

where $(\partial/\partial u_3)\, F(x, t, u_1, u_2, u_3, u_4) \geq \rho > 0,$

and its "hyperbolic" analogue

$$\frac{\partial^2\psi(x, t)}{\partial t^2} - a(x, t)\frac{\partial^2\psi(x, t)}{\partial x^2} = F(x,t,\psi, \frac{\partial\psi}{\partial x}, \frac{\partial\psi}{\partial t}, v(x,t)) \tag{4.13}$$

are considered by Douglas and Jones (1962). Douglas and Jones consider the analogues of various methods for the corresponding partial differential equations and establish theoretical results concerning the convergence of these methods.

References

ADACHI, R. (1956). On the numerical solutions of some integro-differential equations under some conditions. Kumamoto J. Sci. (ser. A) Vol. 2, pp. 322-335.

ALDER, B., FERNBACH, S. and ROTENBERG, M. (1963). Methods in Computational Physics, Vol. 1 - Statistical Physics. Academic Press.

ANSELONE, P.M. (1971). Collectively Compact Operator Approximation Theory, and Applications to Integral Equations. Prentice-Hall.

BENNETT, J.H. (1964). Integral equation methods for transport problems... Some numerical results. Num. Math. Vol. 6, pp. 49-54.

BUCKINGHAM, R.A. and BURKE, P.G. (1960). The solution of integro-differential equations occurring in nuclear collision problems see PICC Proceedings (1960), pp. 458-475.

CLARK, M. and HANSEN, K.F. (1964). Numerical Methods of Reactor Analysis. Academic Press.

C.N.R.S. (1970). Procedures ALGOL en Analyse Numérique. Vol. II. Centre National de la Recherche Scientifique: Paris.

COLLATZ, L. (1960). The Numerical Treatment of Differential Equations (transl. P.G. Williams) Third edition (1966) Springer-Verlag.

CRYER, C.W. (1972). Numerical methods for functional differential equations. Park City Conference on Differential Equations (Proceedings to appear).

DAHLQUIST, G. (1956). Convergence and stability in the numerical integration of ordinary differential equations. Mathematica scand. Vol. 4, pp. 33-53.

DAY, J.T. (1966). Note on the numerical solution of integro-differential equations. Comput. J. Vol. 9, pp. 394-395.

DOUGLAS, J. and JONES, F. (1962). Numerical methods for integro-differential equations of parabolic and hyperbolic type. Num. Math. Vol. 4, pp. 96-102.

ERDOS, P., HALEY, S.B., MARTI, J.T. and MENNIG, J. (1970). A new method for the solution of the transport equation in slab geometry. Jnl. comput. Phys. Vol. 6, pp. 29-55.

FELDSTEIN, A. and SOPKA, J.R. (1973). Numerical methods for nonlinear Volterra integro-differential equations. Report LA-UR-73-85, Los Alamos Scientific Laboratory, Los Alamos (N.M.) USA.

FOX, L. (1962). Numerical Solution of Ordinary and Partial Differential Equations. Pergamon.

KELLER, H.B. and WENDROFF, B. (1957). On the formulation and analysis of numerical methods for time-dependent transport equations. Communs. pure appl. Math. Vol. 10, pp. 567-582.

LAMBERT, J.D. (1973). Computational Methods in Ordinary Differential Equations. Wiley.

LINZ, P. (1969). Linear multistep methods for Volterra integro-differential equations. J. Ass. comput. Mach. Vol. 16, pp. 295-301.

LINZ, P. A method for the approximate solution of linear integro-differential equations. SIAM J. numer. Anal. (to appear).

MOCARSKY, W.L. (1971). Convergence of step-by-step methods for non-linear integro-differential equations. J. Inst. Math. & its Appl. Vol. 8, pp. 235-239.

PHILLIPS, G.M. (1970). Analysis of numerical iterative methods for solving integral and integro-differential equations. Comput. J. Vol. 13, pp. 297-300.

PICC Proceedings (1960). Symposium on the Numerical Treatment of Ordinary Differential Equations, Integral and Integro-differential Equations. Proceedings of the Rome Symposium, Provisional International Computation Centre. Birkhauser-Verlag.

POUZET, P. (1960). Méthode d'integration numérique des équations intégrals et intégro-differentielles du type de Volterra de seconde espéce. Formules de Runge-Kutta. see PICC Proceedings (1960) pp. 362-368.

POUZET, P. (1970). see C.N.R.S. (1970), pp. 201-204.

ROBERTSON, H.H. (1956). Phase calculations for nuclear scattering on the pilot Ace. Proc. Camb. phil. Soc. Vol. 53, pp. 538-545.

RICHTMYER, R.D. and MORTON, K.W. (1967). Difference Methods for Initial Value Problems. Interscience.

SIAM-AMS Proceedings (1969). "Transport Theory" - Volume 1 of SIAM-AMS Proceedings (ed. R. Bellman, G. Birkhoff, and I. Abu-Shumays) American Math. Soc., Providence R.I.

STETTER, H.J. (1973). Analysis of Discretization Methods for Ordinary Differential Equations. Springer-Verlag, Berlin.

TAVERNINI, L. (1969). Numerical methods for Volterra functional differential equations. Ph.D. Thesis, University of Wisconsin.

UNDERHILL, L.H. and RUSSELL, L.M. (1962). The linear transport equation in one and two dimensions. see Fox (1962) pp. 398-422.

VAINIKKO, G.M. (1967) Galerkin's perturbation method and the general theory of approximate methods for non-linear equations. USSR Comp. Math. and Math. Phys. Vol. 7, pp.1-41.

CHAPTER 15 THEORY OF NONLINEAR INTEGRAL EQUATIONS

L.B. Rall
University of Wisconsin

1. The Theory of Nonlinear Integral Equations from the Standpoint of Functional Analysis

To be perfectly correct, there is no such thing as *the* theory of nonlinear integral equations; instead, one has a number of theories which apply more or less completely to special classes of equations. Because of the importance of nonlinear integral equations in applications, it is important for future progress to identify the types of equations of greatest interest, and to develop theories for them if not already available. In the following, attention will be devoted to a description of some of the more elementary theories of nonlinear integral equations. Further developments, proofs of theorems, and applications may be found in the works by Anselone (1964), Collatz (1971), Krasnosel'skii (1963), Krasnosel'skii, Vainikko, Zabreiko, Rutitskii and Stetsenko (1972), Rall (1969), and Vainberg (1964).

If A is a given operator which maps a set X into a set Y, and y in Y is given, then a *theory* of the equation

$$A(x) = y \tag{1.1}$$

will assert either

$$\left.\begin{array}{ll} & \text{(i) existence and uniqueness} \\ \text{or} & \text{(ii) nonexistence or nonuniqueness} \end{array}\right\} \tag{1.2}$$

of solutions x^* in X. For generality, one would want to develop such theories to hold for classes of operators A, and on specified subsets of X and Y. In particular, a characterization of the set of solutions in the case of nonuniqueness is desirable. This alternative structure of the theory of equations is completely developed for nonsingular linear integral equations. For nonlinear integral equations, the discussion will be confined for the most part to circumstances in which alternative (i) of (1.2) can be shown to hold in some subset of X.

In the study of integral equations, the sets X and Y will be taken to be subsets of a Banach space of functions having suitable continuity or integrability properties (see Chapter 4), and A will be an *integral operator* as defined by

$$A(x)(s) = \int_R f(s, t, x(s), x(t))dt, \tag{1.3}$$

where s, t are m-dimensional variables, R is a region in m-dimensional space, and dt is the element of volume. If the region of integration R in (1.3) depends on the space variable s, then equation (1.1) is said to be of <u>Volterra type</u>; otherwise, of <u>Fredholm type</u> (R fixed). In linear spaces, equations of the form (1.1) can be classified as <u>linear</u> or <u>nonlinear</u>, according as A is a linear operator or not. It will be assumed that A is nonlinear for the present purposes. The introduction of functional analysis (that is, analysis on normed linear spaces) allows one to apply general results for equations of the form

$$P(x) = 0, \tag{1.4}$$

or <u>fixed point problems</u>

$$x = F(x) \tag{1.5}$$

directly to integral equations. Many of these results have their counterparts in ordinary scalar analysis, and a number of instances in which the theory of a nonlinear integral equation can be reduced to the theory of a single scalar equation will be cited.

In order to maintain a connection between classical and numerical analysis and functional analysis, attention will be restricted for the most part to operators of <u>second kind</u>,

$$A(x)(s) = x(s) - \int_R f(s, t, x(s), x(t))dt, \tag{1.6}$$

one-dimensional variables s, t, with $R = [0, s]$ in the Volterra case, and $R = [0, 1]$ in the Fredholm case. Furthermore, the Banach spaces considered will be those of continuous functions, with the norm

$$\|x\| = \max_R |x(s)|. \tag{1.7}$$

Particular examples of integral operators of second kind which occur frequently in practice are the <u>Volterra operator</u>

$$V(s)(s) = x(s) - \int_0^s f(s, t, x(t))dt, \tag{1.8}$$

the <u>Urysohn operator</u>,

$$W(x)(s) = x(s) - \int_0^1 f(s, t, x(t))dt, \qquad (1.9)$$

and the Hammerstein operator,

$$H(x)(s) = x(s) - \int_0^1 K(s, t)f(t, x(t))dt. \qquad (1.10)$$

The operator (1.10) is evidently a special case of (1.9). By defining $f(s, t, u) = 0$ for $t > s$, one may also consider the Volterra operator (1.8) to be a special case of the Urysohn operator (1.9). However, the theory and applications of these operators are sufficiently diverse to merit distinction. As the theory of these restricted classes of equations will be discussed in the language of functional analysis, most of the results obtained apply immediately to more general problems.

2. The Classical Schmidt-Lichtenstein Theory

An example of a classical theory of nonlinear integral equations is the one put forward by Schmidt (1907), and further developed and applied by Lichtenstein (1931). From the standpoint of functional analysis, this theory requires $P(x)$ to have a uniformly convergent power series expansion

$$P(x) = P(x_0) + P'(x_0)(x-x_0) + \tfrac{1}{2} P''(x_0)(x-x_0)(x-x_0) + \ldots \qquad (2.1)$$

in some open ball $\mathcal{U}(x_0, r) = \{x: \|x-x_0\| < r\}$ of positive radius. In (2.1), $P^{(k)}(x_0)$ denotes the kth Fréchet derivative of P at x_0; this will be a multilinear operator of order k, and

$$P^{(k)}(x_0)\ \overline{(x-x_0)\ \overset{k}{\ldots}\ (x-x_0)}$$

is the result of operating k times on the point $x-x_0$. Supposing that the equation $P(x) = 0$ has a solution x^* in $\mathcal{U}(x_0, r)$ and the linear operator $P'(x_0)$ has the inverse $[P'(x_0)]^{-1}$, then (2.1) may be written in the form

$$x^*-x_0 = y_0 + A_2(x^*-x_0)(x^*-x_0) + A_3(x^*-x_0)(x^*-x_0)(x^*-x_0) + \ldots, \qquad (2.2)$$

where $y_0 = -[P'(x_0)]^{-1}P(x_0)$, and $A_k = -\frac{1}{k!}[P'(x_0)]^{-1}P^{(k)}(x_0)$, $k = 2, 3, \ldots$. The form of (2.2) suggests the classical technique of inversion of a power series.

Looking upon (2.2) as an expansion of y_0 in terms of x^*-x_0, one tries to find an inverse expansion

$$x^*-x_0 = y_0 + B_2y_0y_0 + B_3y_0y_0y_0 + \dots \tag{2.3}$$

of x^*-x_0 in terms of y_0. Formally, this is done by substituting (2.3) into (2.2) and equating operator coefficients of like order. One obtains

$$B_2 = A_2, \quad B_3 = 2A_2A_2 + A_3, \ \dots \ . \tag{2.4}$$

These expressions become rapidly more complicated, even in the scalar case. However, the important consideration is that if y_0 is in the ball of convergence of the right-hand side of (2.3), and its sum has norm less than r, then the various formal operations used are justified, and the value of x^* obtained is a solution of the equation $P(x) = 0$.

In order to reduce this problem to scalar analysis, one constructs the <u>scalar majorant series</u>

$$\xi = \eta + a_2\xi^2 + a_3\xi^3 + \dots , \tag{2.5}$$

where $\eta \geq \|y_0\|$, $a_k \geq \|A_k\|$, $k = 2, 3, \dots$.

Inversion of (2.5) gives

$$\xi = \eta + b_2\eta^2 + b_3\eta^3 + \dots . \tag{2.6}$$

<u>Theorem 1</u>. If the series (2.6) converges to a point ξ within the radius of convergence of (2.5), then the value of x^* given by (2.3) is a solution of the equation $P(x) = 0$; furthermore, for

$$x_n = x_0 + y_0 + B_2y_0y_0 + \dots + B_n \overline{y_0 \overset{n}{\cdots} y_0}, \tag{2.7}$$

one has

$$\|x^* - x_n\| \leq R_n, \tag{2.8}$$

where

$$\xi = \eta + b_2\eta^2 + \dots + b_n\eta^n + R_n. \tag{2.9}$$

In order to apply Theorem 1 to a Urysohn integral equation

$U(x) = 0$, where U is the operator defined by (1.9), note that $U'(x_0)$ is the linear integral operator such that

$$(U'(x_0)z)(s) = z(s) - \int_0^1 f_x(s, t, x_0(t))z(t)dt, \tag{2.10}$$

and, consequently, $[U'(x_0)]^{-1}$ exists provided the Fredholm resolvent kernel $\Gamma_0(s, t)$ of $f_x(s, t, x_0(t))$ is defined, where

$$([U'(x_0)]^{-1}z)(s) = z(s) + \int_0^1 \Gamma_0(s, t)z(t)dt. \tag{2.11}$$

The higher derivatives of U are simple integral power forms (Schmidt, 1907). For example,

$$(U''(x_0)uv)(s) = - \int_0^1 f_{xx}(s, t, x_0(t))u(t)v(t)dt. \tag{2.12}$$

3. Algebraic Integral Equations

If the power series expansion (2.1) is finite, then the operator P is said to be a polynomial, and the corresponding equation $P(x) = 0$ is called a polynomial, or algebraic equation. For equations of this type, the formation of the inverse series is simpler than in the general case. Also, the series (2.2) will converge for all x, so the only problem in the application of Theorem 1 is the determination of the convergence of the inverse series (2.3). For a quadratic equation, (2.2) takes the form

$$x^*-x_0 = y_0 + A_2(x^*-x_0)(x^*-x_0). \tag{3.1}$$

The inverse series (2.3) may be generated by the relationships

$$y^{(o)} = y_0,\ y^{(k)} = \sum_{i=o}^{k-1} A_2 y^{(i)} y^{(k-1-i)}, \tag{3.2}$$

with $B_k \overbrace{y_0 \cdots y_0}^{k} = y^{(k-1)}$, $k = 2, 3, \ldots$. The scalar majorant series for (3.1) is

$$\xi = \eta + a_2\xi^2. \tag{3.3}$$

The inverse series for (3.3) is the binomial expansion of

$$\xi = \frac{1 - \sqrt{1 - 4\, a_2\eta}}{2a_2}, \tag{3.4}$$

which converges provided

$$a_2\eta \leqq \frac{1}{4}. \tag{3.5}$$

This criterion for the existence of solutions of quadratic integral equations can be found in the book by Lichtenstein (1931). Rall (1961, 1969) showed that it holds for quadratic operator equations in Banach spaces, and used the method of inversion of power series for the solution of the quadratic integral equation of radiative transfer (Chandrasekhar, 1960),

$$H(\mu) = 1 + \tfrac{1}{2}\,\varpi_0\,\mu H(\mu) \int_0^1 \frac{H(\mu')}{\mu + \mu'}\, d\mu'. \tag{3.6}$$

Results similar to (3.5) can also be obtained for cubic and biquadratic equations, using the binomial series expansion of the smallest solution of the scalar majorant equation.

4. Contractive and Monotone Equations of Second Kind

Formulating the integral equation to be solved as a fixed point problem $x = F(x)$ immediately suggests the method of successive substitutions (or iteration), for which one chooses an initial point x_0 and generates the sequence $\{x_n\}$ by means of the relationships

$$x_{n+1} = F(x_n),\ n = 0, 1, 2, \ldots . \tag{4.1}$$

A theory giving conditions for the convergence of $\{x_n\}$ to a fixed point x^* of F is said to be constructive, as it specifies a procedure for obtaining x^* as a limit, and for generating approximations x_n to x^* which will be as close to x^* as desired for n sufficiently large. (The Schmidt-Lichtenstein theory of §2 is thus also constructive.) The utility of a constructive theory is enhanced if it includes a convergence analysis; that is, if it specifies a sequence $\{c_n\}$ of non-negative real numbers such that $\lim_{n\to\infty} c_n = 0$, and

$$\|x^*-x_n\| \leqslant c_n,\ n = 0, 1, 2, \ldots . \tag{4.2}$$

A simple example of a theory of this type for nonlinear integral equations is the classical Picard iteration method

$$x_{n+1}(s) = y(s) + \int_0^s f(s, t, x_n(t))dt,\ n = 0, 1, 2, \ldots, \tag{4.3}$$

for the solution of the Volterra integral equation

$$x(s) = y(s) + \int_0^s f(s, t, x(t))dt. \tag{4.4}$$

One may take $x(s) \equiv 0$ in (4.3) without loss of generality. Under the assumption of <u>Lipschitz continuity</u> of f with respect to x,

$$\max_{0 \leq t \leq s} |f(s, t, x(t)) - f(s, t, z(t))| \leq K \|x-z\| \tag{4.5}$$

for x, z in the closed ball $\bar{\mathcal{U}}_0(r) = \{x: \|x\| \leq r\}$, in operator notation, $\|f(x) - f(z)\| \leq K \|x-z\|$, one has the following result.

<u>Theorem 2</u>. If

$$r \geq e^{Ks} \|x_1\|, \tag{4.6}$$

then the sequence $\{x_n(s)\}$ defined by (4.3) with $x_0(s) \equiv 0$ converges to a solution $x^*(s)$ of (4.4) in $\bar{\mathcal{U}}_0(e^{Ks} \|x_1\|)$ which is unique in $\bar{\mathcal{U}}_0(r)$; furthermore,

$$\|x^*-x_n\| \leq \{e^{Ks} - \sum_{j=o}^{n-1} \frac{(Ks)^j}{j!}\} \|x_1\|. \tag{4.7}$$

Proof: The assertions of this theorem follow directly from the relationships

$$\left.\begin{aligned} &\|x_2 - x_1\| \leq Ks \|x_1 - x_0\| = Ks \|x_1\|, \\ &\|x_3 - x_2\| \leq K \cdot \frac{Ks^2}{2} \|x_1\|, \\ &\|x_4 - x_3\| \leq K \cdot \frac{K^2 s^3}{3!} \|x_1\|, \ \ldots . \end{aligned}\right\} \tag{4.8}$$

Uniqueness holds in $\bar{\mathcal{U}}_0(r)$, as if $x^*(s)$ and $x^{**}(s)$ satisfy (4.4), then

$$\|x^*-x^{**}\| \leq \frac{(Ks)^j}{j!} \|x^*-x^{**}\|, \quad j = 0, 1, 2, \ldots, \tag{4.9}$$

by the same reasoning used to establish (4.8), which leads to a contradiction unless $x^* = x^{**}$.

The inequality (4.6) provides a <u>scalar majorant principle</u> for equation (4.4). For s fixed, the Lipschitz constant K in (4.5) is actually a monotone increasing function $K(r)$ of the radius r of the closed ball $\bar{\mathcal{U}}_0(r)$. If this function is known (or any function not less than it is available), then one may decide if condition (4.6) holds by determining if the scalar equation

$$r = e^{K(r)s} \|x_1\| \tag{4.10}$$

has non-negative solutions.

A similar result holds for the Urysohn equation

$$x(s) = \int_0^1 f(s, t, x(t))dt, \tag{4.11}$$

where one takes $y(s) \equiv 0$ without loss of generality, but the corresponding Lipschitz constant K such that

$$\max_{0 \leq s,t \leq 1} |f(s, t, x(t)) - f(s, t, z(t))| \leq K \|x-z\| \tag{4.12}$$

for x, z in $\bar{\mathcal{U}}_0(r)$ must be less than one. A constructive existence theorem is obtained here by applying Banach's contractive mapping principle (Rall, 1969).

<u>Theorem 3</u>. If $K < 1$ in (4.12) and

$$r \geq \frac{\|x_1\|}{1 - K} \quad , \tag{4.13}$$

then the iteration process

$$x_{n+1}(s) = \int_0^1 f(s, t, x_n(t))dt, \quad n = 0, 1, 2, \ldots, \tag{4.14}$$

starting from $x(s) \equiv 0$, converges to a solution $x^*(s)$ of (4.11) in $\bar{\mathcal{U}}_0 \frac{\|x_1\|}{1 - K}$ which is unique in $\bar{\mathcal{U}}_0(r)$; furthermore,

$$\|x^* - x_n\| \leq \frac{K^n}{1 - K} \|x_1\| , \quad n = 0, 1, 2, \ldots . \tag{4.15}$$

Proof: The relationships corresponding to (4.8) are

$$\left. \begin{aligned} &\|x_2 - x_1\| \leq K\|x_1 - x_0\| = K\|x_1\| , \\ &\|x_3 - x_2\| \leq K^2 \|x_1\| , \\ &\|x_4 - x_3\| \leq K^3 \|x_1\| \quad , \ldots . \end{aligned} \right\} \tag{4.16}$$

and uniqueness follows by assuming that $x^*(s)$, $x^{**}(s)$ are solutions of (4.11) in $\bar{\mathcal{U}}_0(r)$, and noting that

$$\|x^* - x^{**}\| \leq K \|x^* - x^{**}\| , \tag{4.17}$$

a contradiction unless $x^* = x^{**}$.

Given a scalar majorant function $K(r)$ for the Lipschitz constant K in (4.12), one may reduce the study of the existence and uniqueness of solutions of the integral equation (4.11) to finding positive solutions of the scalar equation

$$r = \frac{\|x_1\|}{1 - K(r)} \tag{4.18}$$

for $r \leqq R$, where R is the smallest positive solution of $K(r) = 1$. One can in fact use (4.17) to show that solutions $x^*(s)$ of (4.11) will be unique in the open ball $\bar{\mathcal{U}}_0(R)$.

Another iteration process which leads to constructive existence theorems for nonlinear integral equations is Newton's method. Here, the equation $P(x) = 0$ is transformed into the fixed point problem $x = F(x)$, with

$$F(x) = x - [P'(x)]^{-1}P(x), \tag{4.19}$$

and the sequence $\{x_n\}$ is generated by successive substitutions. One may also write this iteration process as

$$P'(x_n)(x_{n+1} - x_n) = -P(x_n),\ n = 0, 1, 2, \ldots, \tag{4.20}$$

which requires the solution of a linear equation for $x_{n+1} - x_n$, rather then the generally more difficult inversion of a linear operator. As applied to the Urysohn equation (4.11), this method requires the solution of the sequence of linear integral equations

$$z_n(s) - \int_0^1 f_x(s, t, x_n(t))z_n(t)dt = -x_n(s) + \int_0^1 f(s, t, x_n(t))dt, \tag{4.21}$$

where

$$x_{n+1}(s) = x_n(s) + z_n(s),\quad n = 0, 1, 2, \ldots, \tag{4.22}$$

once again starting from $x_0(s) \equiv 0$. Supposing that the Fredholm resolvent kernel $\Gamma_0(s, t)$ of $f_x(s, t, 0)$ exists, and

$$1 + \max_{0 \leqq s \leqq 1} \int_0^1 |\Gamma_0(s, t)|dt \leqq B, \tag{4.23}$$

$\|x_1\| \leqq \eta$, and $f_x(s, t, x(t))$ satisfies a Lipschitz condition of the form (4.12), then the following theorem holds (Tapia, 1971).

Theorem 4 (Kantorovich). If

$$h = BK\eta \leqq \tfrac{1}{2} \tag{4.24}$$

and

$$r \gtrless (1 - \sqrt{1-2h}) \frac{\eta}{h} = r^*, \tag{4.25}$$

then the iteration process defined by (4.21) converges to a solution $x^*(s)$ of (4.11) in the ball $\bar{\mathcal{U}}_0(r^*)$, furthermore,

$$\|x^*-x_n\| \leqslant \frac{\theta^{2^m}}{1-\theta^{2^m}} \cdot 2\sqrt{1-2h}\,\frac{\eta}{h}\ , \quad n = 0, 1, 2, \ldots, \tag{4.26}$$

where

$$\theta = \frac{1-\sqrt{1-2h}}{1+\sqrt{1-2h}} \tag{4.27}$$

If

$$r \gtrless (1 + \sqrt{1-2h}) \frac{\eta}{h} = r^{**}, \tag{4.28}$$

then $x^*(s)$ is unique in the ball $\bar{\mathcal{U}}_0(r^{**})$ if $h < \frac{1}{2}$, or in $\bar{\mathcal{U}}_0(r^{**}) = \bar{\mathcal{U}}_0(2\eta)$ if $h = \frac{1}{2}$.

The inequality (4.26) for $\|x^*-x_n\|$ is due to Gragg and Tapia (1973). An equivalent formulation was given earlier by Ostrowski (1971). This bound is the best possible in general, and shows the rapid convergence of Newton's method for $h < \frac{1}{2}$. The scalar equation corresponding to (4.25) and (4.28) is

$$B\eta\ K(r) = \frac{2\eta(r-\eta)}{r^2} \tag{4.29}$$

(Rall, 1969). If (4.29) has positive solutions $r^* \leqslant 2\eta$ and $r^{**} \geqslant 2\eta$, then the assertions of Theorem 4 concerning the existence and uniqueness of the solution $x^*(s)$ hold.

The operators satisfying the conditions of Theorems 2-4 are said to be contractive as they take some sufficiently small neighbourhood of x^* (or 0) into x^* by repeated application. In order to derive the required scalar majorant functions, note that $K(r) = \sup \|F'(x)\|$ for x in $\bar{\mathcal{U}}_0(r)$ will be a satisfactory Lipschitz constant for F in Theorems 2-3, and one may use $K(r) = \sup \|P''(x)\|$, x in $\bar{\mathcal{U}}_0(r)$ in Theorem 4. If the integral equations considered are algebraic, then these will be simple polynomials, and the analysis of the corresponding conditions will be possible without undue difficulty.

Another approach to the study of the convergence of the method of successive substitutions may be made for operators F of monotone type. There are several ways to generalize the concept of monotonicity to operators. One way to do this is to introduce a partial ordering in the space considered (or a subset of it). The ordering is partial, as not all elements of the space will be comparable. For example, for continuous, real-valued functions on an interval R, $x \geq z$ could be defined by

$$x(s) \geq z(s) \tag{4.30}$$

for all s in R. The iteration operator F will be said to be monotone at x_0 if the sequence $\{x_n\}$ is monotone increasing or decreasing. It is a standard theorem (Collatz, 1971) that if $\{x_n\}$ is monotone increasing and bounded above, then it converges to a fixed point x^* of F, and this is also true for decreasing sequences which are bounded below. If $x_0 \leq z_0$, and F generates a monotone increasing sequence $\{x_n\}$ starting from x_0, and a monotone decreasing sequence $\{z_n\}$ starting from z_0, and

$$z_n \geq x_n, \quad n = 0, 1, 2, \ldots , \tag{4.31}$$

then there is at least one fixed point x^* of F such that

$$z_n \geq x^* \geq x_n, \quad n = 0, 1, 2, \ldots . \tag{4.32}$$

The sequences $\{x_n\}$, $\{z_n\}$ do not necessarily converge to the same fixed point; however, the inequality (4.32) provides upper and lower bounds for any solution of the equation $x=F(x)$ lying in the intervals (4.32). An application of this technique to the integral equation (2.18) was made by Rall (1965).

Another definition of monotonicity for operators in Hilbert spaces (such as $\mathcal{L}_2$) is the one due to Zarantonello and Minty (Collatz, 1971). The operator F is monotone if

$$(F(x_1) - F(x_2), x_1 - x_2) \geq 0 \tag{4.33}$$

for all x_1, x_2, where (,) denotes the inner product. This type of monotonicity has been used extensively in the study of Hammerstein integral equations (Dolph and Minty, 1967; Collatz, 1971). Under certain conditions, the convergence of an iteration process to fixed points of operators of this type can be shown.

5. Bifurcation

Based on experience with nonlinear scalar equations, one would expect to encounter alternative (ii) of (1.2), nonexistence or nonuniqueness of solutions, frequently in the study of nonlinear integral equations. Indeed, the simple Volterra integral equation

$$x(s) = s + \int_0^s x^2(t)dt \tag{5.1}$$

does not have a continuous solution if $s \geq \frac{\pi}{2}$. Another occurrence is the <u>bifurcation</u>, or <u>branching</u>, of solutions, first observed by Schmidt (1907) in connection with the method of inversion of power series described in §2. Writing (2.1) as

$$-P'(x_0)(x^*-x_0) = P(x_0) + \tfrac{1}{2}P''(x_0)(x^*-x_0)(x^*-x_0) + \ldots \tag{5.2}$$

in a neighbourhood of a solution x^* of the equation $P(x) = 0$, one is stuck in trying to form the series (2.2) and (2.3) if $[P'(x_0)]^{-1}$ does not exist. However, if the Fredholm alternative theorem holds for the linear integral operator $P'(x_0)$, then the linear equation $P'(x_0)x = y$ may have a d-parameter family of solutions, and correspondingly, the equation $P(x) = 0$ may have a number of solutions x^* in the neighbourhood of x_0.

To illustrate how this situation could arise, consider the Urysohn integral operator $U(x, \lambda) = x-\lambda F(x)$ depending on the scalar parameter λ, and suppose that $x(s) \equiv 0$ satisfies the equation

$$U(x, \lambda) = x(s) - \lambda \int_0^1 f(s, t, x(t))dt = 0; \tag{5.3}$$

that is, $U(0, \lambda) = 0$, for all λ. The derivative of $U(x, \lambda)$ with respect to x at $x=0$ is the linear integral operator $U'(0, \lambda) = I - \lambda F'(0)$, where I denotes the identity operator, and so

$$(U'(0, \lambda)z)(s) = z(s) - \lambda \int_0^1 f_x(s, t, 0)z(t)dt. \tag{5.4}$$

Now, if λ is <u>not</u> a characteristic value of the kernel $f_x(s, t, 0)$, then $[U'(0, \lambda)]^{-1}$ exists, and

$$([U'(0, \lambda)]^{-1}z)(s) = z(s) + \lambda \int_0^1 \Gamma(s, t; \lambda)z(t)dt, \tag{5.5}$$

where $\Gamma(s, t; \lambda)$ is the Fredholm resolvent kernel of $f_x(s, t, 0)$.

By the implicit function theorem (Hildebrandt and Graves, 1927), the solution $x(s) \equiv 0$ is unique in some neighbourhood of the point $(0, \lambda)$ in the product space $C \times S$ of continuous functions and (complex) scalars. However, if $\lambda = \lambda^*$ is a characteristic value of $f_x(s, t, 0)$, then (5.5) does not hold, and it is possible that (5.3) has nonzero solutions in a neighbourhood of $(0, \lambda^*)$. If this is true, then $(0, \lambda^*)$ is called a bifurcation point for the equation (5.3).

An early application of bifurcation theory was made by Duffing (1918) to the determination of unstable modes of oscillation of massive pendulums. For a modern survey of the subject of bifurcation theory, see Krasnosel'skii et al. (1972).

6. Hammerstein Integral Equations with Green's Function Kernels

An important source of Hammerstein integral equations,

$$x(s) - \int_0^1 K(s, t)\, f(t, x(t))dt = 0, \tag{6.1}$$

is the transformation of boundary value problems for differential equations into integral equations. In this case, the kernel $K(s, t)$ will be a Green's function. For example, from

$$\frac{d^2x}{ds^2} = f(s, x(s)),\ x(0) = x(1) = 0, \tag{6.2}$$

one obtains (6.1) with

$$K(s, t) = \begin{cases} -t(1-s), & 0 \leqq t \leqq s, \\ -s(1-t), & s \leqq t \leqq 1. \end{cases} \tag{6.3}$$

Attention will be confined here to the simple case that $K(s, t)$ has the form

$$K(s, t) = \begin{cases} u(s)v(t), & 0 \leqq t \leqq s, \\ u(t)v(s), & s \leqq t \leqq 1. \end{cases} \tag{6.4}$$

In this case, equation (6.1) may be written as

$$x(s) - \int_0^s [u(s)v(t) - u(t)v(s)]\, f(t, x(t))dt = cv(s), \tag{6.5}$$

where

$$c = \int_0^1 u(t)\, f(t, x(t))dt. \tag{6.6}$$

The system of equations (6.5) - (6.6) may be written as

$$c = \Phi(c), \tag{6.7}$$

a fixed point problem for the single scalar variable c, where Φ is the composition UV of the operator V which takes c into the solution of the Volterra integral equation (6.5), and U is the result of then applying (6.6). For linear functions $f(s, x(s))$, all these equations are linear, and the Fredholm alternative theorem is obtained immediately. In the nonlinear case, this approach has the advantage of reducing the Fredholm equation (6.1) to the Volterra equation (6.5), whose solution may be studied under less restrictive conditions. From a numerical standpoint, this method disposes of the necessity of dealing with the singularities of the derivatives of the Green's function $K(s, t)$ on the line $s = t$. This method may be extended to kernels of the form $a(s)\, G(s, t)\, b(t)$, where $G(s, t)$ is a finite sum of kernels of the form (6.4), and to higher dimensions.

For the differential operator (6.2), equation (6.5) takes the form

$$x(s) - \int_0^s (s-t)\, f(t, x(t))dt = cs, \tag{6.9}$$

and c is simply the unknown value of $x'(0)$, the so-called "shooting" parameter.

For the pendulum problem of Duffing (1918),

$$f(s, x(s)) = \alpha \sin x(s) + \beta \sin s, \quad 0 \leqslant s \leqslant \pi. \tag{6.10}$$

It is interesting to note that Duffing approximated (6.10) by

$$f(s, x(s)) = \alpha(x(s) - \frac{x^3(s)}{3!}) + \beta \sin s, \tag{6.11}$$

which leads to the algebraic integral equation of degree three called the Duffing equation.

7. Acknowledgements

This work was supported by the Science Research Council of Great Britain, and the U.S. Army under Contract No.: DA-31-124-ARO-D-462.

References

ANSELONE, P.M. (1964) (Ed.) Nonlinear Integral Equations, Univ. of Wisconsin Press.

CHANDRASEKHAR, S. (1960) Radiative Transfer, Dover.

COLLATZ, L. (1971) Some applications of functional analysis to analysis, particularly to nonlinear integral equations, Rall (1971), pp. 1-43.

DOLPH, C.L. and MINTY, G.J. (1964) On nonlinear integral equations of the Hammerstein type, Anselone (1964), pp. 99-154.

DUFFING, G. (1918) Erzwungene Schwingungen bei veranderlicher Eigenfrequenz und ihre technische Bedeutung, Vieweg & Sohn.

GRAGG, V.B. and TAPIA, R.A. (1973) Optimal error bounds for the Newton-Kantorovich theorem, SIAM J. numer. Anal. (to appear)

HILDEBRANDT, T.H. and GRAVES, L.M. (1927) Implicit functions and their differentials in general analysis, Trans. Am. math. Soc. Vol. 29, pp. 127-153.

KRASNOSEL'SKII, M.A. (1963) Topological Methods in the Theory of Nonlinear Equations, tr. by A.H. Armstrong, ed. by J. Burlak, Pergamon.

KRASNOSEL'SKII, M.A., VAINIKKO, G.M., ZABREIKO, P.P., RUTITSKII, Ya. B., and STETSENKO, V. Ya. (1972) Approximate Solution of Operator Equations, tr. by D. Louvish, Wolters-Noordhoff.

LICHTENSTEIN, L. (1931) Vorlesungen uber einige Klassen nichtlinearer Integralgleichungen und Integro-Differentialgleichungen nebst Anwendungen, Springer.

OSTROWSKI, A.M. (1971) La méthode de Newton dans les espaces de Banach, C.R. Acad. Sci. Paris Ser. A. Vol. 272, pp. 1251-1253.

RALL, L.B. (1961) Quadratic equations in Banach spaces, Rc. Circ. mat. Palermo Vol. 10, pp. 314-332.

RALL, L.B. (1965) Numerical integrations and the solution of integral equations by the use of Riemann sums, SIAM Rev. Vol. 7, pp. 55-64.

RALL, L.B. (1969) Computational Solution of Nonlinear Operator Equations, Wiley.

RALL, L.B. (1971) (Ed.) Nonlinear Functional Analysis and Applications, Academic Press.

SCHMIDT, E. (1907) Zur Theorie der linearen und nichtlinearen Integralgleichungen, III. Theil. Uber die Auflösung der nichtlinearen Integralgleichungen und die Verzweigung ihrer Lösungen, Math. Annln. Vol. 65, pp. 370-399.

TAPIA, R.A. (1971) The Kantorovich theorem for Newton's method, Am. math. Mon. Vol. 78, pp. 389-392.

VAINBERG, M.M. (1964) Variational Methods for the Study of Nonlinear Operators, tr. by A. Feinstein, Holden-Day.

CHAPTER 16 METHODS FOR NONLINEAR SYSTEMS OF EQUATIONS

L.B. Rall
University of Wisconsin

1. Direct Methods

A method for solving the system of n equations

$$p_i(x_1, x_2, \ldots, x_n) = 0, \qquad i = 1, 2, \ldots, n, \tag{1.1}$$

for the n unknowns $x_1, x_2, \ldots, x_n$ will be said to be direct if it leads to a single equation

$$f(c) = 0 \tag{1.2}$$

which is equivalent to the system (1.1) in the sense that there is a one-to-one correspondence between solution vectors $x^* = (x_1^*, x_2^*, \ldots, x_n^*)$ of (1.1) and solutions c^* of (1.2). In other words, relationships

$$\left.\begin{aligned} x_i^* &= x_i(c^*),\ i = 1, 2, \ldots, n, \\ c^* &= c(x_1^*, x_2^*, \ldots, x_n^*), \end{aligned}\right\} \tag{1.3}$$

exist, and one can obtain information about the existence and uniqueness or multiplicity of solution vectors of (1.1) from the corresponding information about solutions of (1.2). In actual practice it is only rarely that the representations (1.2) and (1.3) can be obtained explicitly, or that formulas are available for solutions c^* of (1.2). However, it is worthwhile to investigate the possibility of applying a direct method to a given system, as theory and technique for solving a single scalar equation (1.2) are in a more advanced state than for systems (1.1) in several variables (Durand, 1960; Householder, 1970; Ostrowski, 1966, Rabinowitz, 1970; Zaguskin, 1961). In particular, if (1.2) is simply a polynomial equation, then the problem may be considered to be solved from a theoretical standpoint (MacDuffee, 1954) and is also well understood computationally (Dejon and Henrici, 1969).

In certain cases, the method of elimination may be applied to (1.1) to obtain the equation (1.2) and the relationships (1.3). For example, if the nth equation of (1.1) can be solved explicitly for x_n,

$$x_n = g_n(x_1, x_2, \ldots, x_{n-1}), \tag{1.4}$$

then this result may be substituted into the first $n - 1$ equations of (1.1) to obtain a new system of $n - 1$ equations

$$p_i^{(1)}(x_1, x_2, \ldots, x_{n-1}) = 0, \quad i = 1, 2, \ldots, n - 1, \tag{1.5}$$

in $n - 1$ unknowns, and so on. If this process can be continued, then the final result will be a single equation

$$p_1^{(n-1)}(x_1) = 0 \tag{1.6}$$

for x_1, which corresponds to (1.2). In the linear case, all of the intermediate systems

$$p_i^{(j)}(x_1, x_2, \ldots, x_{n-j}) = 0, \quad i = 1, 2, \ldots, n - j, \tag{1.7}$$

are linear for $j = 1, 2, \ldots, n - 2$, and the final equation (1.6) is linear, so the method of elimination has not increased the complexity of the problem in terms of the number of arithmetic operations required to evaluate the functions $p_i^{(j)}(x_1, x_2, \ldots, x_{n-j})$ and to calculate $x_k = g_k(x_1, x_2, \ldots, x_{k-1})$, $k = 2, 3, \ldots, n$. For general nonlinear systems, however, the method of elimination may lead to a great increase in the complexity of the functions involved. Consequently, in an actual situation, the advantage (if any) of dealing with a smaller system of equations has to be weighed against the effect of a possible increase in complexity. The method of solution being used and the computational facilities available would have to be taken into account in order to reach a decision.

The possibility of applying a direct method of solution should be kept in mind in the approximate solution of nonlinear integral equations by the use of finite systems of nonlinear equations. For example, it was shown in §6 of Chapter 15 that the Hammerstein integral equation

$$x(s) - \int_0^1 K(x, t)\, f(t, x(t))dt = 0 \tag{1.8}$$

with a Green's function kernel

$$K(s, t) = \begin{cases} u(s)\, v(t), & 0 \leqslant t \leqslant s, \\ u(t)\, v(s), & s \leqslant t \leqslant 1, \end{cases} \tag{1.9}$$

can be replaced by the system

$$\left.\begin{aligned} x(s) - \int_0^s [u(s)\, v(t) - u(t)\, v(s)]\, f(t, x(t))dt &= cv(s), \\ c = \int_0^1 u(t)\, f(t, x(t))dt. & \end{aligned}\right\} \tag{1.10}$$

Choosing a value $h = \frac{1}{n}$ and applying the trapezoidal rule of numerical integration (Milne, 1949) gives the system

$$\begin{cases} x_0 = c\, v_0, \\ x_1 - \frac{h}{2}\left[u_1\, v_0 - u_0\, v_1\right] f_0(x_0) = c\, v_1, \\ \dots \quad \dots \quad \dots \quad \dots \quad \dots \\ x_n - \frac{h}{2}\left[u_n\, v_0 - u_0\, v_n\right] f_0(x_0) - h \sum_{j=1}^{n-1} \left[u_n\, v_j - u_j\, v_n\right] f_j(x_j) = cv_n, \\ c = \frac{h}{2}\left[u_0\, f_0(x_0) + 2 \sum_{j=1}^{n-1} u_j\, f_j(x_j) + u_n\, f_n\, (x_n)\right], \end{cases} \tag{1.11}$$

of $n + 2$ equations for the unknowns $c, x_0, x_1, \dots, x_n$; where

$$v_i = v(ih),\ u_i = u(ih),\ f_i(x_i) = f(ih, x_i),$$

$$i = 0, 1, \dots, n, \tag{1.12}$$

and the numbers $x_0, x_1, \dots, x_n$ are taken as approximations to the unknown values $x(0), x(h), \dots, x(1)$, respectively. Note that the first $n + 1$ equations can be solved explicitly for $x_0, x_1, \dots, x_n$ in terms of c. Substitution of these results into the last equation gives a single equation for c, which can be put in the form (1.2). If $f(t, x(t))$ is a polynomial in $x(t)$, as in the Duffing equation (see equation (6.10) of Chapter 15), then one obtains a polynomial equation for c.

2. Methods based on Functional Analysis

In addition to the absence of direct methods for nonlinear systems in general, it turns out that very few types of systems have been studied in detail, with special methods developed for their solution. The notable exception to this is the algebraic eigenvalue problem, which is equivalent to the system of $n + 1$ equations

$$\begin{aligned} a_{11}\, x_1 + a_{12}\, x_2 + \dots + a_{1n}\, x_n &= \lambda x_1, \\ a_{21}\, x_1 + a_{22}\, x_2 + \dots + a_{2n}\, x_n &= \lambda x_2, \\ \dots \qquad \dots \qquad \dots \qquad \dots \qquad &\dots \\ a_{n1}\, x_1 + a_{n2}\, x_2 + \dots + a_{nn}\, x_n &= \lambda x_n, \\ x_1^2 + x_2^2 \qquad + \dots + x_n^2 \qquad &= 1, \end{aligned} \tag{2.1}$$

for the $n + 1$ unknowns $\lambda, x_1, x_2, \dots, x_n$, the matrix $A = (a_{ij})$ being given. This problem is well understood both theoretically and computationally (Wilkinson, 1965). In view of this lack of theory for the specific problem of solving nonlinear systems of equations, although much can be done by generalization of results for single nonlinear equations (Ostrowski, 1966), it has proved to be valuable to use the methods of functional analysis, and to consider the system (1.1) as a special case of the operator equation

$$P(x) = 0, \tag{2.2}$$

where P maps from a Banach space X into a Banach space Y. This is the point of view adopted in the text by Ortega and Rheinboldt (1970), and allows one to make direct application of the results in the books by Krasnosel'skii, Vainikko, Zabreiko, Rutitskii, and Stetsenko (1972), Rall (1969), and Vainberg (1964) on nonlinear operator equations in general; see also Rabinowitz (1970).

In the setting of nonlinear systems of equations, the spaces X and Y will be spaces of n-dimensional vectors $x = (x_1, x_2, \ldots, x_n)$, with suitable (and perhaps identical) norms. These spaces may be real or complex, but will usually be considered to be real, as complex n-dimensional space can be identified with real 2n-dimensional space. Finite dimensional vector spaces have some useful properties not shared by Banach spaces in general. First of all, any two norms $\|\ \|$, $\|\ \|'$ for an n-dimensional space are <u>equivalent</u> in the sense that positive constants a, b exist such that $a\|x\| \leqslant \|x\|' \leqslant b\|x\|$ for all n-dimensional vectors x. Secondly, all continuous <u>functionals</u> (scalar valued functions) f on x attain their absolute maxima and minima on closed and bounded sets (that is, in finite dimensional spaces, bounded sets are <u>compact</u>). A third useful property is that a linear operator A from an n-dimensional space X into an m-dimensional space Y can be represented uniquely as an $m \times n$ <u>matrix</u> $A = (a_{ij})$, $i = 1, 2, \ldots, m$; $j = 1, 2, \ldots, n$, with

$$y_i = (Ax)_i = \sum_{j=1}^{n} a_{ij} x_j, \quad i = 1, 2, \ldots, m, \tag{2.3}$$

(Taylor, 1958). The set of such matrices A forms a finite-dimensional linear space; the norm

$$\|A\| = \max_{\|x\|=1} \|Ax\| \tag{2.4}$$

is called the <u>operator norm</u> of A.

Representation of derivatives of operators in finite-dimensional spaces is also very simple. If P is the operator with components defined by the functions $p_i(x) = p_i(x_1, x_2, \ldots, x_n)$ and is Fréchet differentiable at x, then its derivative $P'(x)$ is represented uniquely by the Jacobian matrix

$$P'(x) = \left(\frac{\partial p_i}{\partial x_j}\right), \quad i, j = 1, 2, \ldots, n. \tag{2.5}$$

If P is differentiable twice at x, then its second derivative $P''(x)$ is the bilinear (Hessian) operator

$$P''(x) = \left(\frac{\partial^2 p_i}{\partial x_k \partial x_j}\right), \quad i, j, k = 1, 2, \ldots, n. \tag{2.6}$$

Operation by a bilinear operator $B = (b_{ijk})$ on a vector x is defined by

$$(Bx)_{ij} = \sum_{k=1}^{n} b_{ijk}\, x_k, \quad i = 1, 2, \ldots, m;\ j = 1, 2, \ldots, n, \tag{2.7}$$

and on two points x, z by

$$(Bxz)_i = \sum_{j=1}^{n} \sum_{k=1}^{n} b_{ijk}\, x_k\, z_j, \quad i = 1, 2, \ldots, m. \tag{2.8}$$

Consequently, a bilinear operator is linear from X into the space of matrices mapping X into Y. One may define the operator norm (2.4) for B accordingly, and it follows that

$$||Bxz|| \leqq ||B|| \cdot ||x|| \cdot ||z|| \tag{2.9}$$

for all x, z, just as $||Ax|| \leqq ||A|| \cdot ||x||$ for all x for linear A. A bilinear operator $B = (b_{ijk})$ is symmetric if $b_{ijk} = b_{ikj}$; the second derivatives (2.6) are symmetric if they exist. These definitions can be extended inductively to higher order derivatives, which turn out to be symmetric multilinear operators of the same order (Rall, 1969). As in the case of nonlinear integral equations (see §2 of Chapter 15), these analytic tools enable one to contemplate power series expansion of operators, and the solution of equation (2.2) by the inversion of power series. The conditions for the applicability of this method are the same as in Theorem 1 of Chapter 15, and the details follow equations (2.1)-(2.6) of Chapter 15.

The introduction of k-linear operators $A_k = (a_{ij_1 \ldots j_k})$, which will be assumed to be symmetric in the sense that the indices

$j_1, \ldots, j_k$ may be permuted arbitrarily without altering the value of the operator, allows one to define polynomial operators P_d of degree d,

$$P_d(x) = A_d \overbrace{x \ldots x}^{d} + \ldots + A_2\, xx + A_1\, x + A_0, \tag{2.10}$$

where A_k, $k = 2, \ldots, d$, are given k-linear operators, $A_1 = (a_{ij})$ is a given matrix, and A_0 is a given vector. Polynomial operators differentiate according to the same rules as scalar polynomials, for example,

$$P_d'(x) = d\, A_d \overbrace{x \ldots x}^{-(d-1)-} + \ldots + 2A_2\, x + A_1.$$

If $m_k \geqslant \|A_k\|$, $k = 0, 1, \ldots, d$, then

$$p_d(r) = m_d\, r^d + \ldots + m_2\, r^2 + m_1\, r + m_0 \tag{2.11}$$

is a scalar majorant polynomial for $P_d(x)$ in the sense that $\|P_d(x)\| \leqslant p_d\,(\|x\|)$ for all x. It follows that $p_d'(r)$ will be a scalar majorant for $P_d'(x)$, and so on. These considerations make analytic and computational treatment of polynomial systems

$$P_d(x) = 0 \tag{2.12}$$

somewhat easier than that of general nonlinear systems.

The simplest nonlinear polynomial system is, of course, the quadratic system

$$Q(x) \equiv Bxx + Ax + y = 0. \tag{2.13}$$

In a certain sense, it is also the most general, as the following theorem shows.

Theorem 1. Any polynomial system (2.12) of degree $d \geqslant 2$ may be transformed into an equivalent (but usually larger) quadratic system (2.13) which is (i) linear in the coefficients of the original system, and (ii) contains only quadratic terms of the form

$$\xi_i - \xi_j\, \xi_k = 0, \quad i > j,\ i > k \tag{2.14}$$

in the new variables $\xi_1, \xi_2, \ldots, \xi_m$.

Proof: The only operation required to form products of powers of the variables $x_1, x_2, \ldots, x_n$ is multiplication, which is bilinear. Consequently, starting with $\xi_i = x_i$, $i = 1, 2, \ldots, n$, and applying (2.14) as necessary, one obtains a variable ξ_i corresponding to each distinct product $x_1^{\alpha_1} \ldots x_n^{\alpha_n}$ in (2.12). Q.E.D.

For example, the scalar cubic equation

$$ax^3 + bx^2 + cx + d = 0, \tag{2.15}$$

is equivalent to the quadratic system

$$\left.\begin{aligned} \xi_3 - \xi_1 \xi_2 &= 0, \\ \xi_2 - \xi_1^2 &= 0, \\ a\xi_3 + b\xi_2 + c\xi_1 + d &= 0, \end{aligned}\right\} \tag{2.16}$$

with the properties required by Theorem 1. Here, a reduction has been made in the complexity of the equations of the system at the expense of an increase in size. It should also be mentioned that the transformations of the type required in the theorem do not necessarily yield a unique quadratic system. For example, in extending (2.16) to a quartic scalar equation, $\xi_4 = x^4$ could be defined either by $\xi_4 - \xi_2^2 = 0$ or $\xi_4 - \xi_1 \xi_3 = 0$.

3. Iterative Procedures, including Newton's Method

Because of the lack of explicit formulae for the solution of nonlinear systems, including the scalar equations (1.2) obtained by the application of a direct method, one usually is led to some iteration process to find an approximate solution of sufficient accuracy. Here, one reformulates the system (2.2) to be solved as a fixed point problem

$$x = F(x) \tag{3.1}$$

for the iteration operator F. Starting from some $x^{(0)}$, one generates the sequence $x^{(k)}$ by

$$x^{(k+1)} = F(x^{(k)}), \quad k = 0, 1, 2, \ldots \tag{3.2}$$

One would desire that a solution x^* of (3.1) exists, with $\| x^* - x^{(k)} \| \to 0$ as $n \to \infty$. Theorems characterizing the convergence of iterative methods generally may be classified as local, semi-local or global. A local theorem starts from the assumption that x^* exists, and gives convergence (and usually the rate of convergence) of the iteration process if $x^{(0)}$ is sufficiently close to x^*, and the operator F has certain properties in a neighbourhood of x^*. The following theorem is typical of this class.

Theorem 2. Suppose that $x^* = F(x^*)$,

$$F'(x^*) = 0, \ldots, F^{(m)}(x^*) = 0, \tag{3.3}$$

and $F^{(m)}$ satisfies the Lipschitz condition

$$\| F^{(m)}(x) - F^{(m)}(z)\| \leqq (m+1)!\, K \|x - z\|,$$
$$x, z \in \bar{U}(x^{(0)}, \|x^{(0)} - x^*\|). \qquad (3.4)$$

Then, if

$$\theta = K\| x^{(0)} - x^*\|^m < 1, \qquad (3.5)$$

the sequence $x^{(k)}$ defined by (3.2) converges to x^*, with

$$\| x^{(k)} - x^*\| \leqq \theta^{\frac{(m+1)^k - 1}{m}} \|x^{(0)} - x^*\|, \; k = 0, 1, 2, \ldots \qquad (3.6)$$

Proof: Inequality (3.6) follows from the use of (3.2) and Taylor's formula (Rall, 1969), which reduces to

$$x^{(k+1)} - x^* = \int_0^1 \{F^{(m)}(x^* + \lambda(x^{(k)} - x^*)) - F^{(m)}(x^*)\}\, (x^{(k)} - x^*)^m \frac{(1-\lambda)^{m-1}}{(m-1)!}\, d\lambda \qquad (3.7)$$

under the hypothesis (3.3), use of the Lipschitz condition (3.4), followed by (3.5) and mathematical induction, it being assumed that the iterates $x^{(1)}, \ldots, x^{(k)}$ are contained in $\bar{U}(x^{(0)}, \|x^{(0)} - x^*\|)$, which is easily verified for $k = 1$. Q.E.D.

Newton's method for solving the equation $P(x) = 0$ makes use of the iteration operator F defined by

$$F(x) = x - [P'(x)]^{-1} P(x). \qquad (3.8)$$

It is easy to verify that the hypotheses of Theorem 2 are satisfied with $m = 1$ (quadratic convergence) if $[P'(x^*)]^{-1}$ exists and F' is Lipschitz continuous in the required region.

<u>Chebyshev's method</u>, which employs the iteration operator defined by

$$F(x) = x - [P'(x)]^{-1} P(x) - \tfrac{1}{2}[P'(x)]^{-1} P''(x)[P'(x)]^{-1}P(x)[P'(x)]^{-1}P(x), \qquad (3.9)$$

(Rall, 1969) satisfies the conditions of Theorem 2 with $m = 2$ (cubic convergence) under similar assumptions. By taking more terms of the inverse power series for $x^* - x_n$ (see §2 of Chapter 15), one can construct iterative processes of even higher order local convergence. However, as (3.9) shows, the complexity of these formulas is such that their practical utility is doubtful, except perhaps in some special cases.

In general, <u>semi-local</u> theorems assert the existence of x^* and ne convergence of iteration only in a neighbourhood of a suitable nitial approximation $x^{(0)}$. A significant example of a semi-local neorem is the following one on the convergence of Newton's method.

neorem 3. (Kantorovich). If P is Lipschitz continuous with onstant K in $\bar{U}(x^{(0)}, r)$,

$$= \| [P'(x)^{(0)})]^{-1} \| \cdot \| - [P'(x^{(0)})]^{-1} P(x^{(0)}) \| \cdot K \leq \tfrac{1}{2}, \qquad (3.10)$$

nd

$$r \geq (1 - \sqrt{1 - 2h}) \frac{\| -[P'(x^{(0)})]^{-1} P(x^{(0)}) \|}{h} = r^*, \qquad (3.11)$$

hen a solution x^* of the equation $P(x) = 0$ exists in $\bar{U}(x^{(0)}, r^*)$, o which Newton's method converges.

A proof of this theorem may be found in the paper by Tapia (1971), nd various versions exist elsewhere in the literature. A formula for he rate of convergence has already been given in Chapter 15 inequality (4.26) with $\eta = \| -[P'(x^{(0)})]^{-1} P(x^{(0)}) \|$), which shows uadratic convergence if $h < \frac{1}{2}$.

Global theorems assert the existence of x^* and convergence of he iteration process on the basis of properties of F, independently f the choice of $x^{(0)}$. An illustration of this type of theorem is he well-known contraction mapping principle.

heorem 4. (Banach). If

$$\|F(x) - F(z)\| \leq \alpha \|x - z\|, \ \alpha < 1, \qquad (3.12)$$

hen F has a unique fixed point x^* in $\bar{U}\left(x_0, \frac{\| x_0 - F(x_0) \|}{1 - \alpha}\right)$, to hich the sequence $\{x^{(k)}\}$ defined by (3.2) converges, with

$$\|x^* - x^{(k)}\| \leq \frac{\alpha^k}{1-\alpha} \|x^{(0)} - F(x^{(0)})\|, \ k = 0, 1, 2, \ldots \qquad (3.13)$$

The proof of this theorem is standard. It may also be ephrased as a semi-local theorem (Rall, 1969. See also Theorem 3 of hapter 15.).

There are also many other iteration processes that one may onsider; for example, nonstationary processes with

$$x^{(k+1)} = F_k(x^{(k)}), \quad k = 0, 1, 2, \ldots, \qquad (3.14)$$

nd multistep iteration

$$x^{(k+1)} = F_k(x^{(k)}, x^{(k-1)}, \ldots, x^{(k-m+1)}), \ k = m - 1, m, \ldots \qquad (3.15)$$

with $x^{(0)}, \ldots, x^{(m-1)}$ given (Ortega and Rheinboldt, 1970). Generalized secant methods fall into this latter category.

4. Variational Methods

Strictly speaking, variational methods apply to the system of equations $P(x) = 0$ only in case that $P(x)$ is the derivative of a real functional p at x; that is

$$p'(x) = P(x). \tag{4.1}$$

The solutions of (2.2) will be the critical points of p, including the points at which $p(x)$ is a local or global maximum or minimum, if any exist. Consequently, a method for maximizing or minimizing $p(x)$ will give solutions of the system of equations $P(x) = 0$. The well-known necessary and sufficient condition for a differentiable operator P at x to be the derivative of a scalar functional p is that the Jacobian matrix $P'(x)$ is symmetric (Vainberg, 1964); that is

$$\frac{\partial p_i}{\partial x_j} = \frac{\partial p_j}{\partial x_i}, \quad i, j = 1, 2, \ldots, n. \tag{4.2}$$

This condition is very restrictive in practice; applied to the quadratic system

$$Bxx + Ax + y = 0 \tag{4.3}$$

it requires that the matrix A be symmetric, and the bilinear operator $B = (b_{ijk})$ be totally symmetric, so that

$$b_{ijk} = b_{ikj} = b_{jik}, \quad i, j, k = 1, 2, \ldots, m. \tag{4.4}$$

In actual practice, instead of a variational method, one applies a minimization method to solve the equation $P(x) = 0$. Here, one takes a functional p which has a unique minimum at 0 (such as $p(x) = \|x\|$), and minimizes $p(P(x))$. One of the best-known methods of this type is the Davidon-Fletcher-Powell method (Fletcher and Powell, 1963; Rabinowitz, 1970; Ortega and Rheinboldt, 1970). Of course, unless $p(P(x^*)) = 0$ exactly, one cannot conclude for a procedure of this type that a solution x^* of (2.2) exists, or obtain a bound for $\|x^{(0)} - x^*\|$, where $x^{(0)}$ is an approximate minimizer of p. However, the value $x^{(0)}$ obtained in this way could be used in conjunction with semi-local or global theorems of the type discussed in §3 to obtain existence results and error estimates.

5. Continuation Methods

The success and practical usefulness of iteration and also

inimization methods for solving a system of equations of the form 2.2) usually depends on having an initial approximation $x^{(0)}$ which s fairly close to some desired solution x^*. In many cases, the hysical or mathematical basis for the system will suggest a suitable nitial approximation $x^{(0)}$. There is also the possibility, however, hat one is completely in the dark or that the initial approximations t hand are not good enough to ensure convergence of a given process. ne possibility which suggests itself is to embed the operator P into family of operators P_λ, $0 \leqq \lambda \leqq 1$, for which the solution $x^{(0)}$ of

$$P_0(x) = 0 \tag{5.1}$$

s known, and $P_1 = P$. For some value $\lambda_1 > 0$, suppose that $x^{(0)}$ is sufficiently good approximation to the solution $x^{(\lambda_1)}$ of $_{\lambda_1}(x) = 0$, and so on. If $\lambda_{i+1} - \lambda_i \geqslant m$, then this <u>discrete</u> ontinuation process will lead to the solution x^* of $(x) = P_1(x) = 0$ in a finite number of steps. Of course, each <u>ntermediate</u> problem $P_{\lambda_i}(x) = 0$ may require an infinite iteration rocess for its solution. However, it is not necessary to solve each ntermediate problem exactly if one is looking only for a satisfactory nitial approximation to x^*. In certain cases, the parameter λ may ave some significance, so that the intermediate and final problems ould all require accurate solutions. This is called <u>table making</u>. iscrete continuation methods were studied by Ficken (1951), and are lso discussed in the book by Ortega and Rheinboldt (1970).

There is also the possibility of using continuous (or analytic) ontinuation techniques for the solution of systems of equations, as roposed by Davidenko (1953). Starting from

$$P(x^{(0)}) = y^{(0)}, \tag{5.2}$$

ne takes a differentiable function $h(\lambda)$ such that

$$h(0) = y^{(0)},\ h(1) = 0,\ h'(\lambda) = g(\lambda). \tag{5.3}$$

hen,

$$P(x) = h(\lambda), \quad 0 \leqq \lambda \leqq 1, \tag{5.4}$$

efines $x = x(\lambda)$ implicitly, with $x(0) = x^{(0)}$, $x(1) = x^*$, supposing hat this <u>homotopy</u> exists. Differentiating (5.4) yields the initial alue problem

$$\left.\begin{aligned}\frac{dx}{d\lambda} &= [P'(x)]^{-1}\ g(\lambda),\\ x(0) &= x^{(0)},\end{aligned}\right\}\qquad (5.5)$$

a system of n ordinary differential equations in n unknowns. Application of a numerical method for the integration of this system over $0 \leq \lambda \leq 1$ will give an approximation to $x^* = x(1)$. The existence of an integral curve of (5.5) implies the existence of x^*.

The use of the system (5.5) requires the inversion of the Jacobian matrix $P'(x)$ at each evaluation of the right-hand side. It may be noted that the inverse operator

$$\Gamma = [P'(x)]^{-1} \qquad (5.6)$$

satisfies the differential equation

$$\frac{d\Gamma}{d\lambda} = -\ \Gamma\ P''(x)\ \frac{dx}{d\lambda}\ \Gamma, \qquad (5.7)$$

(Rall, 1969), so that (5.5) is also equivalent to the system

$$\left.\begin{aligned}\frac{dx}{d\lambda} &= \Gamma\ g(\lambda), \quad x(0) = x^{(0)},\\ \frac{d\Gamma}{d\lambda} &= -\Gamma\ P''(x)\ \Gamma\ g(\lambda)\Gamma, \quad \Gamma(0) = [P'(x^{(0)})]^{-1},\end{aligned}\right\}\qquad (5.8)$$

of $n^2 + n$ equations, provided $P''(x)$ exists. Although (5.8) is considerably larger than (5.5), only one matrix inversion is required. This approach could be useful in case that $P''(x)$ is sparse (few nonzero entries), or constant, which would be true for quadratic equations.

6. Computational Aspects of the Solution of Nonlinear Systems

Because of the complexity of nonlinear systems, methods for their solution which exhibit rapid convergence and relatively few operations per step appear to be essential. These considerations put Newton's method, which requires only the solution of a linear system at each step to obtain quadratic convergence, and methods related to Newton's method, among the methods of choice for the actual solution of nonlinear systems. The derivatives required may be obtained easily for polynomial systems, in which much common information occurs in the evaluation of $P_d(x)$ and $P_d'(x)$. For systems in which the equations can be written in a common computing language such as FORTRAN, it is possible to write fairly efficient routines to obtain the required derivatives automatically (Kuba and Rall, 1972).

There are other efficient computational methods which avoid

derivatives or approximate them by finite differences. This class of procedures includes secant methods, Newton-like methods, and some minimization methods (Ortega and Rheinboldt, 1970; Rabinowitz, 1970). Many of these converge very rapidly to a vector x^{**} such that $\| P(x^{**}) \|$ is very small, but do not give a guarantee of the existence of a solution x^* or a reliable estimate of $\| x^* - x^{**} \|$. However, if this information is desired, one may take $x^{(0)} = x^{**}$ and apply the conclusions of a semi-local theorem, such as Theorem 3, to obtain the results required. By use of the techniques of interval analysis (Moore, 1966; Hansen, 1969), one may obtain rigorous error bounds which take into account not only truncation and round-off error, but also the effect of errors in the coefficients of the system being solved (Rockne and Lancaster, 1969; Kuba and Rall, 1972).

7. Acknowledgements

This work was supported by the Science Research Council of Great Britain, and the U.S. Army under Contract No. DA-31-124-ARO-D-462.

References

DAVIDENKO, D.F. (1953) On a new method of numerically integrating a system of nonlinear equations (Russian), Dokl. Akad. Nauk SSSR Vol. 88, pp.601-604.

DEJON, B. and HENRICI, P. (1969) (Eds.) Constructive Aspects of the Fundamental Theorem of Algebra, Wiley-Interscience.

DURAND, E. (1960) Solutions Numériques des Équations Algebriques. Tom I: Équations du Type F(x) = 0; Racines d'un Polynôme, Masson et Cie.

FICKEN, F. (1951) The continuation method for functional equations, Communs. pure appl. Math. Vol. 4, pp.435-456.

FLETCHER, R. and POWELL, M.J.D. (1963) A rapidly convergent descent method for minimization, Comput. J., Vol. 6, pp.163-168.

HANSEN, E. (1969) (Ed.) Topics in Interval Analysis, Clarendon Press.

HOUSEHOLDER, A.S. (1970) The Numerical Treatment of a Single Nonlinear Equation, McGraw-Hill.

KRASNOSEL'SKII, M.A., VAINIKKO, G.M., ZABREIKO, P.P., RUTITSKII, Ya. B., and STETSENKO, V. Ya. (1972) Approximate Solution of Operator Equations, Tr. by D. Louvish, Wolters-Noordhoff.

KUBA, D. and RALL, L.B. (1972) A Univac 1108 Program for Obtaining Rigorous Error Estimates for Approximate Solutions of Equations, MRC Tech. Summary Rpt. No. 1168, Univ. of Wisconsin.

MACDUFFEE, C.C. (1954) Theory of Equations, Wiley.

MILNE, W.E. (1949) Numerical Calculus, Princeton.

MOORE, R.E. (1966) Interval Analysis, Prentice-Hall.

ORTEGA, J.M. and RHEINBOLDT, W.C. (1970) Iterative Solution of Nonlinear Equations in Several Variables, Academic Press.

OSTROWSKI, A.M. (1966) Solution of Equations and Systems of Equations 2nd Ed., Academic Press.

RABINOWITZ, P. (1970) (Ed.) Numerical Methods for Nonlinear Algebraic Equations, Gordon and Breach.

RALL, L.B. (1969) Computational Solution of Nonlinear Operator Equations, Wiley.

ROCKNE, J. and LANCASTER, P. (1969) Automatic error bounds for the approximate solution of equations, Computing Vol. 4, pp.294-303.

TAPIA, R.A. (1971) The Kantorovich Theorem for Newton's Method, Am. math. Mon. Vol. 78, pp.389-392.

TAYLOR, A.E. (1958) Introduction to Functional Analysis, Wiley.

VAINBERG, M.M. (1964) Variational Methods for the Study of Nonlinear Operators, Tr. by A. Feinstein, Holden-Day.

WILKINSON, J.H. (1965) The Algebraic Eigenvalue Problem, Prentice-Hall.

ZAGUSKIN, V.L. (1961) Handbook of Numerical Methods for the Solution of Algebraic and Transcendental Equations, Tr. by G.O. Harding, Pergamon.

CHAPTER 17 NUMERICAL ANALYSIS OF METHODS FOR NONLINEAR INTEGRAL EQUATIONS

L.B. Rall

University of Wisconsin

1. Construction and Implementation of Numerical Methods for Nonlinear Integral Equations

When confronted with a mathematical problem, such as the solution of a nonlinear integral equation, the numerical analyst is faced with two tasks: (i) to obtain answers; (ii) to give some estimate of the reliability of the answers obtained. These objectives are usually approached by the construction of an algorithm and an error analysis, respectively. The first of these topics will be examined in this section, it being kept in mind that in producing an algorithm for solving a nonlinear integral equation, one usually tries to make it result in a "good approximation" or a "small error", so that the subjects of algorithm construction and error estimation are rarely independent.

The details of an algorithm will ordinarily depend on the specific integral equation, or type of equation, to be solved. However, it is possible to isolate a few general principles which have proved to be useful in the design of algorithms. Algorithms may be roughly classified into the three categories:

1. Analytic
2. Semi-analytic
3. Approximate.

As a first step, it is usually a good idea to investigate the possibility of solving the given integral equation

$$P(x)\ (s) \equiv \int_R f(s, t, x(s), x(t))\ dt = 0 \qquad (1.1)$$

by an analytic method, such as inversion of power series, successive substitutions, Newton's method, etc. (Anselone, 1964; see also Chapter 15), or if (1.1) has a variational principle (Vainberg, 1964), that is, the solution $x=x(s)$ of (1.1) maximizes (or minimizes) some functional $\phi=\phi(x)$, an algorithm to find the required extremal point of ϕ could then be used; this is called a variational method for solving (1.1). The drawback to analytic and variational methods is the need for the exact evaluation of nonlinear integral transforms, and perhaps also the solution of linear integral equations in their implementation. It is thus not possible in general to carry out the indicated operations explicitly to obtain the solution or an approximate solution in a useful form. However, one should consider the possibility of suitable representations of $x(s)$ and the other functions involved in (1.1) as infinite series

$$x(s) = \sum_{t=0}^{\infty} x_k g_k(s), \tag{1.2}$$

where the <u>basis functions</u> $g_0(s), g_1(s), \ldots$ are such that computer programs can be written to compute the coefficients of the transformed functions. For example, assume that in the Volterra integral equation

$$x(s) - \int_0^s f(s, t, x(t))\, dt = y(s), \tag{1.3}$$

the function $f(s, t, x(t))$ has the power series expansion

$$f(s, t, x(t)) = \sum_{k=0}^{\infty} c_k(s, t)\, x^k(t), \tag{1.4}$$

with coefficients $c_k(s, t)$ also having known power series expansions in s and t. Assuming also that

$$y(s) = \sum_{k=0}^{\infty} y_k s^k, \quad x(s) = \sum_{k=0}^{\infty} x_k s^k, \tag{1.5}$$

the substitution of these power series expansions into (1.3) will lead to an infinite nonlinear system of equations for the coefficients $x_0, x_1, \ldots$. However, although it may be complicated, it is <u>lower triangular</u> in form (Noble, 1964). That is, one finds that

$$x_k = h_k(y_k, x_0, x_1, \ldots, x_{k-1}), \quad k = 0, 1, 2, \ldots, \tag{1.6}$$

with each x_k depending only on the previously found values (x_0 depends only on y_0). As the functions h_k involve only the application of known rules for the manipulation of power series and the integration of monomials, they may be programmed for a digital computer. In practice, the application of a method of this type would necessarily be terminated after the computation of a finite number of coefficients x_t, the contribution of the remaining terms being considered to be "negligible".

If the operations required in an analytic algorithm are carried out only approximately, for example, if one uses numerical integration to valuate integral transforms, then the resulting method will be called <u>semi-analytic</u>. In many cases, semi-analytic methods are equivalent to <u>approximate</u> methods, in which the integral equation (1.1) is replaced by a similar equation

$$S(z) = 0 \tag{1.7}$$

in some Banach space Z, usually finite-dimensional. An approximate algorithm consists of three parts: (i) construction of the approximate equation (1.7); (ii) finding a solution z^* of the approximate equation; and (iii) using the value of z^* to obtain an approximate solution $x^{(0)}(s)$ of the integral equation (1.1).

In order to construct the approximate equation (1.7), two types of method are in common use, both of which lead to finite systems of nonlinear equations. One type is based on <u>numerical integration</u> and <u>collocation</u>. Here, a rule of numerical integration over the region R with <u>weights</u> w_j and <u>nodes</u> t_j, $j = 1, 2, \ldots, n$, is chosen, that is,

$$\int_R f(t)dt = \sum_{j=1}^{n} f(t_j)\, w_j + R(f), \tag{1.8}$$

the error term $R(f)$ is neglected, and by setting $s_i = t_i$, $i = 1, 2, \ldots n$ (collocation), one obtains the finite system of equations

$$\sum_{j=1}^{n} f(s_i, t_j, z_i, z_j)\, w_j = 0, \quad i = 1, 2, \ldots, n, \tag{1.9}$$

where $z_1, z_2, \ldots, z_n$ are the desired approximations to $x(s_1), x(s_2), \ldots, x(s_n)$.

Another class of procedures which lead to finite systems of equations consists of what are called <u>projection</u> methods. Here, one chooses a finite set $y_1(s), y_2(s), \ldots, y_n(s)$ of basis functions, and then attempts to determine constants $z_1, z_2, \ldots, z_n$ such that the function

$$z(s) = z_1\, y_1(s) + z_2\, y_2(s) + \ldots + z_n\, y_n(s), \tag{1.10}$$

belonging to the finite-dimensional space spanned by the basis functions, is considered to be a "good" approximation to a solution $x(s)$ of (1.1). Various criteria of "goodness" may be introduced, leading to different systems of equations for $z_1, z_2, \ldots, z_n$. For example, a <u>Galerkin method</u> is based on the requirement that $P(z)(t)$ be <u>orthogonal</u>

to functions $g_1(t), g_2(t),\ldots, g_n(t)$ with respect to a weight function $w(t)$; that is,

$$\int_R g_i(t)\ [P(z)(t)]\ w(t)\ dt = 0,\ i = 1, 2,\ldots, n. \tag{1.11}$$

It is common to take $g_i(t) = y_i(t)$, $i = 1, 2,\ldots, n$, but this is not essential. Another approach to the choice of the coefficients $z_1, z_2,\ldots, z_n$ in (1.10) is to minimize some functional $\phi(P(z))$, such as the norm of $P(z)$. This leads to variational, or minimization methods for obtaining approximate solutions of integral equations. In Hilbert space, minimizing the norm of $P(z)$ is called a least squares method. This minimization problem could be solved by some direct method, or by finding satisfactory solutions of the variational equations

$$\frac{\partial\phi(P(z))}{\partial z_i} = 0,\quad i = 1, 2,\ldots, n, \tag{1.12}$$

which will again be a system of n nonlinear equations in n unknowns.

The third stage of an approximate method is to use the information obtained, meaning the numbers $z_1, z_2,\ldots, z_n$, to construct an approximate solution $x^{(0)}(s)$ of the integral equation (1.1). In the case of projection methods, this presents no difficulty if, as usual, the functions $y_1(s), y_2(s),\ldots, y_n(s)$ in (1.10) belong to the space in which the solution $x(s)$ is sought. In this case, one simply takes $x^{(0)}(s) = z(s)$. The situation is somewhat different if numerical integration and collocation have been used, as there is no unique function corresponding to the finite set of values $z_1, z_2,\ldots, z_n$. Of course, these could be adequate as approximations to $x(s_i)$, $i = 1, 2,\ldots, n$, but, on the other hand, one may wish approximate values of $x(s)$ at points other than the collocation points, or need a function $x^{(0)}(s)$ as an approximate solution for further analytical work. In this case, a method of interpolation or approximation would be necessary. A function $x^{(0)}(s)$ is said to interpolate $z_1, z_2,\ldots, z_n$ if

$$x^{(0)}(s_i) = z_i,\quad i = 1, 2,\ldots, n. \tag{1.13}$$

There are many methods of constructing interpolating functions, using polynomials, splines, rational functions, and so on. If the integral equation (1.1) can be solved explicitly for $x(s)$, then one may make se of the so-called natural method of interpolation. For example, the rysohn equation

$$x(s) - \int_R f(s, t, x(t))\, dt = 0, \tag{1.14}$$

ay be replaced by the finite system

$$z_i - \sum_{j=1}^{n} f(s_i, t_j, z_j)\, w_j = 0, \ i = 1, 2, \ldots, n. \tag{1.15}$$

iven the solutions $z_1, z_2, \ldots, z_n$ of (1.15), it is easy to see that he function

$$x^{(0)}(s) = \sum_{j=1}^{n} f(s, t_j, z_j)\, w_j \tag{1.16}$$

atisfies the interpolation condition (1.13).

Instead of using interpolation, one may wish to have an approximate olution $x^{(0)}(s)$ having some global property rather than satisfying 1.13) exactly. In this case, one would attempt to find a function of he desired class which would minimize some functional of the eviations.

The central problem of implementation of approximate methods is, of ourse, the ability to solve the system of equations or minimize the unctional which results from the approximation of the integral equation see Chapter 16). One property of the approximating system (1.7) which s often helpful in this respect is consistency of the approximation in he sense that the same method of solution (iteration, etc.) can be hown to be applicable to both equations. Thus, if the integral quation (1.1) has a variational principle, one would also have a ariational principle for the approximate equation (1.7). An even tronger type of consistency is stated in the following definition.

efinition 1. The integral equation (1.1) and the approximating system 1.7) of n equations in n unknowns are said to be n-consistent if or all n sufficiently large, the existence of a solution $x^*(s)$ of 1.1) implies the existence of a solution $z^* = (z_1^*, z_2^*, \ldots, z_n^*)$ of (1.7), and conversely.

One way to demonstrate n-consistency is to show that the values of $x^*(s_i)$, $i = 1, 2, \ldots, n$, are sufficiently good initial approximations o ensure the convergence of an iteration process such as Newton's ethod to z^* (see Chapter 16), and that a function $x^{(0)}(s)$

constructed from $z_1^*, z_2^*, \ldots, z_n^*$ has the same property for the integral equation (see Chapter 15). A theory of n-consistency has been developed by Anselone (1971).

The choice of a computational algorithm is also always constrained by practical considerations of computational facilities and time available.

2. Error Analysis

Error estimation is a topic which permeates numerical analysis (Rall, 1965a). Error estimates are of two types; rigorous, and indicative. A rigorous error analysis produces a number ε and a guarantee that

$$||x^* - x^{(0)}|| \leq \varepsilon, \tag{2.1}$$

where $x^{(0)}$ is an approximate solution. An indicative error estimate is usually derived from order of magnitude arguments which contain constants which cannot be evaluated with precision, but are taken to have certain values. In many cases, rigorous error bounds tend to be pessimistic, with ε much larger than the true error. An indicative error bound may be smaller than the true error, however, it often gives a better idea of the size of the actual error. The choice between these two types of error bounds will depend on a number of factors, including the information available, the requirements of accuracy of the problem being solved, and the time and effort which can be spent on error estimation.

The user of an analytic solution algorithm is ordinarily faced only with truncation error. The algorithm specifies a sequence of functions $x_1(s), x_2(s), \ldots$ which converges to the solution $x^*(s)$. By taking $x^{(0)}(s) = x_n(s)$, one may use

$$||x^*(s) - x_n(s)|| \leq \tau_n \tag{2.2}$$

as the required error estimate. The values of τ_n are obtained from the convergence criteria for the algorithm (see Chapter 15).

In the case of a semi-analytic method, error also arises from the fact that the elements of the sequence $\{x_n(s)\}$ are themselves not calculated exactly. This computational error corresponds to round-off error in ordinary arithmetical calculations. If $y_n(s)$ is the sequence actually computed, and

$$\| x_n - y_n \| \leq \rho_n, \tag{2.3}$$

then one may use

$$\varepsilon = \rho_n + \tau_n \tag{2.4}$$

in (2.1) for the approximate solution $x^{(0)}(s) = y_n(s)$.

In the case of approximate methods, one encounters discretization error. The approximate solution $x^{(0)}(s)$ constructed by a method of this type will belong to a subset (often finite dimensional) of the space in which the solution $x^*(s)$ is sought. Thus, there will usually be a positive minimum for the value of ε in (2.1). Also, the solution of the approximate problem may not be the best possible in the subset considered. Discretization error is often difficult to estimate realistically. A technique which may be helpful is a posteriori error estimation. Here, if the approximate solution $x^{(0)}(s)$, regardless of how obtained, is at all tractable from an analytic standpoint, it may be possible to calculate the necessary quantities to obtain an error bound on the basis of semi-local convergence theorems, such as the ones for simple iteration or Newton's method (see Chapter 15).

Another way to estimate error for approximate methods is to use an inclusion procedure. In addition to solving the approximate problem $S(z) = 0$, one also solves

$$T(w) = 0, \tag{2.5}$$

it being known that

$$z(s) \leq x(s) \leq w(s), \tag{2.6}$$

thus giving error bounds as well as approximate solutions (Collatz, 1970). This type of method requires twice the effort in order to obtain an error estimate as compared to a simple approximate solution of the problem.

By solving more than one approximate problem one may also be able to obtain indicative error bounds in a different way. Suppose, for example, that the discretization error is thought to be of the order n^{-p}, p a positive integer. Conclusions of this type may be based on error terms for numerical integration (Davis and Rabinowitz, 1967), etc. By solving approximate problems for two values of n and comparing the number of digits which agree, one is led to an estimate of the error

(Noble, 1964).

One caution to observe in using numerical integration for the construction of approximate methods and to obtain error estimates is to observe singularities in the integrand or its derivatives. Thus, if the equation involves a Green's function, for example, then one should employ a rule of numerical integration suitable for such integrands, or, perhaps better, transform the problem into one in which there are no singularities in the region of integration (see Equations (1.8)-(1.10) of Chapter 16, also §6 of Chapter 15 for an example of this technique as applied to a Hammerstein integral equation). The following transformation technique also may sometimes be helpful. Suppose that the integrand in (1.1) is singular for $s = t$. Then, if

$$\int_R f(t, s, x(s), x(s))\, dt = g(s, x(s)) \tag{2.7}$$

can be obtained explicitly, one can replace (1.1) by the equation

$$g(s, x(s)) + \int_R \left\{ f(t, s, x(s), x(t)) - f(t, s, x(s), x(s)) \right\} dt = 0 \tag{2.8}$$

in which the integrand will ordinarily be regular at $s = t$.

3. Automatic Methods

The goal of an <u>automatic method</u> for solving a class of integral equations would be to produce computer programs which, given the integrand written in a computer language such as FORTRAN, a description of the region of integration, and other parameters, would give an approximate solution and an error estimate. The present situation is somewhat remote from this objective. However, using available programs for generating Taylor coefficients and the manipulation of power series, the implementation of the techniques suggested in §1 for the Volterra equation (1.3) may be practicable, and a program of this type may also be suitable for Hammerstein equations which can be reduced to Volterra equations. Error estimation could be based on the size of the first few neglected terms in the power series expansion of $x(s)$.

One obstacle to the development of computer programs to implement approximate methods for the solution of nonlinear integral equations is the lack of reliable algorithms for the solution of nonlinear systems of equations. An approach to remedy this defect would be to study the types of systems, such as (1.11) of Chapter 16, which arise specifically rom integral equations, and to develop methods for their solution.

Another approach to automatic methods is to program inclusion lgorithms. An example of this has been given by Rall (1965b). ormulation of inclusion methods in terms of interval analysis (Moore, 966) would give a wider range of applicability. Unfortunately, nclusion techniques tend to be time consuming.

. Examples

Space does not permit the inclusion of specific examples. A number ay be found in the paper by Noble (1964), and elsewhere in the iterature. One nonlinear integral equation which has been investigated ntensively from a numerical standpoint is the equation of radiative ransfer

$$H(\mu) = 1 + \frac{\varpi_0}{2} \mu H(\mu) \int_0^1 \frac{H(\mu')}{\mu+\mu'} d\mu', \qquad (4.1)$$

hich has been solved approximately by iteration (Chandrasekhar, 1960; oble, 1964; Rall, 1969), and also by Newton's method and inversion of ower series (Rall, 1969). An inclusion method bounding the solution of 4.1) rigorously by step functions has also been employed (Rall, 1965b).

. Acknowledgements

This work was supported by the Science Research Council of Great ritain, and the U.S. Army under Contract No. DA-31-124-ARO-D-462.

eferences

NSELONE, P.M. (1964) (Ed.) Nonlinear Integral Equations, Univ. of Wisconsin Press.

NSELONE, P.M. (1971) Collectively Compact Operator Approximation Theory and Applications to Integral Equations, Prentice-Hall.

HANDRASEKHAR, S. (1960) Radiative Transfer, Dover.

OLLATZ, L. (1971) Some applications of functional analysis to analysis, particularly to nonlinear integral equations, Rall (1971), pp. 1-43.

AVIS, P.J. and RABINOWITZ, P. (1967) Numerical Integration, Blaisdell.

OORE, R.E. (1966) Interval Analysis, Prentice-Hall.

OBLE, B. (1964) The numerical solution of nonlinear integral equations and related topics, Anselone (1964), pp. 215-318.

ALL, L.B. (1965a) (Ed.) Error in Digital Computation (2 vols.), Wiley.

ALL, L.B. (1965b) Numerical integration and the solution of integral equations by the use of Riemann sums, SIAM Rev. Vol.7,pp55-64.

RALL, L.B. (1969) Computational Solution of Nonlinear Operator Equations, Wiley.

RALL, L.B. (1971) (Ed.) Nonlinear Functional Analysis and Applications, Academic Press.

VAINBERG, M.M. (1964) Variational Methods for the Study of Nonlinear Operators, Tr. by A. Feinstein, Holden-Day.

CHAPTER 18 PROVISION OF LIBRARY PROGRAMS FOR THE NUMERICAL SOLUTION OF INTEGRAL EQUATIONS

G.F. Miller

National Physical Laboratory

. Introduction

In discussing general numerical methods we have encountered ntegral equations of a great many different types, including linear redholm and Volterra equations of first and second kind, the associated igenvalue problem for Fredholm equations, systems of equations, onlinear equations, equations with singular kernels, equations in which he range (region) of integration is infinite or extends over more than ne dimension, and integro-differential equations.

It is not my intention here to try to sum up what has been said on ach of these topics or to select "best buys". The object will be ather to sketch the situation which now exists with regard to library rograms or algorithms, to discuss what else needs to be done, and to ndicate some of the steps which are being taken to fill in gaps.

We ought also to consider the overall philosophy and strategy of roviding algorithms for integral equations. For which types of quation are algorithms most needed? How general should we seek to ake them? What degree of understanding of the problem should we xpect of the user? Is the theoretical situation in some cases so omplex that it is undesirable to rely on "black boxes"? I am, of ourse, unable to provide definite answers to all of these questions ut will venture a few comments in the hope of provoking discussion and specially of encouraging potential algorithm users to make their needs nown.

There appear to be comparatively few published algorithms for the olution of integral equations. Three which we believe to be among the ost generally useful are described in §§3-5. The shortage may be artly due to the fact that one can manage reasonably well by applying suitable quadrature formula and then solving the resulting algebraic ystem with the aid of one of the many algorithms which exist for that urpose. This is no doubt an over-simplification but at least it can be aid that the problems tend to fall naturally into two or more stages.

In fact there is a good deal to be said for not trying to produce omplicated algorithms in which the various elements are fused together.

It is difficult to allow for every contingency in an algorithm, whereas if suitable methods are to hand it may be comparatively easy to put together programs for the solution of individual problems. Therefore in many cases the user may be best served by providing him with good methods which he can then adapt to meet particular needs. Alternatively, algorithms might be constructed in modular form and combined together as required.

On the other hand there are certain types of integral equation for which the provision of comprehensive library algorithms appears highly desirable. It seems worth remarking, too, that a well-designed algorithm can be much more than just a "black box"; if the associated documentation is well written it can offer real guidance and even stimulate the user to think about his problem.

Let us first briefly examine how these considerations apply to various types of integral equation.

2. Brief Review of Types of Problem

(i) Fredholm equation of second kind and eigenvalue problem. With regard to linear, nonsingular equations over a finite range in one dimension, we are well served by the algorithm of Elliott and Warne (1967), based on the use of Chebyshev series, which is capable of yielding high accuracy (see §3). The only significant difficulty arises when the parameter λ lies near an eigenvalue, when the solution becomes poorly determined. It was suggested by Professor Ben Noble, at a recent seminar at Nottingham, that an algorithm might be made to detect this situation - by examining the successive pivots in Gaussian elimination, for example.

(ii) Volterra equations of second kind. The procedure of Pouzet (described in §4) deals with a very general class of nonlinear equation of Volterra type. However, for this reason it does not appear to be very efficient in application to linear equations, especially those of convolution type. Further provision needs to be made for these and also for several common types of singular equation.

(iii) Equations of first kind. It would be perfectly feasible to write an algorithm to implement, say, the regularization method described in Chapter 13. The algorithm writer would have to decide on a particular choice of the norms and a particular smoothing functional, but otherwise a good deal of latitude could be left to the user. Unfortunately the problems in this category tend to be very diverse and seldom conform to the simple models we assume in theory. Even the most versatile algorithm would probably need to be modified to meet individual circumstances - or the problems manipulated to fit the algorithms.

iv) Systems of integral equations. Methods developed for single equations can usually be generalized to systems. However, an algorithm applicable to systems is likely to be unnecessarily complicated when used to solve a single equation; hence it is desirable to have separate algorithms. There exists an algorithm for the solution of systems of linear Volterra equations, that of Rumyantsev (1965) described in §5.

v) Nonlinear equations. Those of Volterra type are, generally speaking (we disregard possible theoretical complications, instability, etc.), not too difficult to handle since we can proceed one step at a time much as in the solution of initial-value problems for ordinary differential equations. Nonlinear equations with fixed limits are frequently solved by a method of successive approximation involving linearization at every stage, so that algorithms designed for the solution of linear equations may also find a use here. Two ALGOL procedures for equations of Hammerstein type are to be found in C.N.R.S. (1970).

(vi) Singular kernels. Perhaps the most pressing need at the moment is for algorithms for the efficient treatment of singularities of various standard types (e.g. power, logarithmic and Cauchy-type singularities). One approach to this problem is discussed in §7. Consideration may also be given to producing routines which will deal reliably, if less economically, with a broader class of singularities, as discussed in §6.

(vii) Infinite range. This may sometimes be dealt with by truncating at a suitable point, but often there are fundamental difficulties which call for special analysis and methods. Ideally one would like to develop algorithms which automatically detect the rate of convergence and take appropriate action, but these are not yet in sight.

(viii) Multidimensional equations. Such problems are greatly complicated by the need to represent curves, surfaces, etc., and to deal with the singularities of the kernel which are nearly always present. It is not easy to conceive of a library algorithm having wide applicability which is at the same time reasonably concise and convenient to use. This is not, of course, to deny the usefulness of constructing programs for the solution by integral equation methods of particular classes of mathematical problems (e.g. conformal mapping, or certain boundary-value problems).

(ix) Kernels with special structure. When constructing algorithms it is desirable, both on theoretical grounds and in the interests of efficiency, to accord special treatment to equations whose kernels have special properties. Among the most important cases are those in which the kernel is symmetric, of convolution type, or periodic - as when the path of integration is originally a closed contour.

We proceed now to a description of the three published algorithms of Elliott and Warne (1967), Pouzet (1964) and Rumyantsev (1965); all are in ALGOL.

3. Procedure of Elliott and Warne

This solves a linear Fredholm equation of the second kind

$$f(x) - \lambda \int_a^b k(x,y)\ f(y)\ dy = g(x), \quad a \leqslant x \leqslant b, \tag{3.1}$$

or the corresponding Volterra equation in which the upper limit of integration b is replaced by x. The kernel $k(x,y)$ is supposed nonsingular and sufficiently well behaved to permit approximation by a polynomial of moderate degree in y for each x. Exceptionally its derivative with respect to y may possess a simple discontinuity at $y = x$ (see below).

For simplicity of exposition let us assume that the range has been normalized to $(-1, 1)$. The method on which the algorithm is based is then as follows. It is assumed that the solution can be approximated by a finite Chebyshev series

$$f(x) = \sum_{r=0}^{n} a_r\ T_r(x), \quad -1 \leqslant x \leqslant 1. \tag{3.2}$$

Substituting this expression into the integral equation and setting $x = x_i = \cos(i\pi/n)$, $i = 0, 1, \ldots, n$, one obtains a system of $n+1$ linear equations in the $n+1$ unknowns a_j,

$$A\underline{a} = \underline{g} , \quad \text{say},$$

where the $(i + 1, j + 1)$th element of the matrix A is given by

$$T_j(x_i) - \lambda \int_{-1}^{1} k(x_i, y)\, T_j(y)\, dy. \tag{3.3}$$

The integrals in (3.3) are evaluated approximately by expanding $k(x_i, y)$ also as a Chebyshev series and evaluating analytically the integrals of the resulting products $T_j(y)\, T_k(y)$.

For the practical applicability of the method the series for $k(x_i, y)$ should converge reasonably rapidly, and hence the early y-derivatives should be continuous. However, special provision is made for the commonly occurring situation in which the first derivative has a simple discontinuity at $y = x$, and for the case of a Volterra equation. This is achieved by allowing the kernel to have the form

$$k(x, y) = \begin{cases} k_1(x, y), & -1 \leqslant y \leqslant x, \\ k_2(x, y), & x < y \leqslant 1, \end{cases} \tag{3.4}$$

and splitting the integral in (3.3) in the form

$$\int_{-1}^{x_i} k_1(x_i, y)\, f(y)\, dy + \int_{x_i}^{1} k_2(x_i, y)\, f(y)\, dy. \tag{3.5}$$

However, both the functions k_1 and k_2 are now required to be defined, and sufficiently well behaved, on the entire square $-1 \leqslant x, y \leqslant 1$, and this represents something of a limitation.

The algorithm provides no indication of the error; in practice, however, a good estimate may often be obtained by varying the order of the Chebyshev series approximations and observing the rate of decrease of the coefficients a_r.

We note an error in the published algorithm; on p. 220, line 7, the + sign should be changed to − before $C[kp + 1]$.

4. <u>Procedure of Pouzet</u>

This solves a single nonlinear Volterra integral equation of a very general type:

$$\phi(x) = F\left(x, \int_{x_0}^{x} g(x, s, \phi(s))\, ds\right), \tag{4.1}$$

where F and g are well-behaved functions. The algorithm employs two types of method commonly applied to the solution of ordinary differential equations, namely

(a) one-step methods of Runge-Kutta type, and

(b) multi-step methods of Adams type.

Their extension to integral equations is treated in an earlier paper of the author (Pouzet, 1963). The user has the choice of (a) or (b); moreover, the order of the method is at his disposal provided that he can supply the appropriate set of coefficients. The coefficients actually given correspond to order 4 for method (a) and order 5 for method (b). If (b) is selected, (a) must in any case be used to obtain starting values. The user is required to choose a suitable interval of integration and no error estimate is provided.

This procedure also contains an error. On p. 170 on the third line following the comment, N should be replaced by Q.

The algorithm is useful by virtue of its considerable generality; it could readily be adapted to solve systems of equations (see Pouzet, 1963). On the other hand it would appear to be uneconomical in application to linear equations, particularly those of convolution type, and it has the weakness that it applies the same extrapolatory-type formulae over ranges where the integrand is known as over those where it is as yet unknown.

The two volumes of algorithms C.N.R.S. (1967) and C.N.R.S. (1970) include further ALGOL procedures in which similar Runge-Kutta methods are applied to integro-differential equations and, with the use of invariant imbedding techniques, to linear Fredholm equations.

5. Procedure of Rumyantsev

This procedure solves a system of linear Volterra equations of the second kind. The unknown functions are approximated by piecewise quadratic polynomials, so that with the use of a moderate interval the accuracy attained is comparatively modest. Here the user is required to specify accuracy parameters, namely bounds on the residuals for each equation, and the interval is broken down automatically until this specification is met. (We have not yet tested this procedure; I understand from Mr. R.T. Delves of Chelsea College, London, that it too contains errors.)

6. Alternative Methods using Chebyshev Series

There is another method employing Chebyshev series, due to El Gendi (1969), which performs very similarly to that of Elliott and Warne but is considerably simpler to apply. This essentially relates to non-singular equations but may also serve as a starting point for the development of algorithms for particular types of singular equation. It is based on the use of a quadrature formula

$$\int_{-1}^{1} \phi(x)\ dx \simeq \sum_{j=0}^{n} c_j^{(n)}\ \phi(x_j), \tag{6.1}$$

in which the x_j are the Chebyshev points, $\cos(\pi j/n)$, and the coefficients $c_j^{(n)}$ are chosen so that the formula is exact if $\phi(x)$ is a polynomial of degree n. Although the Chebyshev polynomials do not appear here explicitly, the formula is equivalent to the result of expanding $\phi(x)$ as a Chebyshev series and integrating term by term; it is closely related to the Clenshaw-Curtis method of integration.

If we apply the formula (6.1) to the integral equation (3.1), with $x = x_i$, $a = -1$, $b = 1$, we obtain

$$\int_{-1}^{1} k(x_i, y)\ f(y)\ dy \simeq \sum_{j=0}^{n} c_j^{(n)}\ k(x_i, x_j)\ f(x_j). \tag{6.2}$$

Hence the integral equation can be reduced to an approximating system of linear algebraic equations. It is a simple matter to derive the Chebyshev coefficients for $f(x)$, if required, from its values at the points x_i. This method, like that of Elliott and Warne, can be adapted to take account of a simple discontinuity in $k(x, y)$ or its derivative at $y = x$, but again the functions $k_1(x, y)$ and $k_2(x, y)$ of equation (3.4) need to be defined throughout $a \leq x, y \leq b$.

Before proceeding to extend these ideas to singular equations let us note that there are some situations in which the basic strategy of Elliott and Warne may be preferred, notably when the solution itself is known to be well behaved (and hence representable as a finite Chebyshev series) but the kernel has some irregular behaviour which is not of a precisely defined type or which we do not wish to analyse in detail. Their approach has the attractive feature that the moment integrals

$$\int_{-1}^{1} k(x, y)\ T_j(y)\ dy \tag{6.3}$$

are integrals of known functions and so can be evaluated quite independently of the actual process of solution, by whatever method is appropriate. We are not obliged to resort to the method of the published algorithm which assumes that the kernel is well behaved. We could, for example, construct a very general algorithm by use of an adaptive-type quadrature routine.

Here we may also mention an alternative method for treating problems involving a split kernel of the type (3.4), which does not require k_1 and k_2 to be defined over the whole square but only in their appropriate subregions. If we split the range of integration (-1, 1) at x_i and apply formula (6.1) to each of the resulting integrals we obtain, typically,

$$\int_{-1}^{x_i} k(x_i, y)\ f(y)\ dy \simeq \frac{1 + x_i}{2} \sum_{j=0}^{m} c_j^{(m)}\ k(x_i, y_{ij})\ f(y_{ij}), \tag{6.4}$$

where the y_{ij} $(j = 0, 1, \ldots, m)$ are the Chebyshev points for the range $(-1, x_i)$; the integer m may differ from the integer n determining the number of points x_i. Taken as they stand, the set of $n + 1$ equations corresponding to $i = 0, 1, \ldots, n$ would involve a number of unknown function values greatly in excess of $n + 1$, since the points y_{ij} clearly depend on i. However, if we set

$$f(y) = \tfrac{1}{2}a_0 + \sum_{r=1}^{n} a_r\ T_r(y) \tag{6.5}$$

and then express each of the quantities $f(y_{ij})$ in this form and rearrange the resulting double series, we obtain the requisite number o unknowns a_r, and the algebraic linear system can now be solved.

In fact it is convenient to use, instead of equation (6.1), a corresponding "open-type" formula, in which use of the end-points is avoided, since this facilitates the treatment of kernels having, for example, a weak singularity at $y = x$ or $y = \pm 1$. An algorithm along these lines exists at the National Physical Laboratory; further details will be reported elsewhere.

7. Treatment of Singular Kernels

The quadrature formula (6.1) can readily be generalized to cases i which the integrand includes an absolutely integrable weight factor $w(x)$. As shown by Kussmaul (1972), the approximation

$$\int_{-1}^{1} w(x)\ \phi(x)\ dx \simeq \sum_{j=0}^{n}{}'' C_j^{(n)}\ \phi(x_j) \tag{7.1}$$

becomes exact in the case when $\phi(x)$ is a polynomial of degree n if the coefficients are defined by:

$$C_j^{(n)} = \frac{2}{n} \sum_{k=0}^{n}{}'' \left\{ \int_{-1}^{1} w(x)\ T_k(x)\ dx \right\} T_k(x_j). \qquad (7.2)$$

The double prime on the summation symbols here indicates that the irst and last terms are to be halved.) Applying this result to an ntegral equation with kernel $w(y)\ L(x, y)$ we have

$$\int_{-1}^{1} w(y)\ L(x_i, y)\ f(y)\ dy \simeq \sum_{j=0}^{n}{}''\ C_j^{(n)}\ L(x_i, x_j)\ f(x_j), \qquad (7.3)$$

rovided that the product $L(x_i, y)\ f(y)$ behaves like a polynomial of egree n.

However, since the position of a singularity in the kernel almost nvariably depends on x, we need to go a step further and consider eighting functions $w(x, y)$. It is also desirable to take account of ernels which are not integrable in the usual sense; in particular those ontaining a singularity of Cauchy type. These extensions can be fitted nto the same framework, the only essential difference being that we now ave coefficients $C_{ij}^{(n)}$ which depend on i as well as j. We have ritten programs to evaluate the coefficients $C_{ij}^{(n)}$ in the cases where (x, y) has the forms

$$\frac{1}{(1 - y^2)^{\frac{1}{2}}\ (x - y)} \quad \text{and} \quad \frac{\log|x - y|}{(1 - y^2)^{\frac{1}{2}}},$$

hich arise, for example, in certain problems of potential theory and lasticity. The first case is also treated in a slightly different way i.e. using different points x_i) by Erdogan and Gupta (1972).

An attractive feature of the approach here described is that it is asily applicable to (systems of) integral equations exhibiting more than ne type of singularity, since all the formulae involve, for the same hoice of n, the same set of points x_i. Hence algorithms for the olution of a wide variety of equations may readily be constructed.

I am indebted to my colleague, Susan Hill, for her substantial elp in the writing and testing of algorithms.

eferences

.N.R.S. (1967, 1970) Procedures ALGOL en Analyse Numerique. Volumes I and II. Centre National de la Recherche Scientifique: Paris.

L GENDI, S.E. (1969) Chebyshev solution of differential, integral and integro-differential equations. Comput. J. Vol. 12, pp. 282-287.

ELLIOTT, D. and WARNE, W.G. (1967) An algorithm for the numerical solution of linear integral equations. Int. Comput. Cent. Bull. Vol. 6, pp. 207-224.

ERDOGAN, F. and GUPTA, G.D. (1972) On the numerical solution of singular integral equations. Q. appl. Math. Vol. 29, pp. 525-534.

KUSSMAUL, R. (1972) Clenshaw-Curtis quadrature with a weighting function. Computing Vol. 9, pp. 159-164.

POUZET, P. (1963) Etude en vue de leur traitement numérique des integrales de type Volterra. Chiffres Vol. 6, pp. 79-112.

POUZET, P. (1964) Algorithme de résolution des équations integrales de type Volterra par des méthodes par pas. Chiffres Vol. 7, pp. 169-173.

RUMYANTSEV, I.A. (1965) Programme for solving a system of Volterra integral equations (of the second kind). USSR Comput. Math. and Math. Phys. Vol. 5, pp. 218-224.

PART 3

APPLICATIONS

CHAPTER 19 SINGULAR INTEGRALS AND BOUNDARY VALUE PROBLEMS

D. Kershaw

University of Lancaster

1. Singular Integral Equations

In this chapter we shall consider a variety of problems which arise in potential theory. It will be clear by now that there is no lack of methods for solving integral equations and so it is important to be aware of techniques by which they can be found. The most powerful of these is probably the theory of singular integral equations. This theory was developed by Mikhlin, Muskhelishvili and their co-workers for application to the potential problems which arise in hydrodynamics and elasticity theory. The basic references are Mikhlin (1957, 1962), Muskhelishvili (1953a, b); see also the text book by Gakhov (1966).

The integral

$$\int_0^1 \frac{\mu(t)}{t-x}\,dt \tag{1.1}$$

is not defined in the usual sense for $0 < x < 1$, however a meaning can be attached to it by defining it as

$$\lim_{\xi\to 0+}\left\{\int_0^{x-\xi} + \int_{x+\xi}^1 \frac{\mu(t)}{t-x}\,dt\right\}. \tag{1.2}$$

If μ satisfies a Lipschitz condition on $(0, 1)$ then this limit exists. The integral is called a Cauchy principal value integral. For example, if $\mu = 1$, then for $0 < x < 1$

$$\int_0^1 \frac{dt}{t-x} = \lim_{\xi\to 0+}\left\{[\log\xi - \log x] + [\log(1-x) - \log\xi]\right\} = \log(\frac{1-x}{x}).$$

It is a straightforward matter to generalize this concept in order to define the integral

$$\int_C \frac{\mu(\zeta)}{\zeta-z}d\zeta,\ z \in C, \tag{1.3}$$

where C is a smooth (in fact Lipschitz continuous) curve which need not be closed. The principal value of this integral is defined to be

$$\lim_{\xi\to 0} \int_{C_\xi} \frac{\mu(\zeta)}{\zeta-z}\, d\zeta, \qquad (1.4)$$

where C_ξ is C with the part common to C and the circle $|z - \zeta| < \xi$ removed. (If C is not closed then z cannot be an end point of C unless μ vanishes there.)

Suppose now that $z \notin C$ in (1.3) and define

$$w(z) = \frac{1}{\pi i} \int_C \frac{\mu(\zeta)}{\zeta-z}\, d\zeta. \qquad (1.5)$$

The orientation of C is in the usual anti-clockwise direction, and so if C is a closed curve then the interior D of the domain bounded by C will lie to the left as C is traversed in the positive direction. When C is not closed the interior can be thought of as lying to the left as before. (More precise definitions will be found in Gakhov, 1966.)

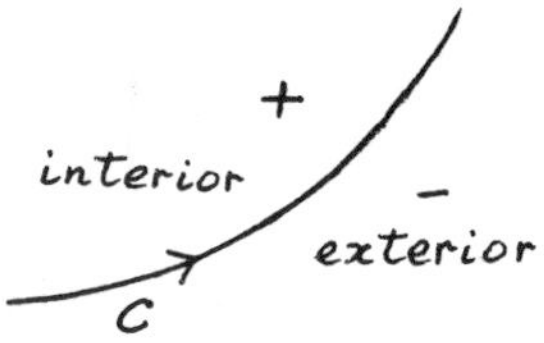

Clearly w as given by (1.5) will be analytic except on C. Designate by w_+ the limiting value of w as z approaches a point of C along a path which lies wholly inside D; w_+ is the <u>interior limit</u> of w. In a similar fashion define w_-, the <u>exterior limit</u>. It can be shown without much difficulty (Gakhov, 1966) that

$$w_+(z) = \mu(z) + \frac{1}{\pi i} \int_C \frac{\mu(\zeta)}{\zeta-z}\, d\zeta, \quad z \in C \qquad (1.6)$$

$$w_-(z) = -\mu(z) + \frac{1}{\pi i} \int_C \frac{\mu(\zeta)}{\zeta-z}\, d\zeta, \quad z \in C, \qquad (1.7)$$

where the integrals are Cauchy principal values.

These formulae are known as <u>Plemelj's formulae</u> (or Sokhotski's formulae according to Gakhov) and are basic to the theory. The <u>density function</u> μ is assumed to satisfy a Lipschitz condition on C.

2. Dirichlet's Problem

As a simple example of the use of Plemelj's formula we shall obtain an integral equation formulation for Dirichlet's problem for Laplace's equation. The problem is as follows.

'Let D be the interior of a simple connected domain in R^2 and let C be its boundary; find a function u which is harmonic in D

and which takes prescribed values on C.'

Let $u = f$ on C and write $u = \mathrm{Re}[w]$ where

$$w(z) = \frac{1}{\pi i} \int_C \frac{\mu(\zeta)}{\zeta - z}\, d\zeta, \quad z \in D, \tag{2.1}$$

the density function μ is assumed to be real (this is always possible without loss of generality). The first of the Plemelj formulae (1.6) gives the limiting value of w on C and so we have

$$f(z) = \mathrm{Re}[w_+(z)] = \mathrm{Re}\left\{\mu(z) + \frac{1}{\pi i} \int_C \frac{\mu(\zeta)}{\zeta - z}\, d\zeta\right\}, \quad z \in C,$$

which, since μ is real, can be written as

$$\begin{aligned} f(z) &= \mu(z) + \frac{1}{\pi} \int_C \mu(\zeta)\, \mathrm{Im}\left(\frac{d\zeta}{\zeta - z}\right) \\ &= \mu(z) + \frac{1}{\pi} \int_C \mu(\zeta)\, \mathrm{Im}\left(\frac{\zeta'}{\zeta - z}\right) d\sigma \end{aligned} \tag{2.2}$$

where σ denotes arc length.

It is not difficult to show (see Mikhlin, 1957) that the kernel of the integral equation (2.2) is continuous in σ except at points of discontinuity of curvature of C. At these points when $\sigma = s$ the kernel suffers a step discontinuity. A proof will be found in Mikhlin that (2.2) has a unique solution μ. Once μ has been found u can be tabulated with the aid of (2.1).

This integral equation can be used to provide a means of calculating Green's function for D. (It will have been noticed that had we supposed u to be the imaginary part of w then we would have arrived at a <u>singular</u> integral equation of the first kind for μ, namely

$$f(z) = -\frac{1}{\pi} \int_C \mu(\zeta)\, \mathrm{Re}\left(\frac{\zeta'}{\zeta - z}\right) d\sigma, \quad z \in C.)$$

3. Two Fredholm Integral Equations of the First Kind

Equations of the first kind and methods for their solution will be dealt with in another chapter, however as an introduction to them and a warning of their frequent ill-posedness we shall derive the explicit solution of the equation

$$u(x) = \frac{1}{\pi} \int_0^1 \frac{(t-x)}{(t-x)^2+a^2} f(t)dt, \; -\infty < x < \infty, \; a > 0. \tag{3.1}$$

We assume that u is given; it is clear from its representation that it is analytic except on the line segments $(\pm ia, 1 \pm ia)$ and that it vanishes at infinity. Thus we have conditions to be satisfied by u before a solution can be contemplated.

This integral equation arises in the problem of finding the current distribution in a thin superconducting strip (Rhoderick and Wilson, 1962). The function u represents the horizontal force induced by the current at a point (x, a).

The equation arises also in geophysics and was considered by Bateman (1956) in the case when the range of integration was $(-\infty, \infty)$, and a solution was given for this case. We shall use Plemelj's formula to provide an alternative derivation of his result.

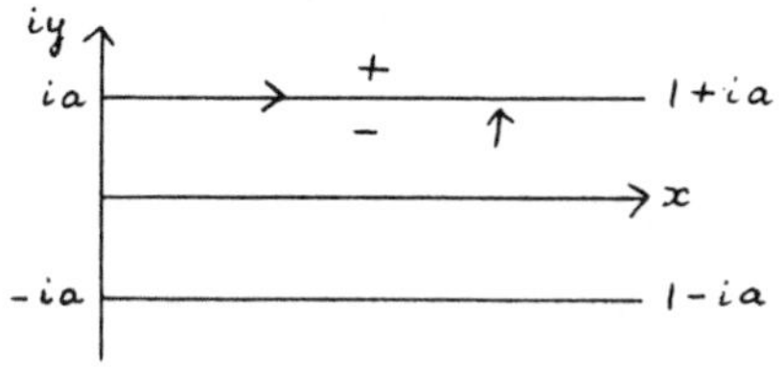

Clearly we can write for $z \notin (\pm ia, 1 \pm ia)$

$$u(z) = \frac{1}{2\pi} \int_0^1 \frac{1}{t-z-ia} f(t)dt + \frac{1}{2\pi} \int_0^1 \frac{1}{t-z+ia} f(t)dt.$$

Take now the exterior limit as z approaches $x + ia$, $0 < x < 1$, then with the aid of (1.7) we have

$$u(x + ia) = \frac{1}{2\pi} \int_0^1 \frac{1}{t-x-2ia} f(t)dt - \frac{i}{2} f(x) + \frac{1}{2\pi} \int_0^1 \frac{f(t)}{t-x} dt, \; 0 < x < 1,$$

the imaginary part of which gives

$$f(x) = -2 \, \text{Im}[u(x+ia)] + \frac{2a}{\pi} \int_0^1 \frac{f(t)}{(t-x)^2+4a^2}, \; 0 < x < 1. \tag{3.2}$$

This is an integral equation of the second kind for f.

However it is easily shown from (3.1) that

$$\int_{-\infty}^{\infty} \frac{(t-x)}{(t-x)^2+a^2} u(t)dt = -\, 2a \int_0^1 \frac{f(t)}{(t-x)^2+4a^2} dt,$$

from which we obtain the solution as

$$f(x) = -2\ \mathrm{Im}\left[u(x+ia)\right] - \frac{1}{\pi}\int_{-\infty}^{\infty} \frac{(t-x)}{(t-x)^2+a^2}\, u(t)dt. \qquad (3.3)$$

Each of the representations requires an analytic continuation of u to the line $(ia, 1 + ia)$, however (3.2) does not use u on the whole line and may be more suitable for computation.

In a similar fashion we can solve the equation

$$\nu(x) = \frac{1}{\pi}\int_0^1 \frac{ag(t)}{(t-x)^2+a^2}\, dt, \quad -\infty < x < \infty. \qquad (3.4)$$

The solution can be shown to satisfy the following equation

$$g(x) = 2\ \mathrm{Re}\left[\nu(x+ia)\right] + \frac{1}{\pi}\int_0^1 \frac{(t-x)g(t)}{(t-x)^2+4a^2}\, dt, \ 0 < x < 1, \qquad (3.5)$$

and it is given explicitly by

$$g(x) = 2\ \mathrm{Re}\left[\nu(x+ia)\right] - \frac{1}{\pi}\int_{-\infty}^{\infty} \frac{a\nu(t)}{(t-x)^2+a^2} dt. \qquad (3.6)$$

The ill-posedness of analytic continuation provides a good reason for the fear that these Fredholm equations are unpleasant. However if a is small then the process might not be too bad - this is reflected in the observation that when a is small the kernels of (3.1) and (3.4) will peak around $t - x = 0$.

4. Flow around a Hydrofoil

Although the representation of an analytic function given in (1.5) is undoubtedly useful in setting up integral equations the well known integral of Cauchy can often be used to produce a similar result.

As the first example of this we consider the problem of finding the flow pattern in a fluid which moves over an infinite cylinder of cross section D and boundary C. The fluid moves steadily at right angles to the generators of the cylinder.

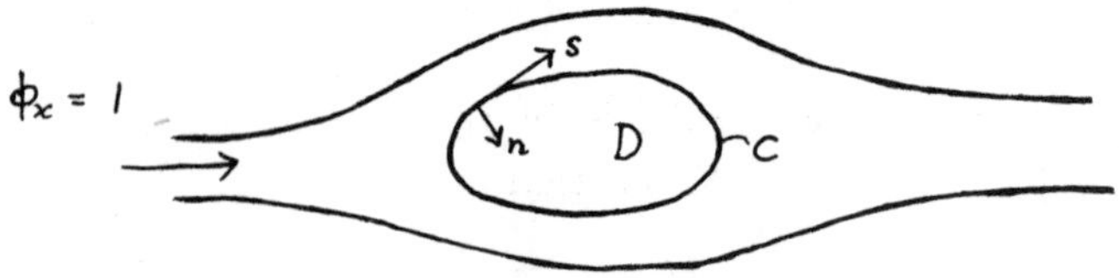

Suppose the fluid moves from left to right, parallel to the x-axis and with unit speed at infinity. The problem is then to find ϕ, harmonic <u>outside</u> D and such that $\phi_x = 1$ at $\pm\infty$ and $\phi_n = 0$ on C (n is in the direction of the normal at a point of C).

Let $\phi + i\psi = w$ where ψ is the conjugate function to ϕ. Then from the Cauchy-Riemann equations, $\phi_n = -\psi_s$. Consequently on C we have $w_s = \phi_s$. It follows that w is to be analytic outside D and such that $w_x = 1$ at $\pm\infty$, $w_s = \phi_s$ on C.

From Cauchy's integral formula

$$w'(z) = 1 + \frac{1}{2\pi i} \int_C \frac{w'(\zeta)}{\zeta - z}\, d\zeta, \quad z \text{ outside } \bar{D}.$$

Let now z tend to a point of C; this will be the interior limit with the orientation of the diagram, then it is easily seen that

$$w'(z) = 2 + \frac{1}{\pi i} \int_C \frac{w'(\zeta)}{\zeta - z} d\zeta, \quad z \in C. \tag{4.1}$$

But $w' = \frac{dw}{dz} = \frac{dw}{ds} \cdot \frac{d\bar{z}}{ds}$, and so on C we have $w'(z) = u(s)\bar{z}'(s)$ where $u = \phi_s$, the speed of flow on C. Hence we obtain the relation which holds on C,

$$u(s) = 2z'(s) + \frac{z'(s)}{\pi i} \int_C u(\sigma) \frac{d\sigma}{\zeta - z}, \quad z \in C. \tag{4.2}$$

The real part of (4.2) gives the following Fredholm equation of the second kind for u

$$u(s) = 2x'(s) + \frac{1}{\pi} \int_C u(\sigma)\, \mathrm{Im}\left(\frac{z'}{\zeta - z}\right) d\sigma. \tag{4.3}$$

The kernel of this integral equation is clearly related to that of (2.2) and indeed the remarks on the latter are valid for the former. However, due to the presence of circulation the solution of the integral equation is not unique (when C is a circle of unit radius the integral equation reduces to

$$u(s) = 2 \sin s + \frac{1}{2\pi} \int_0^{2\pi} u(\sigma)\, d\sigma \Big).$$

When the condition for no circulation is imposed, $\int_C u(\sigma)\, d\sigma = 0$, then the solution will be unique. The flow outside the cylinder can be calculated from (4.1) once u has been found. (This integral equation is known, however a convenient reference to it has not been located.)

We note that as in the Dirichlet problem an integral equation of the first kind can be deduced from (4.2) by taking the imaginary part. The kernel in this case will be singular.

5. The Dock Problem

As the final example of the use of Plemelj's formulae and Cauchy's integral formula we consider the semi-infinite and doubly infinite dock problems which occur in the theory of small amplitude surface waves.

Fluid occupies the lower half plane $y < 0$ and a semi-infinite dock occupies $-\infty < x < 0$, $y = 0$; the positive x-axis is a free surface.

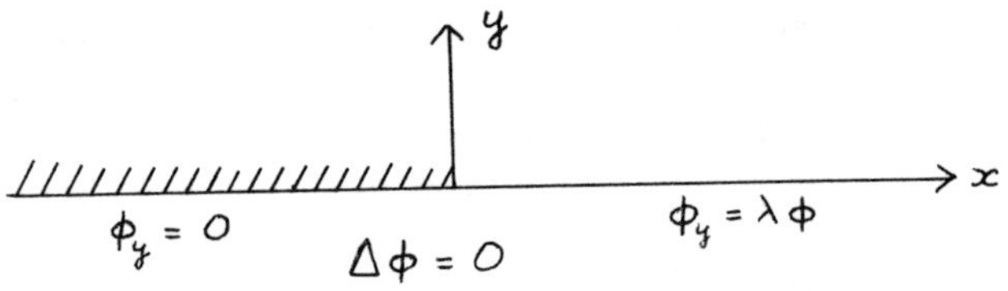

The problem is to find the form of the small amplitude surface waves on $x > 0$, $y = 0$. Friedrichs and Lewy (1948) consider this problem and produce the solutions in explicit form. We shall show that the integral equation formulation is simpler for computation.

Small amplitude waves are considered in Chapter 22 and we quote the boundary value problem without trying to derive it. (See Holford, 1964.)

It is required to find ϕ, harmonic for $y < 0$ and such that

$\phi_y = 0$ for $x < 0$, $y = 0$ (no flow through the dock)

$\phi_y - \lambda\phi = 0$ for $x > 0$, $y = 0$.

Let ψ be the conjugate of ϕ and put $w = \phi + i\psi$. Then

$$w'(z) = \phi_x + i\psi_x = \phi_x - i\phi_y.$$

Consequently on $y = 0$ we have

$$w'(x) = \phi_x, \quad x < 0$$

$$w'(x) = \phi_x - i\lambda\phi, \; x > 0.$$

It follows from Cauchy's integral formula that

$$w'(z) = \frac{1}{2\pi i}\int_{\infty}^{-\infty} \frac{w'(t)}{t-z}\,dt, \text{ where } \operatorname{Im}(z) < 0.$$

Take now the interior limit, this gives easily the result that

$$w'(x) = -\frac{1}{\pi i}\int_{-\infty}^{\infty} \frac{w'(t)}{t-x}\,dt, \; -\infty < x < \infty, \; x \neq 0. \tag{5.1}$$

In more detail we have

$$\phi'(x) = -\frac{1}{\pi i}\int_{-\infty}^{0} \frac{\phi'(t)}{t-x}\,dt - \frac{1}{\pi i}\int_{0}^{\infty} \frac{\phi'(t)-i\lambda\phi(t)}{t-x}\,dt, \; -\infty < x < 0, \tag{5.2}$$

$$\phi'(x)-i\lambda\phi(x) = -\frac{1}{\pi i}\int_{-\infty}^{0} \frac{\phi'(t)}{t-x}\,dt - \frac{1}{\pi i}\int_{0}^{\infty} \frac{\phi'(t)-i\lambda\phi(t)}{t-x}\,dt, \; 0<x<\infty, \tag{5.3}$$

where $\phi(x) = \phi(x, 0)$, $\phi'(x) = \phi_x(x, 0)$, $-\infty < x < \infty$.

The real part of (5.3) gives the result that

$$\phi'(x) = \frac{\lambda}{\pi}\int_{0}^{\infty} \frac{\phi(t)}{t-x}\,dt, \; 0 < x < \infty, \tag{5.4}$$

which is an integro-differential equation. There will be two solutions, one corresponding to $\phi(0) = 0$, and the other to $\phi(0) \neq 0$. Once ϕ is known for $0 < x < \infty$ the form of the surface waves can be found.

<u>Doubly infinite dock</u>. It is easily seen that if the dock occupies the regions $-\infty < x < 0$, $1 < x < \infty$, $y = 0$ then for the waves in $(0, 1)$ we have to solve the equation

$$\phi'(x) = \frac{\lambda}{\pi}\int_{0}^{1} \frac{\phi(t)}{t-x}\,dt, \; 0 < x < 1. \tag{5.5}$$

Indeed the same integro-differential equation arises if the dock occupies $0 < x < 1$, $y = 0$, and the rest of the x-axis is the free surface.

In conclusion we remark that we have clearly only touched on the applications of the theory of singular integral equations to two dimensional potential problems in hydrodynamics. The theory of multidimensional singular integral equations is important for applications to problems in three (or more) dimensions, for this the book by Mikhlin (1962) is of value.

References

BATEMAN, H. (1956) Some integral equations of potential theory. J. app. Phys. Vol. 17, pp.91-102.

FRIEDRICHS, K. and LEVY, H. (1948) The dock problem. Communs pure appl. Math. Vol. 1, pp.135-148.

GAKHOV, F.D. (1966) Boundary Value Problems. Pergamon.

HOLFORD, R.L. (1964) Short waves in the presence of a finite dock. I, II. Proc. Camb. phil. Soc. Vol. 60, pp.957-1012.

MIKHLIN, S.G. (1957) Integral Equations. Pergamon.

MIKHLIN, S.G. (1962) Multidimensional Singular Integral Equations. Pergamon.

MUSKHELISHVILI, N.I. (1953a) Singular Integral Equations. Noordhoff.

MUSKHELISHVILI, N.I. (1953b) Some Basic Problems in the Mathematical Theory of Elasticity. Noordhoff.

RHODERICK, E.H. and WILSON, E.M. (1962) Current distribution in thin superconducting films. Nature Vol. 194, pp.1167-8.

CHAPTER 20 PROBLEMS IN TWO-DIMENSIONAL POTENTIAL THEORY

G. T. Symm
National Physical Laboratory

1. Introduction

Potential theory is essentially the theory of Laplace's equation

$$\nabla^2 f = 0, \tag{1.1}$$

where in two dimensions, in terms of rectangular Cartesian coordinates x and y,

$$\nabla^2 = \partial^2/\partial x^2 + \partial^2/\partial y^2. \tag{1.2}$$

Equation (1.1) governs many physical situations, which give rise to a variety of boundary-value problems. Here we formulate the most common of these problems in terms of integral equations and, using the conformal mapping problem for purposes of illustration, we show how we may thus solve them numerically.

Note that by formulating a boundary-value problem for Laplace's equation as an integral equation one is able effectively to reduce the dimension of the problem by one, by confining attention (at least temporarily) to the boundary of the relevant domain. (Ultimately, of course, one may obtain the solution at points throughout the domain.) This is particularly advantageous for exterior boundary-value problems where the domain in question is unbounded.

2. Boundary-Value Problems for Laplace's Equation

In general (leaving aside free boundary problems) we wish to solve Laplace's equation (1.1) in a given domain D of the x,y-plane. Introducing a vector variable $\underline{p}$ to denote a point of the plane, we wish to solve

$$\nabla^2 f(\underline{p}) = 0, \ \underline{p} = (x,y) \in D, \tag{2.1}$$

where D has a prescribed boundary L. On the boundary L, f must typically satisfy one of the conditions

Dirichlet: $$f(\underline{p}) = h(\underline{p}), \ \underline{p} \in L, \tag{2.2}$$

Neumann: $$\frac{\partial f(\underline{p})}{\partial n} = k(\underline{p}), \ \underline{p} \in L, \tag{2.3}$$

or mixed:

$$\left.\begin{array}{l} f(\underline{p}) = h(\underline{p}),\ \underline{p}\in L_1 \\ \dfrac{\partial f(\underline{p})}{\partial n} = k(\underline{p}),\ \underline{p}\in L_2 \end{array}\right\} \quad L_1 + L_2 = L, \tag{2.4}$$

where h and k are known functions and $\frac{\partial}{\partial n}$ denotes differentiation along the normal to L. Here, we define this normal to be directed <u>into</u> the domain D.

Each of the conditions (2.2) - (2.4) is a particular case of the more general condition

$$\alpha(\underline{p})\ f(\underline{p}) + \beta(\underline{p})\ \frac{\partial f(\underline{p})}{\partial n} = \gamma(\underline{p}),\ \underline{p}\in L, \tag{2.5}$$

in which the functions α, β and γ are all known on the boundary L. Existence and uniqueness theorems for the solution of Laplace's equation subject to (2.5) are available for a fairly wide class of functions α, β and γ provided that L is sufficiently smooth. For details see Kellogg (1929), or Sternberg and Smith (1946).

3. Integral Equation Formulations

There are a number of ways of reducing Laplace's equation to an integral equation. Suppose first that we seek a solution in the form of a simple-layer logarithmic potential

$$f(\underline{p}) = \int_L \log|\underline{p}-\underline{q}|\ \ \sigma(\underline{q})dq,\ \underline{p}\in D, \tag{3.1}$$

where σ represents a continuous simple source density on the boundary L. The representation (3.1) ensures that f satisfies the differential equation (2.1). Then if f satisfies the Dirichlet boundary condition (2.2), we find that σ must satisfy the integral equation

$$\int_L \log|\underline{p}-\underline{q}|\ \ \sigma(\underline{q})dq = h(\underline{p}),\ \underline{p}\in L, \tag{3.2}$$

since (Kellogg (1929)) the potential (3.1) is continuous in $\bar{D}$=D+L. If, alternatively, f satisfies the Neumann condition (2.3), then σ must satisfy the integral equation

$$\int_L \frac{\partial}{\partial n_p}\left\{\log|\underline{p}-\underline{q}|\right\}\sigma(\underline{q})dq + \pi\sigma(\underline{p}) = k(\underline{p}),\ \underline{p}\in L, \tag{3.3}$$

in which $\frac{\partial}{\partial n_p}\log|\underline{p}-\underline{q}|$ denotes the normal derivative of $\log|\underline{p}-\underline{q}|$ at the point $\underline{p}$ assuming $\underline{q}$ fixed. Equation (3.3) results from the fact that the normal derivative of the simple-layer potential (3.1) has a jump discontinuity at the boundary L such that

$$\frac{\partial f(\underline{p})}{\partial n} = \int_L \frac{\partial}{\partial n_p}\left\{\log|\underline{p}-\underline{q}|\right\}\sigma(\underline{q})dq + \pi\sigma(\underline{p}),\underline{p}\in L. \tag{3.4}$$

It follows also from (3.1) and (3.4) that if f satisfies the mixed boundary conditions (2.4) then σ must satisfy the coupled integral equations

$$\left.\begin{aligned}&\int_L \log|\underline{p}-\underline{q}|\ \sigma(\underline{q})dq = h(\underline{p}),\ \underline{p}\in L_1,\\ &\int_L \frac{\partial}{\partial n_p}\left\{\log|\underline{p}-\underline{q}|\right\}\sigma(\underline{q})dq + \pi\sigma(\underline{p}) = k(\underline{p}),\ \underline{p}\in L_2.\end{aligned}\right\} \tag{3.5}$$

Similarly, the general boundary condition (2.5) yields an integral equation for σ whose kernel is a linear combination of $\log|\underline{p}-\underline{q}|$ and $\frac{\partial}{\partial n_p}\log|\underline{p}-\underline{q}|$. In each case, if we can solve the relevant integral equation for σ, then a solution of the associated boundary-value problem is given by formula (3.1).

An alternative formulation of the Dirichlet problem is provided by the double-layer potential representation

$$f(\underline{p}) = \int_L \frac{\partial}{\partial n_q}\left\{\log|\underline{p}-\underline{q}|\right\}\mu(\underline{q})dq,\ \underline{p}\in D, \tag{3.6}$$

in which $\mu(\underline{q})$ denotes a double source density on L. This potential has a discontinuity at the boundary L such that the boundary condition (2.2) yields the integral equation

$$\int_L \frac{\partial}{\partial n_q}\left\{\log|\underline{p}-\underline{q}|\right\}\mu(\underline{q})dq - \pi\mu(\underline{p}) = h(\underline{p}),\ \underline{p}\in L, \tag{3.7}$$

for μ. This formulation, in terms of an integral equation of the second kind, is more common in text-books (Mikhlin (1957); Tricomi (1957)) than the previous one, in terms of an equation, (3.2), of the first kind. For many problems, however, the simple-layer representation has advantages and we shall consider one such problem later.

Further formulations of all the major boundary-value problems for Laplace's equation are provided by Green's formula

$$\int_L \log|\underline{p}-\underline{q}| \frac{\partial}{\partial n_q} f(\underline{q})dq - \int_L \frac{\partial}{\partial n_q}\left\{\log|\underline{p}-\underline{q}|\right\} f(\underline{q})dq = \eta f(\underline{p}), \qquad (3.8)$$

in which

$$\eta = 2\pi, \ \underline{p}\in D, \qquad (3.9)$$

and

$$\eta = \pi, \ \underline{p}\in L. \qquad (3.10)$$

It is assumed here that the boundary L is smooth, in which case the value of η on the boundary, (3.10), follows from its value inside the domain, (3.9), on account of the discontinuity in the second integral in equation (3.8). Inserting (3.10) into (3.8) we have Green's boundary formula

$$\int_L \log|\underline{p}-\underline{q}| \frac{\partial}{\partial n_q} f(\underline{q})dq - \int_L \frac{\partial}{\partial n_q}\left\{\log|\underline{p}-\underline{q}|\right\} f(\underline{q})dq - \pi f(\underline{p}) = 0, \underline{p}\in L, \qquad (3.11)$$

which is a linear relationship between the boundary values of f and $\frac{\partial f}{\partial n}$. Given f or $\frac{\partial f}{\partial n}$ at each point of L, this formula yields an integral equation, or a pair of coupled integral equations, whose solution, together with the given data, can be substituted into the left-hand side of equation (3.8) to give f at any point $\underline{p}\in\overline{D}$.

Note that the integral equations presented above are valid for both interior and exterior boundary-value problems (though slight modifications may be necessary in the latter case to match prescribed behaviour at infinity). Conditions for the existence and uniqueness of solutions to these integral equations are discussed elsewhere (Symm (1964)); here we note only that equation (3.2) has a unique solution for almost all contours L and that any exceptional contours can be avoided by a simple change of scale (Jaswon (1963)).

4. Conformal Mapping

For illustration let us now concentrate upon the particular

problem of conformally mapping a given simply-connected domain D with boundary L, in the z-plane, onto the unit disc $|w| \leq 1$, in the w-plane, in such a way that a particular point $z_0 \in D$ goes into the centre, $w = 0$, of the disc. Without loss of generality we take z_0 to be the origin of coordinates in the z-plane ($z = x + iy$). Then the required mapping function, unique but for an arbitrary rotation, may be written in the form

$$w(z) = u + iv = z \exp(f + ig), \tag{4.1}$$

where f and g are conjugate harmonic functions in D. The requirement that $|w(z)| = 1$ for $z \in L$ implies that f satisfies the Dirichlet boundary condition

$$f(\underline{p}) = -0.5 \log(x^2 + y^2), \ \underline{p} = (x,y) \in L. \tag{4.2}$$

Seeking f in the form of a simple-layer potential (3.1), we now see one of the advantages mentioned earlier; for, the conjugate of this potential is simply

$$g(\underline{p}) = \int_L \theta(\underline{p}-\underline{q}) \ \sigma(\underline{q}) \ dq, \ \underline{p} \in \bar{D}, \tag{4.3}$$

where θ is the angle between the vector $\underline{p}-\underline{q}$ and some fixed direction. Thus, once σ is known, g may be obtained as readily as f and the mapping function w then follows from formula (4.1). (Note that a change in the fixed direction in the definition of θ merely adds a constant to g, corresponding to a rotation of the mapping.)

5. Numerical Approximations

In practical applications, it is almost invariably necessary to solve the relevant integral equations numerically. For most purposes, however, sufficient accuracy can be obtained by relatively simple means, e.g. by approximating the source densities by step-functions and applying collocation (cf. Chapter 7).

Thus, in the case of the conformal mapping problem, we may choose to divide the given boundary L into n intervals and assume that the simple source density σ has a constant value in each interval. Then, denoting these constant values by σ_j, $j=1,2,\ldots,n$, we approximate f and g by

$$\tilde{f}(\underline{p}) = \sum_{j=1}^{n} \sigma_j \int_j \log|\underline{p}-\underline{q}|\, dq \qquad (5.1)$$

and

$$\tilde{g}(\underline{p}) = \sum_{j=1}^{n} \sigma_j \int_j \theta(\underline{p}-\underline{q})dq, \qquad (5.2)$$

where $\int_j$ denotes integration over the jth interval of L. Then, applying the boundary condition (4.2) to f at one point in each interval of L, we obtain, corresponding to the integral equation (3.2) with $h(\underline{p}) = -\log|\underline{p}|$, the set of simultaneous linear equations

$$\sum_{j=1}^{n} \sigma_j \int_j \log|\underline{p}_i-\underline{q}|\, dq = -\log|\underline{p}_i|, \quad i=1,2,\ldots,n, \qquad (5.3)$$

for the σ_j. Because of the weak singularity in the logarithmic kernel at $\underline{p}=\underline{q}$, the diagonal terms are dominant in the matrix of coefficients in (5.3) and no difficulty is encountered in solving these equations directly (cf. Chapter 13). From their solution, $\tilde{f}$ and $\tilde{g}$ are readily computed and hence we obtain an approximation

$$\tilde{w} = z \exp(\tilde{f} + i\tilde{g}) \qquad (5.4)$$

to the mapping function (4.1).

In practice, the coefficients of σ_j in the expressions above can seldom be evaluated analytically and are usually approximated by a simple quadrature formula such as Simpson's 3-point rule. Exceptionally, intervals containing the points $\underline{p}$ or $\underline{p}_i$ are approximated by straight lines along which the integrations can be carried out exactly. We note that if the boundary L is in fact a polygon, or if L is approximated by a polygon, so that every interval is a straight line, then the coefficients of σ_j in (5.1) and (5.3) may all be evaluated analytically, though θ, in (5.2), must still be integrated numerically.

6. Examples

The method of conformal mapping described above was first applied by Symm (1966) using numerical quadrature in the evaluation of the coefficients. A variety of domains D were considered and detailed results were tabulated, including estimates E of the maximum error in $|\tilde{w}|$ on each of the boundaries L. In particular, since $|\tilde{w}| = |w| = 1$ at one point within each interval of L, E was defined as the largest absolute value of $|\tilde{w}| - 1$ computed at the end-points of the boundary intervals. Since, by the maximum modulus theorem for analytic functions, the maximum error in $\tilde{w}$ in $\overline{D}$ occurs on the boundary L, this quantity E provides some measure of the over-all accuracy of the approximate mapping function.

Typically, for an ellipse of axial ratio 2:1 divided into 64 intervals (at equal increments of eccentric angle), we find, if we follow Symm (1966), that

$$E = 0.0005. \tag{6.1}$$

If, alternatively, we replace the ellipse by a polygon, approximating each interval by two chords along which we integrate analytically, we obtain

$$E = 0.0003. \tag{6.2}$$

In each case the centre of the ellipse is mapped into the centre of the disc and the maximum error occurs at the ends of the minor axes , where the intervals over which σ is assumed constant are longest.

Of course, for such an ellipse or for any near circular domain there are many numerical methods (Gaier (1964)) which may be expected to solve this mapping problem with accuracy comparable with (6.1) or (6.2). A much more severe test is provided by a domain with sharp corners, particularly re-entrant corners, such as an L-shaped domain. To conclude this chapter therefore we present a few results obtained recently for the mapping of a domain D formed by removing a unit square from one corner of a square of side two. In this case, dividing the boundary L into n equal intervals and integrating the logarithmic kernel analytically, we find the following results for $\tilde{w}$:

	$\lvert\tilde{w}\rvert$, n = 40	$\lvert\tilde{w}\rvert$, n = 80
At the re-entrant corner	0.933	0.959
one interval from the corner	1.008	1.005
two intervals from the corner	1.000	1.000

Whilst these results indicate maximum errors of several percent at the re-entrant corner, we note that such errors are highly localised. In this example it is the mid-point of the diagonal about which the domain is symmetric which is mapped into the centre of the disc.

References

GAIER, D. (1964) Konstructive Methoden der konformen Abbildung. Springer.

JASWON, M.A. (1963) Integral equation methods in potential theory. I. Proc. R. Soc. vol. A275, pp.23-32.

KELLOGG, O.D. (1929) Foundations of Potential Theory. Springer.

MIKHLIN, S.G. (1957) Integral Equations. Pergamon.

STERNBERG, W.J. and SMITH, T.L. (1946) The Theory of Potential and Spherical Harmonics. University of Toronto.

SYMM, G.T. (1964) Integral equation methods in potential theory and elasticity. NPL Mathematics Report No. 51.

SYMM, G.T. (1966) An integral equation method in conformal mapping. Num. Math. vol. 9, pp. 250-258.

TRICOMI, F.G. (1957) Integral Equations Interscience.

CHAPTER 21 NUMERICAL SOLUTION OF SCALAR DIFFRACTION PROBLEMS

A.J. Burton

National Physical Laboratory

1. Introduction

The phenomenon of diffraction occurs in many branches of physics, but we shall concern ourselves in this chapter only with problems involving the diffraction and scattering of waves by bounded surfaces. In a typical problem the scattering object is immersed in an incident wave field which, for simplicity, will be assumed to have harmonic time dependence. Reflected and diffracted waves are produced which propagate outwards from the scatterer in accordance with the scalar wave equation. Problems of this type are important in acoustics, electromagnetic theory, hydrodynamics and other fields.

The assumption of harmonic time dependence means that once the steady state has been reached all waves present have the same frequency $\frac{c}{\lambda}$ Hz (c is the wave velocity and λ the wavelength of the incident radiation) and can therefore be expressed in the form

$$u(x, y, z)\, e^{-ikct},$$

where $k = \frac{2\pi}{\lambda}$ is the wavenumber and u is a complex function of space variables alone. The function u satisfies the reduced wave equation, or Helmholtz' equation

$$\nabla^2 u + k^2 u = 0 ,$$

and the mathematical problem is to find solutions of this equation in exterior domains (the time factor now being omitted).

In this chapter we carry over some of the ideas of the previous chapter to the formulation of diffraction problems in terms of integral equations, but there are some theoretical difficulties here which did not arise in the Laplace case ($k = 0$). Although the physical problems always have unique solutions the corresponding integral equations may not have this property, and they must be modified if we wish to guarantee reliable numerical solution at all wavenumbers. We include an example of the numerical solution of one such formulation for two-dimensional diffraction problems.

2. Mathematical Statement of the Problem

Let f_{inc} be a given incident wave arriving at a closed smooth scattering surface ∂D. The surface separates the interior region D from the exterior E, and the total wave at any point is the sum of f_{inc} and a scattered wave f_s :

$$f = f_{inc} + f_s .$$

The mathematical problem may be stated as

(i) Solve $\nabla^2 f + k^2 f = 0$ in E

(ii) with one of the following conditions on ∂D

(a) $f = g$ (Dirichlet problem)

(b) $\frac{\partial f}{\partial n} = g$ (Neumann problem)

or (c) $\frac{\partial f}{\partial n} + hf = g$ (Robin problem) ,

where g, h are given boundary functions and n is the outward normal to ∂D.*

(iii) f_s must satisfy the Sommerfeld radiation condition to ensure that the scattered waves are outgoing. If r denotes distance from a fixed origin the condition for three-dimensional problems is

$$\lim_{r \to \infty} \left(\frac{\partial f_s}{\partial r} - ik\, f_s\right) = 0,$$

while in two dimensions it becomes

$$\lim_{r \to \infty} r^{\frac{1}{2}}\left(\frac{\partial f_s}{\partial r} - ik\, f_s\right) = 0 .$$

Existence and uniqueness

Let $k = k_1 + ik_2$ and $k_1 > 0$, $k_2 \geq 0$ with Im (h) ≥ 0, then unique solutions to these boundary value problems exist and are analytic with respect to wavenumber. It will be seen later that we cannot make the same statement about solutions to the classical integral equation formulations of these problems.

Henceforth we assume k to be real (implying a non-dissipative medium) and for ease of presentation we shall only discuss the particular

---

*Readers should note that this convention for the sign of n is the opposite of that used in Chapters 20 and 24. This difference appears naturally because we deal here with exterior problems.

Neumann problem

$$\frac{\partial f}{\partial n} = 0 \quad \text{on} \quad \partial D$$

which occurs, for example, in the scattering of acoustic waves by perfectly hard or rigid obstacles. However, it should be borne in mind that all statements regarding this problem apply practically without change to the general Robin problem, and that there is a similar theory for the Dirichlet case.

Few practical diffraction problems can be solved analytically in closed form, though when k is either large (high frequencies) or small (low frequencies) asymptotic methods are frequently used. In the difficult mid-frequency region where the wavelength is comparable with the size of the scatterer a numerical method will probably be necessary.

Some of the benefits of reformulating boundary value problems as integral equations, especially when solving numerically, have already been mentioned (Chapter 20). Perhaps the most important in the present case is that the original problem in an infinite exterior domain is reduced essentially to one involving only the surface; but a further advantage is that the radiation condition is satisfied automatically.

We may also remark that although we are dealing mainly with the scattering of waves from an external source the theory applies equally well when sources are on the scattering surface, as in acoustic radiation problems for example.

3. Integral Equation Formulations for the Exterior Neumann Problem

There are, in general, two standard ways of formulating diffraction problems as integral equations:

(i) by use of Green's theorem,

(ii) by direct representation of the scattered wave as a Helmholtz potential.

The second of these corresponds to that used in Chapter 20 for the Laplace equation. In both cases the formulation involves the free-space Green's function for Helmholtz' equation

$$G_k(\underline{p},\underline{q}) = \begin{cases} \frac{1}{4\pi r} e^{ikr} & \text{in three dimensions} \\ \frac{i}{4} Ho^{(1)}(kr) & \text{in two dimensions,} \end{cases}$$

where $r = |\underline{p} - \underline{q}|$ and $Ho^{(1)}(x)$ is the Hankel function of the first kind. $G_k(\underline{p},\underline{q})$ is a solution of Helmholtz' equation with respect to either of the variable points $\underline{p},\underline{q}$, except when $\underline{p} = \underline{q}$, and it satisfies the radiation condition. As $r \to 0$

$$G_k(\underline{p},\underline{q}) = \begin{cases} \frac{1}{4\pi r} + O(1) \\ -\frac{1}{2\pi} \ln r + O(1) \end{cases} \tag{3.1}$$

$$= G_0(\underline{p},\underline{q}) + O(1),$$

and we shall see that it plays the same role in diffraction problems as ln r does in two-dimensional potential theory.

(a) Equation from Green's theorem

Since f_{inc} has no singularities within D and f_s obeys the radiation condition it is easily established that f satisfies a modified Green's formula (also known as Helmholtz' formula; see Baker and Copson, (1950)). If $\underline{p}$ is any fixed point and the surface element of integration is at $\underline{q}$, we have

$$f_{inc}(\underline{p}) + \int_{\partial D} \left\{ f(\underline{q}) \frac{\partial}{\partial n_q} G_k(\underline{p},\underline{q}) - G_k(\underline{p},\underline{q}) \frac{\partial f(\underline{q})}{\partial n_q} \right\} dS_q = \begin{cases} f(\underline{p}) & ,\underline{p} \in E \\ \frac{1}{2} f(\underline{p}) & ,\underline{p} \in \partial D \\ 0 & ,\underline{p} \in D \end{cases} \tag{3.2}$$

The first part is an integral representation of f in the exterior region in terms of boundary values of f, $\frac{\partial f}{\partial n}$, while the second is an integral equation connecting boundary values.

The integral terms are generalizations of the layer potentials of classical potential theory, and it will be convenient to define single and double-layer Helmholtz potentials by

$$L_k \sigma \equiv \int_{\partial D} \sigma(\underline{q})\, G_k(\underline{p},\underline{q})\, dS_q ,$$

$$M_k \sigma \equiv \int_{\partial D} \sigma(\underline{q}) \frac{\partial}{\partial n_q} G_k(\underline{p},\underline{q})\, dS_q ,$$

where L_k, M_k are to be interpreted as integral operators acting on a boundary function σ. The jump properties of these potentials as $\underline{p}$

rosses ∂D are the same as for the more familiar Laplace potentials (see 3.1)).

With this notation the integral equation from Green's formula may e expressed as

$$(\tfrac{1}{2} I - M_k) f + L_k \frac{\partial f}{\partial n} = f_{inc}, \quad \underline{p} \in \partial D$$

hich for the Neumann problem $\frac{\partial f}{\partial n} = 0$ reduces to

$$(\tfrac{1}{2} I - M_k) f = f_{inc}, \quad \underline{p} \in \partial D \qquad (3.3)$$

here I is the identity operator. Once we have solved this equation for oundary values of f the first part of Green's formula may be used to btain the exterior field.

b) <u>Equation by the simple source method</u>

Again by analogy with potential problems we may seek a solution to he Neumann problem in the form of a single-layer Helmholtz potential. hus we assume a scattered wave

$$f_s = L_k \sigma \quad , \quad \underline{p} \in E$$

nd aim to choose the layer density σ so that

$$\frac{\partial f_s}{\partial n} = - \frac{\partial f_{inc}}{\partial n} \quad \text{on } \partial D \, .$$

fter taking the exterior limiting value of $\frac{\partial f_s}{\partial n}$ as $\underline{p} \to \partial D$ and using the ump properties of the potential, we have an integral equation for σ :

$$(\tfrac{1}{2} I - M_k^T) \sigma = \frac{\partial f_{inc}}{\partial n} \quad , \quad \underline{p} \in \partial D \qquad (3.4)$$

here M_k^T is the transpose of M_k. (<u>Note</u> $G_k(\underline{p},\underline{q}) = G_k(\underline{q},\underline{p})$.)

. <u>Failure of the Integral Equations at Interior Eigenvalues</u>

Although the integral equation from Green's formula was derived from a boundary value problem which always has a unique solution, it is found hat there are wavenumbers for which its solution is not unique. imilarly in the simple source method we can always find a suitable layer ensity σ except at these special wavenumbers.

These difficulties become apparent by considering the homogeneous quation corresponding to (3.4)

$$(\tfrac{1}{2} I - M_k^T)\,\phi = 0 \,, \tag{4.1}$$

and observing that boundary values of $\frac{\partial v}{\partial n}$ for the following associated interior problem satisfy this equation:

$$\nabla^2 v + k^2 v = 0 \text{ in } D$$
$$v = 0 \text{ on } \partial D \,.$$

Now this problem will normally have only the trivial solution $v \equiv 0$, but at an infinite discrete set of wavenumbers $k \in K_D$ (the interior Dirichlet eigenvalues) both equation (4.1) and its transpose (the homogeneous counterpart of(3.3)) will have non-trivial solutions, and hence even if solutions of (3.3) and (3.4) exist they will be non-unique.

For solutions to exist under these circumstances we know (Chapter 1) that the source term must be orthogonal to all solutions of the transposed homogeneous equation. In the case of equation (3.3) it can be shown by applying Green's theorem to f_{inc}, v in D that this condition is satisfied. The non-uniqueness of the solution when $k \in K_D$ can be shown to be due to the fact that we have failed to impose the boundary condition at these wavenumbers.

However, for the equation (3.4) resulting from the simple source method the orthogonality condition does not usually hold, and we conclude that at these wavenumbers the scattered wave cannot be represented by a single-layer Helmholtz potential. (In general a combination of both single- and double-layer potentials is required - see Kupradze (1965)).

The significance of these failures from the numerical point of view lies in the fact that when the equations have been discretized the related matrices will be nearly singular when $k \in K_D$. But whereas the integral equations break down only at a discrete set of wavenumbers, the approximating system of linear equations will be ill-conditioned even when k lies in the vicinity of these critical values; the numerical problem is therefore more severe for moderate or large k where the density of eigenvalues is high. Since the interior eigenvalues of a general scatterer would not be known in advance there is always the possibility of accidental failure, even at quite low frequencies, unless steps are taken to detect ill-conditioning.

5. Improved Integral Equation Formulations

The difficulties with the standard integral equations have been known for a long time; indeed they were apparently first encountered in attempts to prove the existence of solutions to diffraction problems. But in recent years, with the increasing interest in numerical methods of solution, the question has been re-examined with a view to developing formulations valid for all wavenumbers (see for example Schenck, 1968).

Copley (1967) put forward the idea of using the third part of Green's formula (3.2); that is, instead of taking $\underline{p} \in \partial D$ to give an integral equation for boundary values of f, $\underline{p}$ is taken inside D leading to a functional equation. Provided this equation is satisfied at all points $\underline{p} \in D$ (or, in the case of an axisymmetric body at all points along the axis of symmetry) the solution is unique. In numerical practice, however, the relation will only be imposed at a finite number of points and there is still the likelihood of trouble when $k \in K_D$. Also, being somewhat analogous to an integral equation of the first kind with non-singular kernel, it is not well suited to accurate numerical solution.

A development of Copley's method was introduced by Schenck (1968) who suggested combining the boundary and interior forms of Green's formula:

$$(\tfrac{1}{2} I - M_k) f = f_{inc} \quad , \quad \underline{p} \in \partial D$$

$$- M_k \; f = f_{inc} \quad . \quad \underline{p} \in D$$

Schenck proved that although the solution of the first part is non-unique when $k \in K_D$ only the wanted solution satisfies both equations.

In order to exploit the idea numerically he discretized the first equation in the usual way giving a set of linear equations for sampled boundary values of f, and then supplemented these with further equations from the interior relation. The resulting over-determined system was solved by a least squares procedure. But the effectiveness of the method depends very much on the choice of interior points at which the functional equation is applied (Bolomey and Tabbara, 1973), and the method is still liable to ill-conditioning at higher frequencies even when many interior points are included.

A number of formulations have been proposed which are completely free from interference by interior eigenvalues, though they are more complicated than the two above. One class of methods, which has been used for both Dirichlet and Robin problems, seeks to represent the scattered wave as a mixed potential:

$$f_s = (L_k + \alpha M_k)\,\sigma \quad , \quad \underline{p} \in E$$

where α is a constant which if chosen to be strictly complex when k is real guarantees the existence of a suitable unique layer density σ. (See Brakhage and Werner (1965), Panich (1965), Kussmaul (1969)). The formulation was applied to the numerical solution of two-dimensional Dirichlet problems by Greenspan and Werner (1966).

In a related method given by Burton and Miller (1971) a composite integral equation is constructed from Green's formula; the integral operator of this equation turns out to be just the transpose of that from the mixed potential method. However, there are often clear advantages in working with Green's formula if the boundary condition is homogeneous in f or $\frac{\partial f}{\partial n}$, and our development here will be confined to this approach.

We have already noted that the integral equation from Green's formula has non-unique solutions when $k \in K_D$, but we can derive a further equation by differentiating the first part of (3.2) with respect to the normal at $\underline{p}$ and allowing $\underline{p} \to \partial D$. After inserting the boundary condition we obtain

$$N_k\, f = -\frac{\partial f_{inc}}{\partial n} \quad , \quad \underline{p} \in \partial D \tag{5.1}$$

where the operator N_k is defined as

$$N_k\,\sigma \equiv \frac{\partial}{\partial n_p} \int_{\partial D} \sigma(\underline{q})\, \frac{\partial}{\partial n_q}\, G_k(\underline{p},\underline{q})\; dS_q . \tag{5.2}$$

The solution of this equation is also non-unique at a critical set of wavenumbers, this time whenever the associated interior problem

$$\nabla^2 v + k^2 v = 0 \quad \text{in} \quad D$$
$$\frac{\partial v}{\partial n} = 0 \quad \text{on } \partial D$$

has eigenfunctions. The infinite set of wavenumbers at which this occurs ($k \in K_N$) will be termed the interior Neumann eigenvalues.

We now have two integral equations for the problem:

$$(\tfrac{1}{2} I - M_k) f = f_{inc} \qquad (A)$$

$$N_k \; f = - \frac{\partial f_{inc}}{\partial n}, \qquad (B)$$

he first having non-unique solutions when $k \in K_D$ and the second when $\in K_N$. Nonetheless the two equations always have only one solution in ommon even when $k \in K_D \cap K_N$.

We construct the following composite integral equation and show hat for a suitable choice of the coupling constant α its solution is nique:

$$(\tfrac{1}{2} I - M_k + \alpha \, N_k) f = f_{inc} - \alpha \frac{\partial f_{inc}}{\partial n}, \; \underline{p} \in \partial D. \qquad (5.3)$$

niqueness will be assured if the homogeneous equation

$$(\tfrac{1}{2} I - M_k + \alpha \, N_k) \phi = 0 \qquad (5.4)$$

as only the trivial solution $\phi = 0$.

Consider the double-layer Helmholtz potential with layer density ϕ

$$v(\underline{p}) = M_k \phi \quad , \quad \text{where } \; \underline{p} \in D + \partial D + E.$$

n view of the jump relations for the potential, equation (5.4) shows hat interior boundary values of v, $\frac{\partial v}{\partial n}$, are related by

$$v_{int} - \alpha \left(\frac{\partial v}{\partial n}\right)_{int} = 0. \qquad (5.5)$$

herefore applying Green's second identity to v, $\bar{v}$ in D, we find that

$$\int_{\partial D} \left(v \frac{\partial \bar{v}}{\partial n} - \bar{v} \frac{\partial v}{\partial n}\right) dS = 2i \; \mathrm{Im}(\alpha) \int_{\partial D} \left|\frac{\partial v}{\partial n}\right|^2 dS$$

$$= 4i \; \mathrm{Re}(k) \; \mathrm{Im}(k) \int_D |v|^2 \; dV.$$

Hence for real k the choice $\mathrm{Im}(\alpha) \neq 0$ ensures that $\left(\frac{\partial v}{\partial n}\right)_{int} = 0$ and y (5.5) $v_{int} = 0$ as well. Then from the jump relations and the niqueness of the exterior Neumann problem we conclude that $\phi = 0$, and the solution of the composite equation (5.3) is unique.

6. Transformation of the Operator N_k

One of the drawbacks of both the mixed potential method and the composite Green's formula from a numerical point of view is that in the case of the Robin problem (including the present Neumann problem) the integral equations contain the highly singular operator N_k. If the indicated differentiation in (5.2) is carried through, the integral becomes divergent when $\underline{p} \in \partial D$. N_k does not occur in equations for the Dirichlet problem and their numerical solution is consequently much easier.

In the mixed potential methods a theoretical difficulty is also introduced because N_k is not completely continuous. Thus while it is not difficult to confirm the uniqueness of appropriate layer densities we cannot immediately verify their existence since ordinary Fredholm theory is not applicable. Both Panich and Kussmaul overcame this problem by working with integral equations that had been regularized.

By regularization in this context we mean that given an operator equation

$$K f = g \,, \tag{6.1}$$

in which K is not completely continuous, we attempt to find a regularizing operator K_1 such that

$$K_1 \, K f = (I - K_2) \, f = K_1 \, g \,, \tag{6.2}$$

where K_2 is completely continuous. The regularization may be carried out either by pre- or post-multiplication, and we say it is equivalent if the solutions of (6.2) are the same as those of (6.1) - in particular there must be no gain or loss in number.

We give an example of equivalent regularization with the integral equation from the composite Green's formula. It depends on the following operator relation

$$L_k \, N_k \, \sigma = (M_k + \tfrac{1}{2} I) \, (M_k - \tfrac{1}{2} I)\sigma \quad , \quad \underline{p} \in \partial D \tag{6.3}$$

which is easily proved from Green's theorem for $G_k(\underline{p},\underline{q})$ and $M_k \, \sigma$ in D. The relation used by Panich is the transpose of this with k = 0.

In the composite integral equation (5.3) we write

$$N_k = (N_k - N_o) + N_o$$

and pre-multiply the equation by L_o. Then using (6.3) with $k = 0$ we transform the term $L_o N_o$ into the product of two weakly singular operators; $(N_k - N_o)$ is also only weakly singular, and the final equation becomes

$$L_o(\tfrac{1}{2} I - M_k + \alpha(N_k - N_o)) f + \alpha(M_o + \tfrac{1}{2} I)(M_o - \tfrac{1}{2} I) f$$
$$= L_o(f_{inc} - \alpha \frac{\partial f_{inc}}{\partial n}), \quad \underline{p} \in \partial D$$

Note: Pre-multiplication with L_k rather than L_o would not be an equivalent regularization when $k \in K_D$, also if the idea is used in two dimensions L_o may need to be modified slightly (for details see Burton and Miller, 1971). The elegant method of Kussmaul (1969) for two-dimensional problems uses a different regularizing operator but is more expensive in computing time.

Another approach for two-dimensional problems which may be more economical (in that it does not require full matrix multiplications) is to transform N_k directly using a vector relationship given by Maue (1949):

$$N_k f = \int_{\partial D} \left\{ (\underline{n}_q \times \nabla_q f) \cdot (\underline{n}_p \times \nabla_p G_k) + k^2 \underline{n}_p \cdot \underline{n}_q f(\underline{q}) G_k(\underline{p},\underline{q}) \right\} dS_q, \quad \underline{p} \in \partial D$$

where $\underline{n}_p$, $\underline{n}_q$ are unit outward normal vectors at $\underline{p}$, $\underline{q}$, and ∇_p, ∇_q are the gradient operators at these points.

Specialising to two dimensions the relation is simply

$$N_k f = \int_{\partial D} \left\{ \frac{\partial f}{\partial t_q} \frac{\partial G_k}{\partial t_p} + k^2 \underline{n}_p \cdot \underline{n}_q f(\underline{q}) G_k \right\} ds_q, \qquad (6.4)$$

in which t_p, t_q are the positive tangent directions at $\underline{p}$ and $\underline{q}$. Now while both terms under the integral are still singular ($\frac{\partial G_k}{\partial t_p}$ has a Cauchy singularity and G_k a logarithmic singularity when $\underline{q} = \underline{p}$), satisfactory numerical approximations can be constructed as we show in the next section.

7. Numerical Approximations

We shall outline one method for the numerical solution of the composite integral equation (5.3) in two dimensions when the operator N_k has been transformed using (6.4). The boundary ∂D is a closed curve (assumed smooth) and we can approximate the various integrals in the usual way with quadrature formulae after the boundary has been divided into N arc intervals. The integrals are replaced by weighted sums of boundary functions sampled at the end-points of the intervals, and an approximate solution of the integral equation is found by collocation at the same set of N points.

Consider first the operator M_k in which the kernel $\frac{\partial G_k}{\partial n_q}$ is non-singular for two-dimensional problems. In fact

$$\lim_{\underline{q} \to \underline{p}} \frac{\partial}{\partial n_q} G_k(\underline{p},\underline{q}) = -\tfrac{1}{2}\,(\text{Curvature at } \underline{p});$$

but to avoid computing curvatures it is helpful to write $M_k f$ in the form

$$M_k f = \int_{\partial D} \left\{ f(\underline{q}) \frac{\partial}{\partial n_q} G_k(\underline{p},\underline{q}) - f(\underline{p}) \frac{\partial}{\partial n_q} G_o(\underline{p},\underline{q}) \right\} ds - \tfrac{1}{2} f(\underline{p}), \qquad \underline{p} \in \partial D$$

where we have used Gauss' formula to integrate the subtracted term (Green's second theorem for G_o and 1 in D). Let the end-points of the intervals be $\underline{q}_1, \underline{q}_2, \ldots, \underline{q}_N$ and the corresponding integration weights $w_1, w_2, \ldots, w_N$; then if $\underline{p} = \underline{q}_i$ we have the approximation

$$M_k f \doteq \sum_{\substack{j=1 \\ j\neq i}}^{N} w_j\, f(\underline{q}_j) \frac{\partial}{\partial n_j} G_k(\underline{q}_i,\underline{q}_j) - f(\underline{q}_i) \left\{ \tfrac{1}{2} + \sum_{\substack{j=1 \\ j\neq i}}^{N} w_j \frac{\partial}{\partial n_j} G_o(\underline{q}_i,\underline{q}_j) \right\} .$$

The representation of $N_k f$ is considered in two parts: firstly there is the term

$$I_1 = \int_{\partial D} \frac{\partial f}{\partial t_q} \frac{\partial}{\partial t_p} G_k(\underline{p},\underline{q})\, ds ,$$

which has a Cauchy singularity at $\underline{q} = \underline{p}$, and secondly

$$I_2 = k^2 \int_{\partial D} \underline{n}_p \cdot \underline{n}_q\, f(\underline{q})\, G_k(\underline{p},\underline{q})\, ds$$

with a logarithmic singularity. In the case of I_1 we effectively cancel the singularity by writing

$$I_1 = \int_{\partial D} \left\{ \frac{\partial f(\underline{q})}{\partial t_q} \frac{\partial}{\partial t_p} G_k(\underline{p},\underline{q}) + \frac{\partial f(\underline{p})}{\partial t_p} \frac{\partial}{\partial t_q} G_k(\underline{p},\underline{q}) \right\} ds ,$$

noting that the integral of the added part vanishes. With this modification

$$\lim_{\underline{q} \to \underline{p}} \text{(Integrand)} = \frac{1}{2\pi} \frac{\partial^2 f(\underline{p})}{\partial t_p^{\;2}} ,$$

and the approximation is carried out by first discretizing the integral as before and then replacing derivatives by finite differences. (Note $\frac{\partial f}{\partial t} = \frac{\partial f}{\partial s}$, and since $\frac{\partial f}{\partial n} = 0$ for this problem $\frac{\partial^2 f}{\partial t^2} = \frac{\partial^2 f}{\partial s^2}$.)

For the second term I_2 we use the method given by Kussmaul (1969) which was in part suggested by the work of Atkinson (1966). After subtracting out the logarithmic component of $G_k(\underline{p},\underline{q})$ the remaining integral is easily discretized, and the main problem is therefore to represent integrals of the type

$$J = \int_{\partial D} v(\underline{q}) \ln r \, ds ,$$

where $v(\underline{q})$ is well-behaved and $r = |\underline{p}-\underline{q}|$. Kussmaul writes this as

$$J = \int_{\partial D} v(\underline{q}) \ln \left(\frac{r}{\sin \frac{\pi s}{c}}\right) ds + \int_{\partial D} v(\underline{q}) \ln \left(\sin \frac{\pi s}{c}\right) ds,$$

in which s is the arc distance of $\underline{q}$ from $\underline{p}$ and $c = \int_{\partial D} ds$.
The first integral is now well-behaved and can be approximated in the usual way, while the second is handled by a special integration formula obtained by first expanding $v(\underline{q})$ in a finite Fourier series and then integrating analytically; the final expression is rearranged as a weighted sum of the $v(\underline{q}_j)$.

As an example of the performance of the method we consider the diffraction of plane waves by an infinite circular cylinder of unit radius and compare the results with the exact solution. The boundary is divided into equal arc intervals and the trapezoidal rule used to approximate the well-behaved integrals (particularly effective here as the integrands are periodic). Derivatives in the representation of

N_k were replaced by five-point finite-difference formulae.

A convenient physical quantity with which to make numerical comparisons is the plane-wave scattering cross-section of the obstacle τ; in some problems it is not necessary to know the exterior wave field in complete detail and this measure suffices. For our purposes we simply regard it as a number that characterizes the scattering properties of the surface at a particular frequency and which can be expressed as

$$\tau = \frac{1}{k} \operatorname{Im} \int_{\partial D} f(\underline{q}) \frac{\partial \bar{f}_{inc}(\underline{q})}{\partial n_q} ds. \tag{7.1}$$

The table shows the relative error in computed values of τ after a straightforward discretization of (7.1). Approximate surface values of f were first obtained at a non-critical wavenumber (k = 4) and results are given for each of the basic integral equations (A) and (B) and also for the composite equation (A) + 0.1 i (B). It is clear that the first equation is solved more accurately than the second, as would be expected, but ultimately the errors decrease at about the same rate.

N	% Relative error in τ at k = 4		
	Equation (A)	Equation (B)	(A) + 0.1 i (B)
24	0.227	6.538	0.853
48	0.026	0.435	0.072
96	0.003	0.032	0.006

However, if we take k = 3.8317, which is very close to an element of K_D (and K_N also), the scattering cross-section computed from the solution of the standard equation (A) has errors of 50-60% whereas the corresponding errors from the composite equation are slightly less than those listed in the last column of the table. Fortunately only a small complex value of α is necessary to ensure freedom from ill-conditioning, which mitigates the poorer approximation of the second integral equation (B).

Finally we would mention that the theoretical aspects of this chapter are discussed in much greater detail in a recent report (Burton (1973)).

References

ATKINSON, K.E. (1966) Extensions of the Nyström method for the numerical solution of linear integral equations of the second kind. Thesis, University of Wisconsin. (Section 2.6).

BAKER, B.B. and COPSON, E.T. (1950) The mathematical theory of Huygens' principle. 2nd Ed., Oxford. (Chapter 1).

BOLOMEY, J-C. and TABBARA, W. (1973) Numerical aspects on coupling between complementary boundary value problems. IEEE Trans AP-21 No.3, pp 356-363.

BRAKHAGE, H. and WERNER, P. (1965) Über das Dirichletsche Außenraumproblem für die Helmholtzsche Schwingungsgleichung. Arch. Math. vol. 16, pp 325-329.

BURTON, A.J. and MILLER, G.F. (1971) The application of integral equation methods to the numerical solution of some exterior boundary-value problems. Proc. R. Soc. vol. A 323, pp 201-210.

BURTON, A.J. (1973) The solution of Helmholtz' equation in exterior domains using integral equations, NPL Mathematics Report NAC 30.

COPLEY, L.G. (1967) Integral equation method for radiation from vibrating surfaces. J. Acoust. Soc. Am. vol. 41, pp 807-816.

GREENSPAN, D. and WERNER, P. (1966) A numerical method for the exterior Dirichlet problem for the reduced wave equation. Archs. ration. Mech. Analysis, vol. 23, pp 288-316.

KUPRADZE, V.D. (1965) Potential methods in the theory of elasticity. Israel Program for Scientific Translations, Jerusalem. (Chapter 6).

KUSSMAUL, R. (1969) Ein numerisches Verfahren zur Lösung der Neumannschen Außenraumproblems für die Helmholtzsche Schwingungsgleichung. Computing vol. 4, pp 246-273.

MAUE, A.W. (1949) Zur formulierung eines allgemeinen Beugungsproblems durch eine Integralgleichung. Z. Phys. vol. 126, pp 601-618.

PANICH, O.I. (1965) On the question of the solubility of the exterior boundary problem for the wave equation and Maxwell's equation

(Russian). _Usp. Mat. Nauk._ (in Russian) vol. _20_ pp 221-226.

SCHENCK, H.A. (1968) Improved integral formulation for acoustic radiation problems. _J. Acoust. Soc. Am._ vol. _44_, pp 41-58.

CHAPTER 22 A PROBLEM IN THE THEORY OF WATER WAVES

F. Ursell

University of Manchester

1. Introduction

The problem which I shall discuss in the present chapter arises in the linearized theory of water waves.

Let us consider a horizontal circular cylinder which in its mean position is half-immersed in an inviscid fluid and which oscillates vertically with small amplitude and constant frequency about this mean position. Let the origin of rectangular coordinates be taken at the mean position of the centre, with the x-axis horizontal and normal to the axis of the cylinder, and the y-axis vertical (y increasing with depth). We also write $x = r \sin \theta$, $y = r \cos \theta$. By symmetry it is sufficient to consider the quadrant $0 \leqslant \theta \leqslant \frac{1}{2}\pi$.

The velocity potential $\phi(x, y)e^{-i\sigma t}$ satisfies

$$\frac{\partial^2 \phi}{\partial x^2} + \frac{\partial^2 \phi}{\partial y^2} = 0 \tag{1.1}$$

in the region $r > a$, $0 \leqslant \theta \leqslant \frac{1}{2}\pi$; the boundary conditions are

$$K\phi + \frac{\partial \phi}{\partial y} = 0 \quad \text{on the free surface} \quad y = 0, \ |x| > a, \tag{1.2}$$

where $K = \sigma^2/g$, and $2\pi/\sigma$ is the period;

$$\frac{\partial \phi}{\partial r} = V_0 \cos \theta \quad \text{on the quadrant} \quad r = a, \ 0 \leqslant \theta \leqslant \tfrac{1}{2}\pi; \tag{1.3}$$

and $$\frac{\partial \phi}{\partial \theta} = 0 \quad \text{on the line of symmetry} \quad \theta = 0. \tag{1.4}$$

Also, the waves travel outwards towards infinity:

$$\frac{\partial \phi}{\partial r} - iK\phi \to 0 \quad \text{as} \quad r \to \infty. \tag{1.5}$$

I shall consider two methods, using series expansions and integral equations respectively.

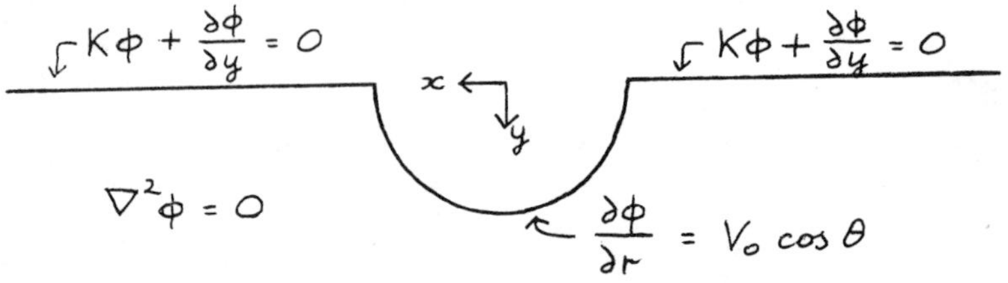

2. Series Expansions

We write

$$\phi(x, y) = \phi(r \sin\theta, r\cos\theta)$$

$$= C\left\{\Phi_0(Kr;\ \theta) + \sum_{m=1}^{\infty} p_{2m}(Ka)a^{2m}\left(\frac{\cos 2m\ \theta}{r^{2m}} + \frac{K}{2m-1}\frac{\cos(2m-1)\theta}{r^{2m-1}}\right)\right\} \tag{2.1}$$

where $\Phi_0(Kr;\ \theta) = \int_0^{\infty} e^{-kr\cos\theta}\cos(kr\sin\theta)\frac{dk}{k-K}$ is the potential of a wave source at the origin (a known but complicated function), and where the coefficients C, p_2, p_4,... are to be determined. We shall first assume, and then verify, that $p_{2m} = O(1/m^3)$. Then term-by-term differentiation is allowed. The expansion (2.1) satisfies all the equations and boundary conditions except the condition (1.3) on the semi-circle, and this also is satisfied if

$$\left\langle a\frac{\partial\Phi_0}{\partial r}(Kr;\ \theta)\right\rangle_{r=a} - \sum_1^{\infty} 2m.p_{2m}\left\{\cos 2m\ \theta + \frac{Ka}{2m}\cos(2m-1)\theta\right\} = \frac{aV_0}{C}\cos\theta \quad \text{when} \quad 0 \leqslant \theta \leqslant \tfrac{1}{2}\pi. \tag{2.2}$$

When $Ka = 0$, the series is a Fourier cosine series, with one term missing but with the additional term $\cos\theta$ on the right-hand side. This system can obviously be solved (when $Ka = 0$) by taking the Fourier components of (2.2) with respect to the complete set of functions $\cos 2s\theta$ $(s = 0, 1, 2,\ldots)$. The same idea will now be applied when $Ka > 0$. It is convenient to begin by eliminating V_0. By integrating (2.2) from 0 to $\frac{1}{2}\pi$ we find that

$$\int_0^{\frac{1}{2}\pi}\left\langle a\frac{\partial}{\partial r}\Phi_0(Kr,\ \theta')\right\rangle_{r=a}d\theta' - \sum_1^{\infty} 2m.p_{2m}\frac{Ka.\ (-1)^{m-1}}{2m(2m-1)} = a\frac{V_0}{C}. \tag{2.3}$$

When this is substituted in (2.2) we see that

$$F(\theta, Ka) \equiv \left\langle a\frac{\partial}{\partial r}\Phi_0(Kr;\ \theta)\right\rangle_{r=a} - \cos\theta\int_0^{\frac{1}{2}\pi}\left\langle a\frac{\partial}{\partial r}\Phi_0(Kr;\ \theta')\right\rangle d\theta'$$

$$= \sum_1^\infty(2mp_{2m}(Ka))\left[\cos\ 2m\theta + \frac{Ka}{2m}\left\{\cos(2m-1)\theta - \frac{(-1)^{m-1}}{2m-1}\cos\ \theta\right\}\right] \tag{2.4}$$

in the range $0 \leqslant \theta \leqslant \frac{1}{2}\pi$. In other words the functions $[\ldots]$ on the right-hand side (which satisfy $\int_0^{\frac{1}{2}\pi}[\ldots]d\theta = 0$) are to be combined to give the function $F(\theta, Ka)$ on the left-hand side (which also satisfies $\int_0^{\frac{1}{2}\pi}F(\theta)d\theta = 0$).

The expansion (2.4) is now transformed into an equivalent infinite system of equations by application of the operators $\int_0^{\frac{1}{2}\pi}\ldots\cos 2s\theta\, d\theta$ $(s = 0, 1, 2,\ldots)$; equivalent because the set $\{\cos 2s\theta\}$ is complete over $(0, \frac{1}{2}\pi)$. When $s = 0$ we obtain $0 = 0$; when $s \geqslant 1$, we obtain

$$c_{2s}(Ka) \equiv \frac{4}{\pi}\int_0^{\frac{1}{2}\pi}F(\theta, Ka)\cos 2s\theta\, d\theta = (2s.p_{2s})$$
$$+ \frac{4}{\pi}Ka\sum_{m=1}^{\infty}(2m.p_{2m}).\frac{1}{2m}\int_0^{\frac{1}{2}\pi}\cos 2s\theta\left\{\cos(2m-1)\theta - \frac{(-1)^{m-1}}{2m-1}\cos\theta\right\}d\theta, \tag{2.5}$$

where the integrals are elementary. We find that

$$c_{2s} = 2s.p_{2s} + Ka\sum_{m=1}^{\infty}(2m.p_{2m})b_{sm}, \tag{2.6}$$

$$\text{where}\quad b_{sm} = \frac{32}{\pi}(-1)^{m+s}\frac{s^2(m-1)}{(4s^2-1)(2m-1)(2s+2m-1)(2s-2m+1)} \tag{2.7}$$

and where c_{2s} is given by (2.5). This is the system which we wish to study. We observe that the unknown p_{2s} occurs outside as well as inside the sign of summation. Systems of this form have a theory analogous to the Fredholm theory of integral equations. (If the abstract theory of compact operators is applied to the Hilbert space L_2 we obtain the latter theory; if to the Hilbert space ℓ_2 we obtain the former theory.) We have the following

Theorem 1.

In the system

$$C_m = X_m + \sum_{n=1}^{\infty} A_{mn} X_n, \quad m = 1, 2, \ldots,$$

suppose that $\Sigma|C_m|^2 < \infty$, $\Sigma\Sigma|A_{mn}|^2 < \infty$. Then we have the Fredholm Alternative:

Either $\det(\delta_{mn} + A_{mn}) \neq 0$, then for given C_m there exists a unique solution X_m such that $\Sigma|X_m|^2 < \infty$.

Or $\det(\delta_{mn} + A_{mn}) = 0$, then there exists a solution only if C_m is orthogonal to all the solutions of the homogeneous transposed system. (We shall not discuss this case which can occur at most at a discrete set of values of Ka at which the wave amplitude vanishes at infinity, and at which the term Φ_0 must be omitted from the expansion (2.1). No serious difficulty arises.)

If Theorem 1 is to be directly applicable to the system (2.6) we must verify that $\Sigma|c_{2s}|^2 < \infty$, and that $\Sigma\Sigma|b_{sm}|^2 < \infty$, and we can then conclude that there is a solution $\{2s.p_{2s}\}$ such that $\Sigma|2s.p_{2s}|^2 < \infty$. This result is correct but is not sufficient for the physical problem which requires something like $\Sigma|2s.p_{2s}| < \infty$. Accordingly we rewrite (2.6) in the form

$$2s.c_{2s} = (2s)^2.p_{2s} + \text{Ka} \sum (2m)^2 p_{2m}.\left(\frac{sb_{sm}}{m}\right), \tag{2.8}$$

and we shall verify that $\sum|2s.c_{2s}|^2 < \infty$ and $\sum\sum\left|\frac{sb_{sm}}{m}\right|^2 < \infty$. We can conclude that $\sum(2s)^4|p_{2s}|^2 < \infty$ from Theorem 1. It will finally be shown that $p_{2s} = O(1/s^3)$.

To obtain a bound for c_{2s} we observe that

$$\tfrac{1}{4}\pi c_{2s} = \int F.\cos 2s\theta \; d\theta = -\int F \frac{d^2}{d\theta^2} \frac{\cos 2s\theta}{s^2} d\theta$$

$$= -\int F'' \frac{\cos 2s\theta}{s^2} d\theta + \left[F' \frac{\cos 2s\theta}{s^2} - F \frac{d}{d\theta} \frac{\cos 2s\theta}{s^2}\right]_{\theta=0}^{\frac{1}{2}\pi}, \tag{2.9}$$

by integration by parts. The last term vanishes when $\theta = 0$ and $\theta = \frac{1}{2}\pi$, all the other terms are evidently $O(1/s^2)$, (the leading term is actually $F'(\frac{1}{2}\pi)\cos s\pi . s^{-2})$. Thus

$$|s^2 c_{2s}(Ka)| < A(Ka). \tag{2.10}$$

It follows that

$$\sum s^2 |c_{2s}|^2 < \infty. \tag{2.11}$$

Also, from (2.7), we see that

$$\left|\frac{sb_{sm}}{m}\right| < A \frac{s}{m(2s + 2m - 1)|2s - 2m + 1|} .$$

It follows that $\sum_s \sum_m \left|\frac{s}{m} b_{sm}\right|^2 < \infty$; for

$$\sum_s \sum_m \frac{s^2}{m^2} |b_{sm}|^2 \leqslant \sum_{s=1}^{\infty} \left(\sum_{m=1}^{[\frac{1}{2}s]} + \sum_{[\frac{1}{2}s]}^{[\frac{3}{2}s]} + \sum_{[\frac{3}{2}s]}^{\infty} \right)$$

$$= \sum_s (S_1 + S_2 + S_3), \text{ say,} \tag{2.12}$$

where $$S_1 = s^2 \sum_{m=1}^{[\frac{1}{2}s]} \frac{1}{m^2(2s + 2m - 1)^2(2s - 2m + 1)^2}$$

$$\leqslant \frac{As^2}{s^2 . s^2} \sum_{1}^{[\frac{1}{2}s]} \frac{1}{m^2} < \frac{A}{s^2} ;$$

$$S_2 \leqslant \frac{As^2}{s . s^2} \sum_{[\frac{1}{2}s]}^{[\frac{3}{2}s]} \frac{1}{(2s - 2m + 1)^2} < \frac{A}{s^2} ;$$

$$S_3 \leqslant \frac{As^2}{s^2 . s^2} \sum_{[\frac{3}{2}s]}^{\infty} \frac{1}{m^2} < \frac{A}{s^2} .$$

Thus, from (2.12), we see that

$$\sum_m \frac{s^2}{m^2} |b_{sm}|^2 < \frac{A}{s^2} , \tag{2.13}$$

and $$\sum\sum \frac{s^2}{m^2} |b_{sm}|^2 \leqslant A \sum_s \frac{1}{s^2} < \infty. \tag{2.14}$$

The inequalities (2.13) and (2.14) show that Theorem 1 is applicable to the system (2.8), which therefore has a unique solution p_{2s}, such that

$$\sum s^4 |p_{2s}|^2 < \infty \tag{2.15}$$

(except possibly for a discrete set of values of Ka).

It only remains to show that the solution $\{p_{2s}\}$ satisfies the stronger inequality $|p_{2s}| < A(Ka)s^{-3}$. This is done by iterating the system (2.8) once. We have

$$\begin{aligned}|2s.c_{2s} - 4s^2 p_{2s}|^2 &= (Ka)^2 \left| \sum_m (2m^2 p_{2m}).(\tfrac{s}{m})\ b_{sm}\right|^2 \\ &\leqslant 4(Ka)^2 (\sum |m^2 p_{2m}|^2)(\sum_m \frac{s^2}{m^2}\ |b_{sm}|^2),\end{aligned} \tag{2.16}$$

by Schwarz's inequality. The first series on the right-hand side is convergent, from (2.15); the second series is bounded by As^{-2}, see (2.13) above. Thus, from (2.16) we have

$$|2s.c_{2s} - 4s^2 p_{2s}|^2 \leqslant A.(Ka)^2.s^{-2},$$

i.e. $4s^2 p_{2s} - 2sc_{2s} = O(1/s)$,

i.e. $4s^2 p_{2s} = O(1/s)$, from (2.10).

Thus the required precise order of magnitude of the unknown p_{2s} is given by

$$p_{2s} = O(1/s^3). \tag{2.17}$$

On substituting this result back in (2.8) we can show, with some difficulty, that

$$(-1)^s\ s^3\ p_{2s} \to \text{a limit depending on } Ka.$$

(Here we use the result $(-1)^s\ s^2 c_{2s} \to -F'(\tfrac{1}{2}\pi)$ which, as we have seen, follows easily from (2.9).)

The original computation was slightly different, instead of (2.4) it used the expansion which is obtained by integrating (2.4) between 0 and θ, i.e.

$$F_1(\theta) \equiv \int_0^\theta F(\theta')d\theta'$$

$$= \sum_1^\infty p_{2m}(Ka)\left[\sin 2m\theta + \frac{Ka}{2m-1}\ (\sin(2m-1)\theta - (-1)^{m-1}\sin\theta)\right],$$

$$\text{when } 0 \leqslant \theta \leqslant \tfrac{1}{2}\pi. \tag{2.18}$$

We observe that both sides vanish when $\theta = \frac{1}{2}\pi$. We can use this expansion, instead of (2.4), to show that $p_{2m} = O(1/m^3)$. In the original computation a large integer M was chosen, only the first $M - 1$ terms on the right were retained, and the equation (2.18) was assumed to hold at the points $\theta_k = \frac{k\pi}{2M}$, $k = 0, 1, \ldots, M$; this gives $M - 1$ equations since the equations for $k = 0$ and $k = M$ are trivial. Let the resulting coefficients be denoted by $p_{2m}(M)$, $m = 1, \ldots, M - 1$. It can be shown that $p_{2m}(M) \to p_{2m}$ as $M \to \infty$ and that $\sum_{m=1}^{M-1} m^4 |p_{2m}(M) - p_{2m}|^2 \to 0$ as $M \to \infty$.

If we try to apply a similar method to the earlier expansion (2.4) we must then apply the integral side condition $\int_0^{\frac{1}{2}\pi} F(\theta')d\theta' = 0$. It was found that this must be replaced by a suitable sum condition if equivalent convergence is to be obtained.

In the hydrodynamical application we are concerned with the total vertical force, which (apart from a constant factor) is

$$\int_0^{\frac{1}{2}\pi} \phi(a \sin \theta, a \cos \theta)\cos \theta \, d\theta = C\left\{\int_0^{\frac{1}{2}\pi} \Phi_0(Ka;\, \theta)\cos \theta \, d\theta \right. \tag{2.19}$$

$$\left. + \tfrac{1}{4}\pi Ka . p_2(Ka) + \sum_{m=1}^{\infty} (-1)^{m-1} \frac{p_{2m}(Ka)}{4m^2 - 1} \right\}, \tag{2.20}$$

from (2.1), by an elementary integration, where C is given by (2.3). The expression for C involves the series $\sum (-1)^{m-1} \frac{p_{2m}(Ka)}{2m - 1}$, in which terms are seen to be ultimately comparable with $1/m^4$; while (2.20) involves the series $\sum (-1)^{m-1} \frac{p_{2m}(Ka)}{4m^2 - 1}$, in which the terms are ultimately comparable with $1/m^5$. The convergence in both cases was found to be adequate for numerical purposes.

3. Another Approach, by Integral Equations

The method of the previous section can be applied in principle for all values of Ka, but in practice only for small and moderate values of Ka. For large values of Ka a method based on integral equations has been used to obtain analytical results. Let Green's theorem

$$\oint \left(\phi \frac{\partial G}{\partial n} - G \frac{\partial \phi}{\partial n}\right) ds = 0 \tag{3.1}$$

be applied to the potential $\phi(x, y)$ and to the fundamental solution $G(x, y; \xi, \eta)$

$$= \tfrac{1}{2} \log \frac{(x - \xi)^2 + (y - \eta)^2}{(x - \xi)^2 + (y + \eta)^2} - 2 \int_0^\infty e^{-k(y+\eta)} \cos k(x - \xi) \frac{dk}{k - K}$$

$$= \tfrac{1}{2} \log \frac{(x - \xi)^2 + (y - \eta)^2}{(x - \xi)^2 + (y + \eta)^2} + G_1(x, y; \xi, \eta),$$

and let the integration be taken along the boundary of the fluid closed by a large semi-circle at infinity. The functions ϕ and G are both harmonic and satisfy the boundary conditions on the free surface and at infinity. Let $(\xi, \eta) = (a \sin \alpha, a \cos \alpha)$ be taken on the semi-circle, and excluded from the region of integration by a small indentation. Then the only non-vanishing contributions to the left-hand side of (3.1) come from the circular arc and from the small indentation near (ξ, η). Let the value of $\phi(a \sin \theta, a \cos \theta)$ be denoted by $\phi(\theta)$. Then we find that

$$\pi\phi(\alpha) + a \int_0^{\frac{1}{2}\pi} \phi(\theta) \left\langle \frac{\partial G_1(\theta, \alpha)}{\partial r} + \frac{\partial G_1(-\theta, \alpha)}{\partial r} \right\rangle_{r=a} d\theta$$

$$= - \pi a V_0 \cos \alpha + a V_0 \int_0^{\frac{1}{2}\pi} \cos \theta \left\langle G_1(\theta, \alpha) + G_1(-\theta, \alpha) \right\rangle d\theta \tag{3.2}$$

in an obvious notation. (The functions ϕ and G depend also on the parameter Ka.) This is an integral equation of the second kind with a complicated kernel. An existence theory can be based on this; a numerical treatment should also be possible but has not been carried out. Difficulties may once again be expected for large Ka, when the kernel becomes large near the free surface, i.e., when θ and α are both near $\frac{1}{2}\pi$.

The integral equation (3.2) has however been used to obtain asymptotic results for large Ka (Ursell, 1953). We observe that, although $\phi(\alpha)$ is uniquely defined by the integral equation, the integral equation is not unique. For instance, if we apply Green's theorem to $\phi(x, y)$ and to $\Phi_0 = G(x, y; 0, 0)$, we obtain the integral relation

$$aA_0(\alpha)\int_0^{\frac{1}{2}\pi}\phi(\theta)\left\langle\frac{\partial\Phi_0}{\partial r}(Kr;\theta)\right\rangle_{r=a}d\theta = aV_0A_0(\alpha)\int_0^{\frac{1}{2}\pi}\cos\theta\left\langle\Phi_0(Ka,\ \theta)\right\rangle d\theta \quad (3.3)$$

for any arbitrary function $A_0(\alpha)$. Let (3.3) be added to the integral equation (3.2); in this way we obtain a new integral equation of the second kind, still with solution $\phi(\alpha)$. For a general choice of $A_0(\alpha)$ the kernel still remains large for large Ka, but there are physical arguments which suggest that $A_0(\alpha)$ can be chosen in such a way that the kernel of the resulting integral equation tends to 0 as Ka tends to ∞. This was verified by a detailed analytical examination (Ursell, 1953) which is too long to be given here. The new equation can therefore be solved by iteration when Ka is large enough, and the iterative solution is both asymptotic and convergent. In this way it was shown rigorously that

$$\phi(\alpha) + a\,V_0 \cos\alpha \to 0 \quad \text{as } Ka \to \infty.$$

Higher approximations were obtained from which the behaviour of the real and imaginary parts of the force coefficient (2.19) could be inferred for large Ka. (These are proportional to the added-mass and damping coefficients familiar in ship hydrodynamics.) The analytical results for large Ka joined up smoothly with the computations which had earlier been made for small and moderate values of Ka (see §2 above) and in this way the real and imaginary parts of the force coefficient were obtained in the whole range $0 < Ka < \infty$; see Ursell (1957).

References

URSELL, F. (1949) On the heaving motion of a circular cylinder on the surface of a fluid. Q. Jl. Mech. appl. Math. Vol. 2, pp.218-231.

URSELL, F. (1953) Short surface waves due to an oscillating immersed body. Proc. Roy. Soc. A, Vol. 220, pp.90-103.

URSELL, F. (1957) On the virtual mass and damping of ships at zero speed ahead. Proc. Symp. Behaviour of Ships in a Seaway. Wageningen pp.374-387.

CHAPTER 23 USE OF FINITE ELEMENTS IN MULTIDIMENSIONAL PROBLEMS IN PRACTICE

R. Wait

University of Liverpool

1. Introduction

Integral equations are widely used to solve boundary value problems in mathematical physics, as the use of such methods often reduces the dimensionality of the problem by one. This reduction is particularly significant in exterior boundary value problems and this type of solution is generally preferred to the use of finite difference or finite element methods for the differential form of the equations as in these latter methods it is necessary to introduce a false boundary at a suitable distance from the interior surface and such a procedure is difficult to perform successfully.

An exterior boundary value problem can be formulated as an integral equation in two ways.

i) We can assume that the solution to the boundary value problem, which we denote by $f(\underset{\sim}{q})$, is derived from a particular surface potential, which we denote by $\sigma(\underset{\sim}{q})$, as

$$f(\underset{\sim}{p}) = \iint_S \sigma(\underset{\sim}{q}) . K(\underset{\sim}{q},\underset{\sim}{p}) \, S_q, \tag{1.1}$$

where the kernel $K(\underset{\sim}{q},\underset{\sim}{p})$ is defined by the differential operator involved. If we consider this definition of $f(\underset{\sim}{p})$ as the point $\underset{\sim}{p}$ tends towards the boundary, we can apply the appropriate boundary condition and derive an integral equation for the potential $\sigma(\underset{\sim}{q})$. The form of the integral equation depends on both the boundary conditions (Neumann or Dirichlet?) and the type of potential represented by $\sigma(\underset{\sim}{q})$ (single or double layer potential?), and it can be an integral equation of the first or second kind (see Chapters 21, 22 and 24, and Hess and Smith (1967); Copley (1968)). Once we have obtained $\sigma(\underset{\sim}{q})$, the value of $f(\underset{\sim}{p})$ at any point can be found from (1.1) directly.

ii) An alternative approach is to discard the potential $\sigma(\underset{\sim}{q})$ and consider the appropriate Green's formula

$$\iint_S \left(f(\underset{\sim}{q}) . \frac{\partial}{\partial n_q}\left\{G(\underset{\sim}{q},\underset{\sim}{p})\right\} - G(\underset{\sim}{q},\underset{\sim}{p}) . \frac{\partial f(\underset{\sim}{q})}{\partial n_q}\right) dS_q = \lambda f(\underset{\sim}{p}) \tag{1.2}$$

where $\lambda = 1, \frac{1}{2}$ or 0 depending on the position of the point $\underset{\sim}{p}$, $G(\underset{\sim}{q},\underset{\sim}{p})$ denotes the appropriate Green's function and $\frac{\partial}{\partial n_q}$ denotes differentiation along the outward normal to S at $\underset{\sim}{q}$. If Neumann boundary conditions are given, that is if

$$\frac{\partial f(\underset{\sim}{q})}{\partial n} = g(\underset{\sim}{q}) \tag{1.3}$$

on the surface S, we can apply (1.2) with the point $\underset{\sim}{p}$ on the boundary to derive an integral equation for the value of $f(\underset{\sim}{q})$ on the surface S. Once the surface values of $f(\underset{\sim}{q})$ have been obtained, the value at any exterior point $\underset{\sim}{p}$ can be obtained directly from (1.2).

The exterior problem for the Helmholtz equation has received a certain amount of theoretical attention because the reduction of the problem to an integral equation by either of the above methods can lead to difficulties of non-uniqueness not inherent in the original problem (see Chapter 21). Various strategies which have been used to overcome this difficulty are discussed in Chapter 21, and a good survey of all available methods has recently been given by Burton (1973). The object of this chapter is to discuss a simple method for implementing one such strategy on a computer.

In any implementation of the integral equation formulation of a boundary value problem, by far the largest part of the computation will be the numerical evaluation of the integrals. For it is only in very exceptional circumstances that integration can be performed analytically. The most popular approach is to define the surface S as being made up of a collection of small surface elements or shells, thus

$$S = \sum_{j=1}^{N} S_j. \tag{1.4}$$

Then, on any element S_j, the solution is approximated by a constant value f_j, and then the integrals of the form

$$\iint_S f(\underset{\sim}{q}).\ K(\underset{\sim}{q},\underset{\sim}{p})\ dS_q \tag{1.5}$$

become

$$\sum_{j=1}^{N} f_j \iint_{S_j} K(\underset{\sim}{q},\underset{\sim}{p})\ dS_q \tag{1.6}$$

which can be evaluated numerically. If in each surface element S_j, a representative point p_j is selected, it follows from the form of the approximate solution that we assume

$$f(p_j) = f_j. \qquad (1.7)$$

Then if we evaluate the integral equation at the points p_j, $j=1,2,\ldots,N$ using the approximations given by (1.6) and (1.7) we have a system of linear equations in the unknown values f_j. (See Schenck, (1968) or Hess and Smith (1967).)

Although most authors choose to represent the surface as in (1.4), there is a notable division in the definition of the partition.

i) The first method is restricted to problems in which the surface S is well defined as a piecewise smooth surface such that each surface element S_j is defined by local co-ordinates (u_j, v_j) which are analytic functions of position. For this class of problems, it is possible to store a list of standard types of surfaces such as flat plates, cylindrical shells and spherical shells. Then each surface element S_j, is taken to be defined in terms of one of these standard geometries, (see Barach and Schenck (1969)).

This approach has been tested on a number of standard problems and found to be extremely efficient (see Schenck (1968); Van Buren (1970); Liu and Martenson (1969)). The major drawback of this method is in defining and updating a list of geometries that is sufficient to take account of surfaces encountered in practical problems (e.g. Submarine hulls, aircraft fuselages and whales!). See for example Hess (1971).

ii) In order to overcome the restrictive conditions of the above approach various authors have suggested that the surface S be replaced by a piecewise planar approximation $\hat{S}$, made up of surface elements S_j that are either triangular flat plates (see Chen and Schweikert (1963)) or quadrilateral flat plates (see Hess and Smith (1967)). This style of surface approximation compares badly with the first approach in those model problems for which either method is possible, see for example the results quoted by Liu and Martenson (1969). However, it has the great advantage that it can be applied to all surfaces, and in particular it can be applied to structural surfaces that are defined solely in terms of physical measurements at various points on the surface, as such points could be used to define the vertices of such a piecewise partition directly.

2. The Finite-Element-Type Solution of Integral Equations

Approximate solutions of integral equations, based on a representation of the surface by means of planar triangles resemble, in some ways, the approximate solutions of two dimensional boundary value problems by finite element methods, (see for example Zlamal (1968); and the references therein). The major difference is clearly that the domain involved is a closed surface and not planar.

In finite element approximations, it is usual to define the unknown solution in terms of parameters that correspond to function values at points on the boundary of the triangles, in particular to the values at the vertices. For example, using triangular elements we define points (p_i, q_i) $i=1,2,\ldots,6$ on each triangle where p and q are the co-ordinates of the planar element, as shown in Figure 1.

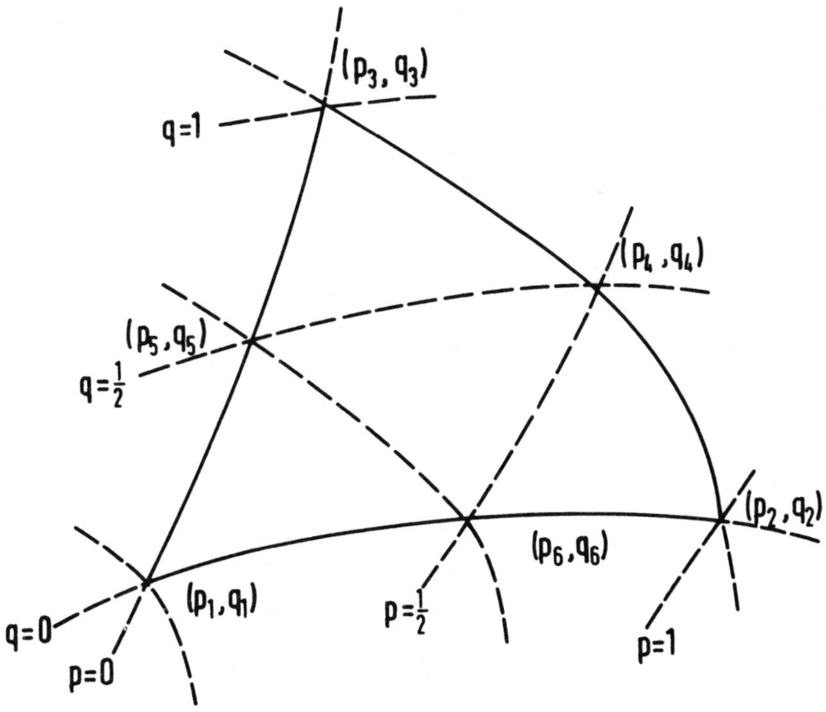

Figure 1 Partition of a triangular surface element in terms of local variables p and q.

Then possible approximations $\hat{f}$ to f are

$$\hat{f}(p,q) = 1/3\ f(p_1,q_1) + 1/3\ f(p_2,q_2) + 1/3\ f(p_3,q_3), \qquad (2.1)$$

if a piecewise constant approximation is still adequate, or

$$\hat{f}(p,q) = (1-p-q).f(p_1,q_1) + p\ f(p_2,q_2) + q\ f(p_3,q_3) \qquad (2.2)$$

which defines a piecewise linear approximation. Alternatively, by introducing the function values at the side mid-points, p_4,p_5 and p_6, it is possible to define a piecewise quadratic approximation

$$\begin{aligned}\hat{f}(p,q) = {} & (1-p-q)(1-2p-2q)\ f(p_1,q_1) + p(2p-1)\ f(p_2,q_2) \\ & + q(2q-1)\ f(p_3,q_3) + 4pq\ f(p_4,q_4) \\ & + 4q(1-p-q)\ f(p_5,q_5) + 4p(1-p-q)\ f(p_6,q_6). \qquad (2.3)\end{aligned}$$

The amount of computation is directly proportional to the number of surface points at which the integral equation is evaluated, and in a general triangular partition there are approximately twice as many triangles as vertices (each triangle has 3 vertices but most vertices are in 6 triangles). The form of piecewise constant approximation given by (2.1) therefore leads to a shorter computation than if one representative point is selected for each triangle as is done by most other authors (see Schenck (1968); Hess (1971); Hess and Smith (1967)).

The actual position of the surface can be defined in a similar manner in terms of p and q if we replace $\hat{f}(p,q)$ by $x = x(p,q)$, $y = y(p,q)$ and $z = z(p,q)$. The piecewise linear surface is defined over each triangle by equation (1.5) as

$$\begin{aligned} x &= (1-p-q)\ x_1 + px_2 + qx_3, \\ y &= (1-p-q)\ y_1 + py_2 + qy_3, \\ \text{and} \quad z &= (1-p-q)\ z_1 + pz_2 + qz_3. \qquad (2.4)\end{aligned}$$

It is equally possible to define a piecewise quadratic surface parametrically by equation (1.6) as

$$\begin{aligned} x = {} & (1-p-q)(1-2p-2q)\ x_1 + p(2p-1)\ x_2 + q(2q-1)x_3 \\ & + 4pqx_4 + 4q(1-p-q)\ x_5 + 4p(1-p-q)\ x_6, \qquad (2.5)\end{aligned}$$

together with similar expressions for y and z.

Clearly we can define a function $f(p,q)$ of the form (2.1), (2.2) or (2.3) over a surface defined by either (2.4) or (2.5). For the numerical results given in Section 4, we have taken approximations (2.1) or (2.2) defined on a surface given by (2.4), and (2.2) defined on a surface given by (2.5).

If the function $f(\underset{\sim}{q})$ is a piecewise smooth function on the surface S defined locally by (2.1), (2.2) or (2.3), it is possible to represent $f(\underset{\sim}{q})$ in the global form

$$f(\underset{\sim}{q}) = \sum_{l=1}^{m} f_l \, \phi_l(\underset{\sim}{q}), \tag{2.6}$$

where m is the number of nodes used in the approximation, f_l is the value of f at the node $\underset{\sim}{p}_l$ and the functions $\phi_l(\underset{\sim}{q})$ are piecewise smooth functions of position on the surface S, also defined locally by (2.1), (2.2) or (2.3), but such that

i) $\phi_l(\underset{\sim}{q}) \equiv 0$ in triangles <u>not</u> not containing the node $\underset{\sim}{p}_l$, and, with the exception of the simple piecewise constant approximation

ii) $\phi_l(\underset{\sim}{q}) = \delta_{jl}$ for $l,j=1,2,\ldots,m$.

It is not necessary to have the same degree of approximation for both the surface and the solution, but it would seem sensible from a purely empirical point of view for the degrees of approximation to be roughly equal.

3. Applications of the Method

Let us consider the application of the method to the exterior Helmholtz equation

$$\nabla^2 f + k^2 f = 0,$$

subject to

$$\frac{\partial f}{\partial n} = g$$

on the surface S, together with the radiation condition at infinity (see Chapter 21). The Green's function for this problem is

$$G(\underset{\sim}{p},\underset{\sim}{q}) = \left\{ \frac{e^{ik|\underset{\sim}{p}-\underset{\sim}{q}|}}{|\underset{\sim}{p}-\underset{\sim}{q}|} \right\}.$$

The method suggested by Schenck (1968) for solving the exterior

Helmholtz equation is to calculate $f(\underset{\sim}{q})$ on the surface by means of the Green's formula, which for this problem is the surface Helmholtz integral equation. The non-uniqueness is removed by constructing an overdetermined system by adding the interior Helmholtz integral equation evaluated at selected internal points. If $\underset{\sim}{p}$ is any point on the surface S the appropriate Helmholtz integral equation is

$$f(\underset{\sim}{p}) - \frac{1}{2\pi}\iint_S f(\underset{\sim}{q}) . \frac{\partial}{\partial n_q}\left\{\frac{e^{ik|\underset{\sim}{p}-\underset{\sim}{q}|}}{|\underset{\sim}{p}-\underset{\sim}{q}|}\right\} dS_q$$

$$= -\frac{1}{2\pi}\iint_S g(\underset{\sim}{q}) . \frac{e^{-ik|\underset{\sim}{p}-\underset{\sim}{q}|}}{|\underset{\sim}{p}-\underset{\sim}{q}|} dS_q, \qquad (3.1)$$

where $\underset{\sim}{q}$ varies over the whole surface and $|\underset{\sim}{p}-\underset{\sim}{q}|$ is the Euclidean distance between the points $\underset{\sim}{p}$ and $\underset{\sim}{q}$. The additional integral equation corresponding to any strictly internal point $\underset{\sim}{x}$ is

$$\iint_S f(\underset{\sim}{q}) . \frac{\partial}{\partial n_q}\left\{\frac{e^{ik|\underset{\sim}{x}-\underset{\sim}{q}|}}{|\underset{\sim}{x}-\underset{\sim}{q}|}\right\} dS_q$$

$$= -\iint_S g(\underset{\sim}{q}) . \frac{e^{-ik|\underset{\sim}{x}-\underset{\sim}{q}|}}{|\underset{\sim}{x}-\underset{\sim}{q}|} dS_q. \qquad (3.2)$$

We can apply the finite element approximation to the surface S and the unknown solution f and then evaluate equation (3.1) at each vertex of the surface approximation, together with equation (3.2) evaluated at a small number of internal points to give a set of algebraic equations of the form

$$A_h^* \underset{\sim}{f} = \underset{\sim}{g}_h. \qquad (3.3)$$

Where $\underset{\sim}{f}$ contains the values of the approximate solution at the vertices of the surface triangles and the components of A_h^* and $\underset{\sim}{g}_h$ are obtained by evaluating the integrals in (3.1) and (3.2).

If we evaluate equation (3.1) at each of the nodes of the surface partition, we obtain m equations of the system denoted by (3.3). If we assume that the approximate solution is given by (2.6) it follows that the entries of the lth row ($l \leq m$) of A_h^* are of the form

$$\delta_{jl} - \frac{1}{2\pi}\iint_S \phi_j(\underset{\sim}{q}) . \frac{\partial}{\partial n_q}\left\{\frac{e^{-ik|\underset{\sim}{p}-\underset{\sim}{q}|}}{|\underset{\sim}{p}-\underset{\sim}{q}|}\right\} dS_q \qquad (3.4)$$

and the lth ($l \leq m$) components of g_h is of the form

$$\frac{1}{2\pi}\iint_S g(\underset{\sim}{q}) . \frac{e^{-ik|\underset{\sim}{p}-\underset{\sim}{q}|}}{|\underset{\sim}{p}-\underset{\sim}{q}|} \; dS_q \qquad (3.5)$$

The entries in the additional m′ rows, corresponding to the m′ interior points are of the form

$$\frac{1}{2\pi}\iint_S \phi_j(\underset{\sim}{q}) \frac{\partial}{\partial n_q}\left\{\frac{e^{-ik|\underset{\sim}{x}-\underset{\sim}{q}|}}{|\underset{\sim}{x}-\underset{\sim}{q}|}\right\} dS_q, \qquad (3.6)$$

for $j=1,2,\ldots,m$ and

$$\frac{1}{2\pi}\iint_S g(\underset{\sim}{q}) \frac{e^{-ik|\underset{\sim}{x}-\underset{\sim}{q}|}}{|\underset{\sim}{x}-\underset{\sim}{q}|} \; dS_q. \qquad (3.7)$$

The algebraic system thus defined is then solved by some least squares or l_1 approximation procedure.

4. Numerical Results

As a test for the accuracy of this approach, we apply the method to the model problem of a uniformly vibrating sphere. This problem has been solved by Schenck (1968), using an exact representation of the surface, that is, using the alternative approach, set out in Section 1, in which the surface elements are taken to be parts of spherical shells etc. As the surface pressure is constant for this model problem, a piecewise constant approximation is capable of providing the exact solution if there is no error in the surface representation. Thus in the numerical calculations used by Schenck (1968) and Van Buren (1970) the only errors are those introduced by the numerical integration, and results are obtained which are correct to 0.3%.

An exact representation is not however possible with the finite element type of surface, since we have introduced errors into the surface representation. Tables I and II give the results obtained with one simple triangulation, in which we first divide the sphere into n "lemon slices" and then cut these into sections at constant latitude difference. The resulting quadrilaterals are then halved, and the curvilinear triangles approximated linearly (see Figure 2a) or quadratically (Figure 2b). The Tables give the results for both piecewise constant, and piecewise linear approximations of the solutions, as the two sets of computed values for this model problem only differed in the fourth and fifth significant figure. As can be seen from Table I the errors introduced by the piecewise planar surface approximation are quite large, and the convergence of the solution, as the number of triangles is increased, is very slow. The results of Table II are clearly much better.

TABLE I Piecewise linear surface, piecewise constant/linear approximate solution

No. of triangles	least sq. error
64	11.5%
120	11 %
144	10.7%
224	10.5%

We can understand these results if we assume that the solution converges in a way analogous to the convergence of the finite element solution of two-dimensional boundary value problems (see Zlamal (1968)), then

$$||\text{error}|| = c. \frac{h^s_{max}}{\sin \theta_{min}}$$

where h_{max} is the largest side of any triangle in the partition and θ_{min} is the smallest angle of any triangle in the partition; and c and s are constants independent of the partition. Such an assumption is consistent with the results if we observe that the partition illustrated in Figure 2 leads to a sequence of θ_{min} (at the poles) which tend to zero. The results of Table I suggest that s=1 in the case of a piecewise planar partition, while in the case of the piecewise quadratic surface a value of $s \approx 2$ would lead to results consistent with those given in Table II. The results may also be improved by a different partition such as the regular polyhedra used by Chen and Schweikert (1963) for example, for which none of the triangles becomes pathologically thin.

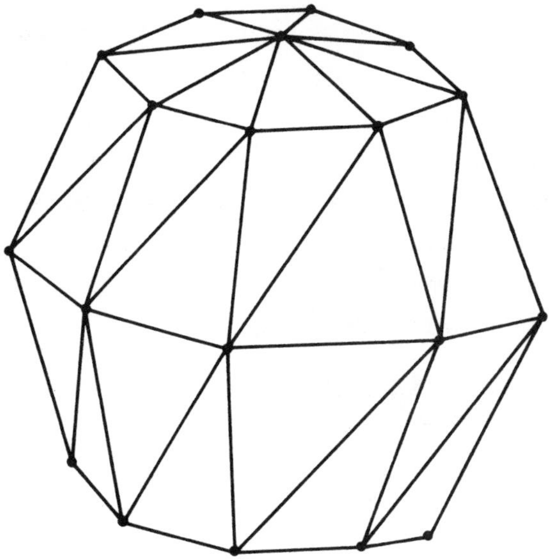

Figure 2A Example of a piecewise linear approximation to a spherical surface.

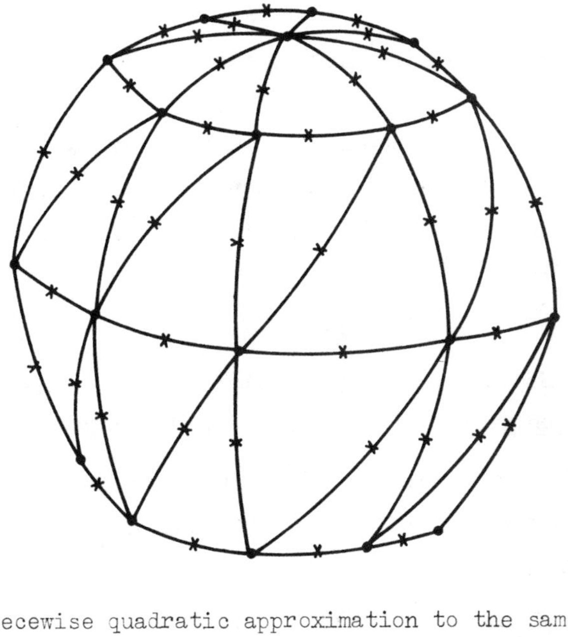

Figure 2B A piecewise quadratic approximation to the same sphere.

Another way of computing the solution would be to collocate equation (3.1) at m points that are not necessarily the vertices of the partition; alternatively it might improve the solution if (3.1) was evaluated at n(> m) points and then the (l_2 or l_1) solution of the resulting n+m' equations determined. Both of these alternatives introduce additional difficulties in evaluating the integrals, as the points at which the integrand becomes singular would no longer be the vertices of the surface triangles.

TABLE II Piecewise quadratic surface, piecewise constant/linear approximate solution

No. of triangles	least sq. error
24	8.8%
64	2.1%
120	1.3%
224	.6%

The results in Table I were obtained on the ICL 4130 at University of Dundee, whereas those in Table II were obtained either on a Modular 1 at University of Liverpool or on the C.D.C. 7600 at U.M.R.C.C.

Acknowledgements

This work was financed in part by the Ministry of Defence, with an A.U.W.E. Research Fellowship.

This paper contains material previously included in a Ministry of Defence report written by the present author (Wait, (1972)), and is published here by permission of the Director, Admiralty Underwater Weapons Establishment, Portland, Dorset.

References

BARACK, D. and SCHENCK, H.A. (1969). CHIEF computer program description, Mart. Contract Ref. 66, N.U.S.W.; San Diego, Calif.

BURTON, A.J. (1963). The solution of Helmholtz' equation in exterior domains using integral equations, N.P.L. Report NAC 30, Teddington.

BURTON, A.J. and MILLER, G.F. (1970). The application of integral equation methods to the numerical solution of some exterior boundary value problems, Proc. R. Soc. vol. A 323, pp 201-210.

CHEN, L.H. and SCHWEIKERT, D.G. (1963). Sound radiation from an arbitrary body, J. Acoust. Soc. Am., vol. 35, pp 1626-1632.

CHERTOCK, G. (1964). Sound radiation from vibrating surfaces, J. Acoust. Soc. Am., vol. 36, pp 1305-1313.

COPLEY, L.G. (1968). Fundamental results concerning integral representations in acoustic radiation, J. Acoust. Soc. Am., vol. 44, pp 28-32.

HESS, J.L. and SMITH, A.M.O. (1964). Calculation of non lifting potential flow about arbitrary three dimensional bodies, J. Ship. Res., vol. 8, pp 22-44.

HESS, J.L. and SMITH, A.M.O. (1967). Calculation of potential flow about arbitrary bodies, Prog. Aeronaut. Sci., vol. 8, pp 1-138.

HESS, J.L. (1971). Numerical solution of the integral equation for the Neumann problem with application to aircraft and ships, presented at S.I.A.M. Symp. Numerical solution of integral equations with physical applications, Univ. of Wisconsin.

LIU, H.K. and MARTENSON, A.J. (1969). A numerical technique for solving acoustic radiation problems, Unpublished Ms.

PANICH, O.I. (1965). On the question of the solvability of the exterior boundary value problem for the wave equation and Maxwell's equations, Russ. Math. Surv., vol. 20, pp 221-226.

SCHENCK, H.A. (1968), Improved integral formulation for acoustic radiation problems, J. Acoust. Soc. Am., vol. 44, pp 41-58

VAN BUREN, A.L. (1970). A test of the capabilities of CHIEF in the numerical calculation of acoustic radiation from arbitrary surfaces, N.R.L. Report 7160; Washington.

WAIT, R. (1972). Final report of the Ministry of Defence Fellow, Univ. Dundee.

ZLAMAL, M. (1968). On the finite element method, Num. Math., vol. 12, pp 394-409.

CHAPTER 24 POTENTIAL PROBLEMS IN THREE DIMENSIONS

G. T. Symm

National Physical Laboratory

1. Introduction

Many problems of potential theory are governed by Poisson's equation

$$\nabla^2 f = -4\pi\rho, \qquad (1.1)$$

in which ρ represents a known function and, in three dimensions, in terms of rectangular Cartesian coordinates x, y and z,

$$\nabla^2 = \partial^2/\partial x^2 + \partial^2/\partial y^2 + \partial^2/\partial z^2. \qquad (1.2)$$

If $\rho\equiv 0$, equation (1.1) reduces to Laplace's equation, in which case associated boundary-value problems may be formulated in terms of integral equations analogous to those in two dimensions (described in Chapter 20) by replacing $\log|\underline{p}-\underline{q}|$ and π by $1/|\underline{p}-\underline{q}|$ and -2π respectively. The vector variables $\underline{p}$ and $\underline{q}$ now denote points in three-dimensional space and we suppose that, corresponding to the plane domain D with boundary L, we have a volume V bounded by a closed surface S.

In this chapter we shall assume that $\rho\neq 0$ and consider the problem of solving

$$\nabla^2 f(\underline{p}) = -4\pi\rho(\underline{p}), \ \underline{p}=(x,y,z)\in V, \qquad (1.3)$$

subject to the boundary condition

$$\frac{\partial f(\underline{p})}{\partial n} = 0, \ \underline{p}\in S, \qquad (1.4)$$

where $\frac{\partial}{\partial n}$ denotes differentiation along the normal to the surface S directed into the volume V. We shall formulate this boundary-value problem in terms of integral equations in two ways and we shall discretise these equations in a manner suggested by Lynn and Timlake (1968). This discretisation is designed to preserve as far as possible the properties of the integral operators involved.

Note that, from (1.3),

$$\int_V \nabla^2 f(\underline{p})dV_p = -4\pi\int_V \rho(\underline{p})dV_p, \qquad (1.5)$$

where dV_p denotes a volume element at $\underline{p}$, whilst, by Gauss's divergence theorem,

$$\int_V \nabla^2 f(\underline{p}) dV_p = -\int_S \frac{\partial f(\underline{p})}{\partial n} dS_p \qquad (1.6)$$

where dS_p represents an element of surface area at $\underline{p}$. Consequently, any solution of equation (1.3) must satisfy

$$\int_S \frac{\partial f(\underline{p})}{\partial n} dS_p = 4\pi \int_V \rho(\underline{p}) dV_p \qquad (1.7)$$

and, in particular, for the problem posed above to have a solution, we must have

$$\int_V \rho(\underline{p}) dV_p = 0 \; ; \qquad (1.8)$$

we shall therefore assume this condition to hold.

2. First Formulation

Assuming that the function ρ is differentiable in V, a particular integral of equation (1.3) is given by

$$g(\underline{p}) = \int_V (1/|\underline{p}-\underline{q}|)\, \rho(\underline{q}) dV_q. \qquad (2.1)$$

If then we let

$$f(\underline{p}) = g(\underline{p}) + h(\underline{p}), \qquad (2.2)$$

the function h satisfies Laplace's equation in V with the boundary condition

$$\frac{\partial h(\underline{p})}{\partial n} = -\frac{\partial g(\underline{p})}{\partial n}, \; \underline{p} \in S. \qquad (2.3)$$

Thus, representing h as a simple-layer potential

$$h(\underline{p}) = \int_S (1/|\underline{p}-\underline{q}|)\, \sigma(\underline{q}) dS_q, \; \underline{p} \in V, \qquad (2.4)$$

we obtain, corresponding to equation (3.3) of Chapter 20, the integral equation

$$\int_S K(\underline{p},\underline{q})\, \sigma(\underline{q}) dS_q - \sigma(\underline{p}) = k(\underline{p}),\ \underline{p}\in S, \tag{2.5}$$

where

$$K(\underline{p},\underline{q}) = [\partial(1/|\underline{p}-\underline{q}|)/\partial n_p]/2\pi, \tag{2.6}$$

$\partial/\partial n_p$ denoting the inward normal derivative at $\underline{p}$, and

$$k(\underline{p}) = -\frac{1}{2\pi}\frac{\partial g(\underline{p})}{\partial n}\ . \tag{2.7}$$

If we can solve equation (2.5) for σ (see Section 4), then h is given by (2.4) and hence, from (2.2), we obtain the required solution

$$f(\underline{p}) = g(\underline{p}) + \int_S (1/|\underline{p}-\underline{q}|)\sigma(\underline{q})dS_q,\ \ \underline{p}\in\overline{V}\ , \tag{2.8}$$

where $\overline{V}$=V+S. Note that, since any constant satisfies the homogeneous interior Neumann problem, h and hence f are unique only to within an arbitrary additive constant.

3. Second Formulation

If we apply Green's formula, sometimes called the third identity (Kellogg (1929)), to the function f on the surface S, we obtain, corresponding to equation (3.11) of Chapter 20,

$$\int_S (1/|\underline{p}-\underline{q}|)\frac{\partial f(\underline{q})}{\partial n_q} dS_q - 2\pi\int_S K(\underline{q},\underline{p})\ f(\underline{q})dS_q +$$

$$\int_V (1/|\underline{p}-\underline{q}|)\ \nabla^2 f(\underline{q})dV_q + 2\pi f(\underline{p}) = 0,\ \ \underline{p}\in S. \tag{3.1}$$

Hence, substituting from (1.3), (1.4) and (2.1), we have the integral equation

$$f(\underline{p}) - \int_S K(\underline{q},\underline{p})\ f(\underline{q})dS_q = 2g(\underline{p}),\ \underline{p}\in S. \tag{3.2}$$

If we can solve this equation for f (see again Section 4), we have immediately the surface values of the required solution and we may then determine f within V from the Green's formula obtained by replacing the term $2\pi f(\underline{p})$ in (3.1) by $4\pi f(\underline{p})$, $\underline{p}\in V$.

Note that from the definition of K, (2.6), it follows that

$$\int_S K(\underline{q},\underline{p})dS_q = 1, \ \underline{p}\in S. \tag{3.3}$$

Hence the homogeneous form of equation (3.2) is satisfied by any constant and, as before, f is unique only to within an arbitrary additive constant.

4. Existence of Solutions

Result (3.3) implies that the integral operator appearing in equation (3.2) has a unit eigenvalue and a corresponding eigenfunction $e(\underline{p}) = 1$, $\underline{p}\in S$. From the Fredholm theory therefore, the associated integral operator, which appears in equation (2.5), also has a unit eigenvalue to which there corresponds some eigenfunction $\eta(\underline{p})$ satisfying the homogeneous equation

$$\int_S K(\underline{p},\underline{q})\,\eta(\underline{q})dS_q - \eta(\underline{p}) = 0, \quad \underline{p}\in S. \tag{4.1}$$

It follows, on again applying the Fredholm theory, that the integral equation (2.5) has a solution only if

$$\int_S k(\underline{p})dS_p = 0, \tag{4.2}$$

i.e., from (2.7), only if

$$\int_S \frac{\partial g(\underline{p})}{\partial n}\, dS_p = 0, \tag{4.3}$$

which, on account of (2.3), is the Gauss condition for the interior harmonic function h. Now, from (2.1) and (2.6),

$$\frac{\partial g(\underline{p})}{\partial n} = 2\pi \int_V K(\underline{p},\underline{q})\rho(\underline{q})dV_q, \ \underline{p}\in S \tag{4.4}$$

whence, integrating with respect to $\underline{p}$ over the surface S and interchanging the order of the double integration, justified by Fubini's theorem (Petrovsky (1971)), we have

$$\int_S \frac{\partial g(\underline{p})}{\partial n}\, dS_p = 2\pi \int_V \rho(\underline{q}) \int_S K(\underline{p},\underline{q})dS_p\, dV_q. \tag{4.5}$$

However, analogously with (3.3),

$$\int_S K(\underline{p},\underline{q})dS_p = 2, \ \underline{q}\epsilon V, \tag{4.6}$$

and hence, from (4.5),

$$\int_S \frac{\partial g(\underline{p})}{\partial n} dS_p = 4\pi \int_V \rho(\underline{q})dV_q. \tag{4.7}$$

It follows that condition (4.3) is an immediate consequence of the assumption (1.8).

Similarly, the integral equation (3.2) has a solution only if

$$\int_S g(\underline{p})\ \eta(\underline{p})dS_p = 0 \tag{4.8}$$

and this condition too follows directly from (1.8). Multiplying (2.1) by $\eta(\underline{p})$, integrating over S, and reversing the order of the double integration as above, we obtain

$$\int_S g(\underline{p})\ \eta(\underline{p})dS_p = \int_V \rho(\underline{q}) \int_S (1/|\underline{p}-\underline{q}|)\ \eta(\underline{p})dS_p\ dV_q, \tag{4.9}$$

whence we deduce, from (4.1), that

$$\int_S g(\underline{p})\ \eta(\underline{p})dS_p = m \int_V \rho(\underline{q})dV_q, \tag{4.10}$$

where m is an arbitrary constant (satisfying the homogeneous interior Neumann problem).

5. Discretisation

Dividing the surface S into n smooth elements, with areas μ_j, j=1,2,...,n, we define

$$\sigma_j = \int_j \sigma(\underline{p})dS_p, \tag{5.1}$$

$$f_j = \int_j f(\underline{p})dS_p/\mu_j, \tag{5.2}$$

$$k_j = \int_j k(\underline{p})dS_p, \tag{5.3}$$

$$g_j = 2\int_j g(\underline{p})dS_p/\mu_j, \tag{5.4}$$

and

$$\Omega_{j,q} = \int_j K(\underline{p},\underline{q})dS_p, \tag{5.5}$$

where $\int_j$ denotes integration over the jth element of S. Note that $2\pi\Omega_{j,q}$ represents the solid angle subtended at the point $\underline{q}$ by the jth element of S, from which fact (3.3) follows. Note also that, from (4.2) and (5.3),

$$\underline{e}^* \, \underline{k} = 0, \tag{5.6}$$

where $\underline{e}^* = (1,1,\ldots,1)$ and $\underline{k}^* = (k_1,k_2,\ldots,k_n)$.

Now, integrating equation (2.5) over the ith element of S, we have

$$\sigma_i = \int_i \Big[\sum_{j=1}^{n} \int_j K(\underline{p},\underline{q})\, \sigma(\underline{q})dS_q\Big]dS_p - k_i, \tag{5.7}$$

from which, applying Fubini's theorem, we obtain

$$\sigma_i = \sum_{j=1}^{n} \int_j \Omega_{i,q}\, \sigma(\underline{q})dS_q - k_i. \tag{5.8}$$

Following Lynn and Timlake (1968), we thus obtain

$$\underline{\sigma} = \underline{B}\,\underline{\sigma} - \underline{k} + \underline{\zeta}, \tag{5.9}$$

where $\underline{B} = (b_{ij})$ is an n×n matrix with elements

$$b_{ij} = \int_j \Omega_{i,p}\, dS_p/\mu_j, \quad i,j=1,2,\ldots,n, \tag{5.10}$$

and $\underline{\zeta}$ is the vector of corresponding 'remainder' terms.

Similarly, integrating equation (3.2) over the ith element of S, and dividing by μ_i for each i=1,2,...,n, we obtain

$$\underline{f} = \underline{B}^*\underline{f} + \underline{g} + \underline{\xi}, \tag{5.11}$$

where $\underline{B}^*$ is the transpose of $\underline{B}$ and $\underline{\xi}$ is a vector of 'remainder' terms.

At this point we note that, from (5.5) and (5.10), we have,

corresponding to (3.3),

$$\sum_{i=1}^{n} b_{ij} = 1, \; j=1,2,\ldots,n, \tag{5.12}$$

or, in matrix notation,

$$\underline{e}^* \, \underline{B} = \underline{e}^*. \tag{5.13}$$

Hence, $\underline{I}-\underline{B}$ is singular ($\underline{B}$ has a unit eigenvalue) and we cannot immediately drop the remainder terms $\underline{\zeta}$ and $\underline{\xi}$ from equations (5.9) and (5.11) since solutions of the resulting linear algebraic systems may not exist. However, in appropriate circumstances, e.g. when the volume V is convex, the unit eigenvalue of $\underline{B}$ is simple and in this case, to which we restrict our attention from now onwards, it is permissible to drop the term $\underline{\zeta}$ from equation (5.9) since the resulting approximating system does have a solution in view of (5.6). However, we still cannot drop the term $\underline{\xi}$ from equation (5.11), for if $\underline{\eta}$ is the eigenvector of B satisfying

$$\underline{B} \, \underline{\eta} = \underline{\eta} \tag{5.14}$$

there is no guarantee that $\underline{\eta}^* \, \underline{g} = 0$ (although $\underline{\eta}^*(\underline{g}+\underline{\xi}) = 0$ for the continuous problem to have a solution). We must therefore proceed somewhat differently and we do so using the concept of a Wielandt deflation (Wilkinson (1965)).

Introducing a vector $\underline{y}$ such that

$$\underline{y}^* \, \underline{e} = 1, \tag{5.15}$$

we define

$$\underline{C} = \underline{C}(\underline{y}) = \underline{B} - \underline{y} \, \underline{e}^*. \tag{5.16}$$

Then $\underline{C}$ is an n×n matrix with the same eigenvalues as $\underline{B}$ except that the unit eigenvalue of $\underline{B}$ is replaced by an eigenvalue of $\underline{C}$ at the origin, so that $\underline{I}-\underline{C}$ is non-singular. If then we seek that solution of equation (2.5) which satisfies

$$\int_S \sigma(\underline{q}) dS_q = 0, \tag{5.17}$$

or equivalently $\underline{e}^*\underline{\sigma}=0$, we may replace equation (5.9) by

$$\underline{\sigma} = \underline{C} \, \underline{\sigma} - \underline{k} + \underline{\zeta}. \tag{5.18}$$

Similarly, if we seek that solution of equation (3.2) corresponding to which $\underline{y}^*\underline{f}=0$, we obtain, equivalent to (5.11), the equation

$$\underline{f}_y = \underline{C}^* \underline{f}_y + \underline{g} + \underline{\xi}, \tag{5.19}$$

in which the subscript indicates that the solution is dependent upon the vector $\underline{y}$.

Our final discretisations of the integral equations (2.5) and (3.2) are now derived by dropping the terms $\underline{\zeta}$ and $\underline{\xi}$ from (5.18) and (5.19). Thus we obtain

$$\hat{\underline{\sigma}} = \underline{C}\,\hat{\underline{\sigma}} - \underline{k} \tag{5.20}$$

and

$$\hat{\underline{f}}_y = \underline{C}^* \hat{\underline{f}}_y + \underline{g}. \tag{5.21}$$

A similar discretisation of equation (2.8) gives

$$\hat{\underline{f}} = \underline{R}\,\hat{\underline{\sigma}} + 0.5\,\underline{g}, \tag{5.22}$$

where $\underline{R} = (r_{ij})$ and

$$r_{ij} = \int_j \int_i (1/|\underline{p}-\underline{q}|)\,dS_p\,dS_q/\mu_i\mu_j \;. \tag{5.23}$$

Hence, by solving equations (5.20) and (5.21), we may determine two approximations $\hat{\underline{f}}$ and $\hat{\underline{f}}_y$ to the vector $\underline{f}$.

Lynn and Timlake (1968) go on to consider the effects on $\hat{\underline{f}}_y$ of different choices of the vector $\underline{y}$ (denoted by $\underline{p}$ in their notation) and to obtain bounds on the errors in the approximate solutions $\hat{\underline{f}}$ and $\hat{\underline{f}}_y$. Further error analysis is carried out by Ikebe, Lynn and Timlake (1969) and we note, in conclusion, that Ikebe (1972) reports that the approximating systems of equations (5.20) and (5.21) have been solved (by iterative methods) for values of n up to 2500.

References

IKEBE, Y. (1972) The Galerkin method for the numerical solution of Fredholm integral equations of the second kind. SIAM Review vol. 14, pp.465-491.

IKEBE, Y., LYNN, M.S. and TIMLAKE, W.P. (1969) The numerical solution of the integral equation formulation of the single interface Neumann problem. SIAM J. Numer. Anal. vol. 6, pp.334-346.

KELLOGG, O.D. (1929) Foundations of Potential Theory. Springer.

LYNN, M.S. and TIMLAKE, W.P. (1968) The numerical solution of singular integral equations of potential theory. Num. Math. vol. 11, pp. 77-98.

PETROVSKY, I.G. (1971) Lectures on the Theory of Integral Equations. MIR, Moscow.

WILKINSON, J.H. (1965) The Algebraic Eigenvalue Problem. Clarendon.

CHAPTER 25 SCATTERING PROBLEMS IN QUANTUM MECHANICS

L.M. Delves

University of Liverpool

1. Introduction

The quantum mechanical scattering of one particle fron another - an electron from a proton, or a neutron from a uranium nucleus, say, is a wave phenomenon. As such, it is described by equations which have some points of similarity with the diffraction theory equations discussed in Chapter 21. We give here a brief discussion of the equations which arise in the scattering of two and three particles, and look at their particular difficulties, and how these have been overcome numerically and analytically.

The equations all follow from the basic energy equation, which has the familiar form

$$T + V = E \tag{1.1}$$

where T is the kinetic energy, V the potential energy, and E the (conserved) total energy of the system. In nonrelativistic quantum mechanics, T and V are operators in a Hilbert space whose elements are the state vectors of the system. If we choose a particular coordinate system these operators can be written down explicitly. Suppose that the system contains N unit mass particles interacting in pairs. Then choosing as coordinates the position vectors $\underline{r}_i$ of these particles, (1.1) takes on the form

$$[-\sum_{i=1}^{N} \nabla_i^2 + \sum_{i<j} V_{ij}]\, \psi(\underline{r}_1 \dots \underline{r}_N) = E\,\psi \tag{1.2}$$

where the second sum is over all pairs of particles, and ψ, the wave-function of the system, describes the motion of the particles in a probabilistic sense.

The first term represents a sum over the kinetic energies of the particles. The individual potential terms V_{ij} may take a number of forms depending on the system considered; the simplest form is termed a local potential:-

$$V_{ij} = V(|r_i - r_j|) \qquad \text{(local potential)} \tag{1.3}$$

2. Collision Between Two Particles

The simplest scattering system has only two particles. Then, there is only one potential term V_{12}. Moreover, as in classical mechanics, we can separate out the centre of mass motion of the system by introducing the relative coordinates $\underline{r} = \underline{r}_1 - \underline{r}_2$, $\underline{R} = (\underline{r}_1 + r_2)/2$. The resulting equation involves only $\underline{r}$, and has the form

$$[-\tfrac{1}{2}\nabla^2 + V_{12} - E]\,\psi(\underline{r}) = 0 \tag{2.1}$$

To this differential equation we must add the boundary conditions. If we specify the initial condition: particle one coming from some direction (which we take as the z-axis) with initial momentum $\underline{p}$, say; then in other directions the solution is required to be "wholly outgoing" at infinity (cf the radiation condition imposed in chapter 21). For a given $\underline{p}$, we make the expansion

$$\psi = \sum_q t(\underline{p},\underline{q},E)\,e^{i\,\underline{q}\cdot\underline{r}} \tag{2.2}$$

Formally, we can then derive an integral equation for the so-called "t-matrix" t. Introducing the Green's function G_0:

$$G_0 = [-\tfrac{1}{2}\nabla^2 - E]^{-1}$$

we find

$$t = V - VG_0\,t \tag{2.3}$$

We look at the structure of this equation more closely.

We notice that if V_{12} is spherically symmetric, equation (2.1) <u>separates</u> in polar coordinates. We therefore write

$$t(\underline{p},\underline{q},E) = \sum_{l=0}^{\infty} (2l+1)\, t_l(p,q,E)\, P_l(\cos\theta) \tag{2.4}$$

where $P_l(x)$ is a **Legendre polynomial.**
We then find that (2.3) has the explicit form (called in this field the "Lippmann-Schwinger" equation):

$$t_l(p,q,E+i\epsilon)=v_l(p,q)-\lambda\int_0^\infty \frac{r^2\,dr}{r^2-E-i\epsilon}\,v_l(p,r)t_l(r,q,E+i\epsilon) \qquad (2.5)$$

In this equation, $v_l(p,q)$ is the coefficient derived from the Fourier transform of V_{12} by making an expansion of the type (2.4); it may be assumed as given. For a "local" potential (1.3) it has the form:

$$v_l(p,q) = v_l(p-q) \qquad (2.6)$$

The constant λ depends on the mass of the particles involved. Finally, we note the occurrence of $(E + i\epsilon)$ rather than E. This stems from the Green's function G_0, equation (2.3); the integral in (2.5) is in fact a contour integral, and the $i\epsilon$ is conventionally inserted to define the contour to be used in the neighbourhood of the singularity at $r^2 = E$.

In the next two sections we discuss the numerical solution of (2.5).

3. Solution for Negative E

The physical derivation of (2.5) suggests that p,q and E are real, and that since p is the momentum of the incoming particle we should set $\frac{p^2}{2m} = E$. In fact, however, it turns out that the solution is of interest for all complex values of p, q and E. We limit ourselves to real values here, and consider first the case $E<0$. Then the kernel of (2.5) is regular and we may drop the term $i\epsilon$. We make some general comments on the resulting equation:-

1) The variable q appears only as a parameter; we may solve for arbitrary fixed q.

2) Although (2.5) is formally singular, due to the infinite range of integration, the function v_l is in practice $L_2(0,\infty)$ and the infinite interval is not a real difficulty.

Since the kernel is L_2, any of the standard methods of solution are applicable. The most straightforward is to use a quadrature method, and this procedure converges very rapidly provided that the quadrature rule used reflects the rather slow decay to zero of the integrals, since typically we have for fixed q and large p

$$t_1(p,q),\ v_1(p,q) = O(p^{-2})$$

An appropriate rule is the "Gauss-Rational" rule (Stroud and Secrest (1966), p.52). However, this procedure has two disadvantages:-

a) It produces a solution only for fixed q. We can obtain the q-dependence by repeated solution, but the matrices involved are functions of q so that this procedure is expensive.

b) Moreover, it can be shown that because $v_1(p,q)=v_1(q,p)$ is symmetric, so is the solution $t_1(p,q)$. It would be nice if the approximate solution retained this exact symmetry, but the quadrature procedure does not.

There are two ways to retain this symmetry. The first, and most commonly used, is to introduce a <u>separable expansion</u> of the function $v_1(p,q)$. If we have an expansion of the form

$$v_1(p,q) = \sum_{i=1}^{\infty} b_i\, l_i(p) l_i(q) \tag{3.1}$$

which we truncate after N terms, then an extension of the discussion of Chapter 1 on degenerate (finite rank) kernels shows that an <u>exact</u> solution of the <u>truncated</u> equation is

$$t_1^{(N)}(p,q) = \sum_{i=1}^{N} c_i\, l_i(p)\, l_i\,(q) \tag{3.2}$$

where the c_i may be found as the solution of a set of N linear algebraic equations. The form (3.2) automatically preserves the symmetry of t_1.

For an arbitrary (but even smooth) choice of $v_1(p,q)$, it is well known that eigenfunction expansions of the form (3.1) may converge only slowly; hence the required value of N may be quite high. In practice, N=4 is considered high by workers in this field, and N=1 is a very common choice (that is, the potential $v_1(p,q)$ is arbitrarily assumed separable from the beginning). Of course, such a choice makes the solution of (2.5) completely trivial; but see section 5 below.

An alternative approach is to solve (2.5) by a Ritz-Galerkin procedure (Delves 1973). For such a procedure we would normally choose an expansion set $\left\{h_i(p)\right\}$ and write

$$t_1(p,q,E) \approx \sum_{i=1}^{N} a_i(q,E)\, h_i(p) \qquad (3.3)$$

deriving by the Ritz procedure a set of equations for the a_i for each fixed E,q. Let us instead write the double expansion

$$t_1(p,q,E) \approx \sum_{i=1}^{N} \sum_{j=1}^{N} a_{ij}(E)\, h_i(p)\, h_j(q) \qquad (3.4)$$

Equation (3.3) would seem less efficient than (3.2): it contains terms with $i \neq j$. However, the underlying expansion (3.1) is not valid for arbitrary $l_i(p)$; rather, the eigenfunctions l_i must be computed one by one, from the given $v_1(p,q)$, and this is an expensive process. But the $h_i(p)$ are completely at our disposal, and can in practice be chosen so that (3.3) converges extremely rapidly.

We now have N^2 coefficients to compute in (3.3); and we would like to ensure that the matrix $A=(a_{ij})$ is symmetric. We outline one procedure (see Delves (1973)) which is both efficient and stable. We write

$$a_i(q) = \sum_{j=1}^{N} a_{ij}\, h_j(q) \qquad (3.5)$$

thus re-setting (3.4) into the form (3.3).

We then choose a weight function w(q) and impose the integral condition

$$\int_0^\infty h_k(q)a_i(q)w(q)dq = \sum_{j=1}^{N} a_{ij} \int_0^\infty h_k(q)h_j(q)w(q)dq \quad k=1,\ldots,N \qquad (3.6)$$

which is an identity if (3.5) holds. Now the Galerkin equations for the coefficients $a_i(q)$ are

$$\sum_{j=1}^{N} L_{ij}\, a_j(q) = g_i(q) \qquad (3.7)$$

where

$$L_{ij}=\int_0^\infty w(p)h_i(p)h_j(p)dp+\lambda \int_0^\infty dr \int_0^\infty dp w(p) \frac{r^2}{r^2-E} v_1(p,r)h_i(p)h_j(r) \qquad (3.8a)$$

and

$$g_i= \int_0^\infty w(p)\; v_1(p,q)\; h_i(p)\; dp. \qquad (3.8b)$$

If we solve (3.7) as a function of q, we can insert the $a_i(q)$ into (3.6) to obtain a set of N linear equations for the N coefficients a_{ij}, j=1,...,N, for each i=1,...,N. We make the following comments on this procedure.

1) In practice each integration is performed with, say, a P-point Gauss quadrature rule. We need therefore solve (3.7) only for each of the P quadrature points $q_1,\ldots,q_P$.

2) In these P solutions only the vector $\underline{g}$ depends on q; the matrix L does not. We can therefore compute, and triangulate, L <u>once only</u>. The time taken to produce these P solutions is then, since in practice $N< \frac{1}{2}P$, rather less than twice the time for a solution at a single value of q.

3) The matrix of the linear equations (3.6) for the a_{ij} has already been computed for the solution of (3.7) (see 3.8a). It too is independent of q, and hence the solution of the N sets of N equations for the a_{ij} is also very fast.

4) If we make the natural choice $w(q)=\frac{q^2}{q^2-E}$, it can be shown that the resultant coefficient matrix a_{ij} is symmetric, as desired.

4. Solution for Positive E

If E>0 equation (2.5) is singular, and the numerical procedures above fail. As in all such cases, the most appropriate course of action is to remove the singularity before solution, and this is achieved by the following decomposition (Osborne 1968). We set $E=k^2>0$, and introduce the auxiliary function $f_1(p,E)$ satisfying the integral equation

$$f_1(p,E) = \frac{v_1(p,k)}{v_1(k,k)} + \lambda \int_0^\infty \Lambda_1(p,r,E)\ f_1(r,E)\ \frac{r^2\,dr}{r^2-E-i\epsilon} \qquad (4.1)$$

where $$\Lambda_1(p,q,E)=\frac{v_1(p,k)\ v_1(k,q)}{v_1(k,k)} - v_1(p,q) \qquad (4.2)$$

We see that

$$\Lambda_1(k,q,E) = 0 \text{ and hence}$$
$$f_1(k,E) = 1 \qquad (4.3)$$

We now write $t_1(p,q,E+i\epsilon)$ in the form:-

$$t_1(p,q,E+i\epsilon) = f_1(p,E)t_1(k,k,E+i\epsilon)\ f_1(q,E)+ \lambda H_1(p,q,E) \qquad (4.4)$$

and find for H_1 the integral equation

$$H_1(p,q,E)= -\Lambda_1(p,q,E)+\lambda \int_0^\infty \Lambda_1(p,r,E)H_1(r,q,E)\ \frac{r^2}{r^2-E}\,dr \qquad (4.5)$$

while it can be shown that $t_1(k,k,E+i\epsilon)$ satisfies the relation

$$t_1(k,k,E+i\epsilon)=\frac{v_1(k,k)}{1+\lambda I+ikv_1(k,k)} \tag{4.6}$$

$$I=\int_0^\infty \frac{q^2}{q^2-E-i\epsilon}\; v_1(k,q)\; f_1(q,E)\; dq \tag{4.7}$$

We now sort out what we have achieved. First, since $\Lambda_1(p,k,E)=0$ the kernels of the two integral equations (4.1) and (4.5) are <u>regular</u> at $r^2=E$. Hence the term $i\epsilon$ is irrelevant: the equations are nonsingular, and may be treated by standard techniques. In particular, H_1 is symmetric in p,q and symmetric approximations may be derived in the same manner as for $E<0$. The singularity has not vanished completely; it resides in the principal value integral (4.7) for I. Many simple numerical techniques exist for handling this equation, and these techniques in general involve no complex arithmetic. Although $t_1(p,q)$ is complex, the complex part resides wholly in (4.6) and apart from this all of the calculations can be carried out with real arithmetic.

5. Three-Body Scattering

When we move on to consider the mutual scattering of three particles, we find a very large increase in difficulty both formally and numerically. Equation (1.2) now has three potential terms V_{12}, V_{23}, V_{31} and, after we eliminate the centre of mass motion, six space dimensions. Of these, it is possible by invoking the rotational symmetry of the system to separate out a further three; so that the final equations to be solved will be in three space dimensions. If we proceed as for two particles we will define

$$V = V_{12} + V_{23} + V_{31} = V_3 + V_1 + V_2 \quad \text{in an obvious notation}$$

$$G_0 = (-\nabla_1^2 - \nabla_2^2 - \nabla_3^2 - E)^{-1} \tag{5.1}$$

and introduce a three-body t-matrix T satisfying the operator equation

$$T = V - VG_0T \tag{5.2}$$

which when we introduce the relevant variables explicitly, is a multi-dimensional integral equation. Unfortunately, the kernel of this equation is highly singular; it is not $\mathcal{L}^2$, nor can it be made so by any finite number of iterations. The standard procedure is therefore to make the following simple manipulations (Faddeev (1961)). We split T into three parts:

$$T = \sum_{\alpha=1}^{3} T^{(\alpha)} \qquad (5.3)$$

where the $T^{(\alpha)}$ satisfy the equation

$$T^{(\alpha)} = V_\alpha - V_\alpha G_0 T \qquad \alpha=1,2,3 \qquad (5.4)$$

This split is clearly permissible from (5.1). It is then quite simple to show that $T^{(\alpha)}$ satisfies the equation

$$T^{(\alpha)} = t_\alpha - t_\alpha G_0 \sum_{\beta\neq\alpha} T^{(\beta)} \qquad (5.5)$$

where $t^{(\alpha)}$ is the two body t matrix associated with the potential V_α and satisfies (2.3).

Proof

Using (2.3) we have from (5.5)

$$T^{(\alpha)} = V_\alpha - V_\alpha G_0 t_\alpha - (V_\alpha - V_\alpha G_0 t_\alpha) G_0 \sum_{\beta\neq\alpha} T^{(\beta)}$$

$$= V_\alpha - V_\alpha G_0 \left\{ \sum_{\beta\neq\alpha} T^{(\beta)} + T_\alpha - T_\alpha G_0 \sum_{\beta\neq\alpha} T^{(\beta)} \right\}$$

$$= V_\alpha - V_\alpha G_0 \left\{ \sum_{\beta\neq\alpha} T^{(\beta)} + T^{(\alpha)} \right\}$$

$$= V_\alpha - V_\alpha G_0 T \text{ in agreement with (5.4)}$$

Equation (5.5) is referred to as the "Faddeev equation" for the system. It is perhaps not very transparent in its operator form, so let us introduce the relevant variables and display it in its simplest form as an integral equation. We explicitly remove the centre of mass motion of the three particles; we define relative momenta $\underline{q}_i = \frac{1}{2}(\underline{p}_j - \underline{p}_k)$, where (i,j,k) is a cyclic permutation of (1,2,3); a vector $\underline{p}$ with components $(|p_1|, |p_2|, |p_3|)$; with similar definitions for primed and double primed variables. Then for a spherically symmetric system, (5.5) becomes explicitly

$$T^{(1)}(\underline{p}, \underline{p}', E) = \langle \underline{p} | t_1 | \underline{p}' \rangle$$

$$-\lambda \int_\Delta \frac{1}{p_1'^2} \frac{p_1'' p_2'' p_3'' \, dp_1'' \, dp_2'' \, dp_3'' \, \delta(p_1'' - p_1')}{p_1''^2 + p_2''^2 + p_3''^2 - \frac{1}{2} E}$$

$$\times \sum_{l=0}^{\infty} (2l+1) \, p_l(\cos \gamma_1') \, p_l(\cos \gamma_2') \, t_l(q_1', q_1'', E - \tfrac{3}{2} p_1'^2)$$

$$\times \left[T^{(2)}(\underline{p}'', \underline{p}; E) + T^{(3)}(p'', p; E) \right] \qquad (5.6)$$

where $\cos \gamma_i = \hat{p}_i \cdot \hat{q}_i$

and $$\langle \underline{p} | t_1 | \underline{p}' \rangle = \frac{\delta(p_1' - p_1)}{p_1'^2} \sum_{l=0}^{\infty} (2l+1) \, p_l(\cos \gamma_1) \, p_l(\cos \gamma_1')$$

$$\times \, t_l(q_1, q_1', E - \tfrac{3}{2} p_1^2) \qquad (5.7)$$

The region Δ is that region of the space R^3 such that the three variables p_1'', p_2'', p_3'' can form a physical triangle.

We notice several features of this equation:

1) It is rather messy. Numerically, this is not very relevant; the structure of the equation is quite straightforward.

2) Both the inhomogeneous term and the kernel still contain a δ function. This δ-function may however be removed by iterating the equations sufficiently often (five times!) The iterated kernel is then L^2. This implies in practice that the δ functions are not hard to handle, and the equation is usually solved as it stands (uniterated).

3) If the three particles are identical $T^{(1)} = T^{(2)} = T^{(3)}$ and the system of coupled equations reduces to a single integral equation.

4) The equations require as input the two-body $t_\alpha(p,q,E')$ for <u>all</u> positive momenta p,q and also for all energies $E' \epsilon (-\infty, E]$. This explains why solutions of (2.5) were required as a function of the second variable q, and suggest that a compact scheme for expressing the energy dependence would also be desirable (see Delves (1973)).

6. Numerical Solution of the Faddeev Equations

Equation (5.6) represents a non-trivial computing problem even for three identical particles, and many attempts have been made to find efficient methods of solution. The most popular start from the separable expansion (2.7) of the two-body potential. If we truncate the sums over l in (5.7) at l=L, and truncate the expansion (3.1) at i=N, it can be shown that (5.6) reduces to an $N\times(L+1)$ set of coupled <u>one-dimensional</u> integral equations, from which all δ-functions have been removed explicitly. For $N(L+1)$ small, this set is comparatively simple to solve. The most popular choice is L=0, N=1; unfortunately, the numerical difficulties rise very sharply indeed as L and/or N increase, and 5 or 6 coupled equations represents about the largest set that has been solved directly in this way.

Very many attempts have therefore been made to find alternative approaches; and it is probably true to say that almost any technique ever suggested for integral equations, has been tested on the Faddeev equation with more, or sometimes less, success. If we eschew the separable expansions, an L-truncation (which seems inevitable) yields (L+1) coupled two-dimensional integral equations. These have been solved by variational/Galerkin methods (Delves (1973)): by quadrature methods of a special type (Osborne (1968)) and using product integration (Kim (1969)); by perturbation theory starting from an initial separable calculation (Afnan and Read (1973)); and even by such unlikely approaches as iteration (Laverne and Grignoux (1972)) and Bateman's method (we quote no names here!). Many of these calculations involve a preliminary rewriting of the equations to put them into a more tractable form; but we cannot discuss these further here.

References

AFNAN, I.R., and READ, J.M. (1973) T matrix Perturbation Theory: the Three-Nucleon Bound State II. Flinders University Preprint.

DELVES, L.M. (1973) Variational Techniques in the Nucleon Three-Body Problem: a Review. Advances in Nuclear Physics Vol. 5, pp.1.-223. Plenum Press.

FADDEEV, L.D. (1961) Scattering Theory for a Three-Particle System Soviet Phys. JETP Vol. 12, pp.1024.

KIM, Y.E. (1969) Solution of the Faddeev Equation for Local Potential by Approximate Product Integration. J. math. Phys. vol. 10, pp.1491-1503

LAVERNE, A., and GIGNOUX, C. (1972) Detailed Analysis of Faddeev equations in Configuration Space and their Solution for Bound States, Internal Report I SN-72-11, University of Grenoble.

OSBORNE, T.A. (1968) The Faddeev Equation for Local Potentials Ph.D. Thesis, Stanford University.

STROUD, A.H., and SECREST, D. (1966) Gaussian Quadrature Formulae, Prentice-Hall.

INDEX